physica status solidi (b)

ISSN 0370-1972 · SUBJECT AND AUTHOR INDEX (b) VOLUMES 109 TO 114 · 1982

ISSN 0031-8965 · SUBJECT AND AUTHOR INDEX (a) VOLUMES 69 TO 74 · 1982

physica status solidi (a)

THEORIA CUM PRAXI

AKADEMIE-VERLAG · BERLIN

Classification Scheme

1. Structure of Crystalline Solids
   1.1 Perfectly Periodic Structure
   1.2 Solid-State Phase Transformations
   1.3 Alloys. Metallurgy
   1.4 Microstructure (Magnetic Domains See 18; Ferroelectric Domains See 14.4.1)
   1.5 Films
   1.6 Surfaces
2. Non-Crystalline State
3. Crystal Growth
4. Bonding Properties
5. Mössbauer Spectroscopy
6. Lattice Dynamics. Phonons
7. Acoustic Properties
8. Thermal Properties
9. Diffusion
10. Defect Properties (Irradiation Defects See 11)
    10.1 Metals
    10.2 Non-Metals
11. Irradiation Effects (X-Ray Diffraction Investigations See 1 and 10)
12. Mechanical Properties (Plastic Deformations See 10)
    12.1 Metals
    12.2 Non-Metals
13. Electron States
    13.1 Band Structure
    13.2 Fermi Surfaces
    13.3 Surface and Interface States
    13.4 Impurity and Defect States
    13.5 Elementary Excitations (Phonons See 6)
        13.5.1 Excitons
        13.5.2 Plasmons
        13.5.3 Polarons
        13.5.4 Magnons
14. Electrical Properties. Transport Phenomena
    14.1 Metals. Semi-Metals
    14.2 Superconductivity. Superconducting Materials and Devices
    14.3 Semiconductors
        14.3.1 Films
        14.3.2 Surfaces and Interfaces
        14.3.3 Devices. Junctions (Contact Problems See 14.3.4)
        14.3.4 High-Field Phenomena, Space-Charge Effects, Inhomogeneities, Injected Carriers (Electroluminescence See 20.3; Junctions See 14.3.3)
    14.4 Dielectrics
        14.4.1 Ferroelectrics
15. Thermoelectric and Thermomagnetic Properties
16. Photoconductivity. Photovoltaic Effects
17. Emission of Electrons and Ions
    17.1 Field Emission Microscope Investigations
18. Magnetic Properties
    18.1 Paramagnetic Properties
    18.2 Ferromagnetic Properties
        18.2.1 Ferromagnetic Films
    18.3 Ferrimagnetic Properties
    18.4 Antiferromagnetic Properties

(Continued on cover three)

physica status solidi (b)

basic research

physica status solidi (a)

applied research

SUBJECT AND AUTHOR INDEX

Volumes 109 to 114, 1982 — Volumes 69 to 74, 1982

ISSN 0370-1972/0031-8965

Edited by

E. BORCHARDT

AKADEMIE-VERLAG · BERLIN

Subscriptions and orders for this Subject and Author Index
or single copies of physica status solidi (b) and physica status solidi (a) should be directed
in the GDR:

to the Postzeitungsvertrieb or
to the Akademie-Verlag, DDR-1086 Berlin, Leipziger Str. 3 – 4;

in the other socialist countries:
to a book-shop for foreign language literature or
to the competent news-distributing agency:

in the FRG and BERLIN (WEST):
to a book-shop or
to the wholesale distributing agency Kunst und Wissen, Erich Bieber OHG, Wilhelmstr. 4 – 6, D-7000 Stuttgart 1;

in the other Western European countries:
to Kunst und Wissen Erich Bieber GmbH, Dufourstr. 51, CH-8008 Zürich;

in USA and CANADA:
to Verlag Chemie International Inc., 303 N.W. 2th Avenue, Deerfield Beach, FL 33441, USA;

in other countries:
to the international book- and journal-selling trade,
to Buchexport, Volkseigener Außenhandelsbetrieb der Deutschen Demokratischen Republik, DDR-7010 Leipzig, Postfach 160, or
to the Akademie-Verlag, DDR-1086 Berlin, Leipziger Straße 3 – 4.

Schriftleiter und verantwortlich für den Inhalt der Zeitschriften:
Professor Dr. Dr. h. c. P. Görlich, DDR-1020 Berlin, Neue Schönhauser Str. 20 bzw. DDR-6900 Jena, Schillbachstr. 24.

Herausgeber und verantwortlich für den Inhalt des Registers:
Dipl.-Phys. E. Borchardt, DDR-1020 Berlin, Neue Schönhauser Str. 20.

Verlag:
Akademie-Verlag, DDR-1086 Berlin, Leipziger Str. 3–4; Fernruf: 2236221 und 2236229; Telex-Nr.: 114420;
Bank: Staatsbank der DDR, Berlin, Kto.-Nr.: 6836-26-20712.

Veröffentlicht unter der Lizenznummer 1310 und 1620 des Presseamtes beim Vorsitzenden des Ministerrates der Deutschen Demokratischen Republik.

Gesamtherstellung:
VEB Druckerei „Thomas Müntzer", DDR-5820 Bad Langensalza.

Bestellnummer des Registers 1068/85 R15.

Printed in the German Democratic Republic.
AN (EDV) 20831/20735

Preface

The fifteenth index volume contains volumes 109 to 114 of "physica status solidi (b) — basic research" as well as volumes 69 to 74 of "physica status solidi (a) — applied research", i.e. all volumes having appeared in 1982 containing 1509 papers.

The general scheme of last index volume was maintained. The basis for quoting is the subject classification given every papers in the respective issue. Nevertheless, before taking over these subject classifications into the index volume they were once more carefully revised and in some cases improved and completed in order to get a most uniform and clear classification for all papers. The average number of quotations per paper in this index volume is nearly the same as in the last one, namely about 3.0.

Again I have to express my thanks to the readers for their interest, and I will always readily take up any critical remarks and proposals for improvements. I am especially obliged to Prof. Dr. P. Görlich, Prof. Dr. E. Gutsche, Dr. H.-J. Hänsch, Dr. H. Lange, Dr. S. Oberländer, and Dipl. Phys. K. Müller for giving me advice and unselfish help. Furthermore I have to thank Mrs. E. Schwan for her exact and reliable secretary's work at the manuscript.

Berlin, January 1983

Eva Borchardt

Classification Scheme

1. Structure of Crystalline Solids
   1.1 Perfectly Periodic Structure
   1.2 Solid-State Phase Transformations
   1.3 Alloys. Metallurgy
   1.4 Microstructure (Magnetic Domains See 18; Ferroelectric Domains See 14.4.1)
   1.5 Films
   1.6 Surfaces
2. Non-Crystalline State
3. Crystal Growth
4. Bonding Properties
5. Mössbauer Spectroscopy
6. Lattice Dynamics. Phonons
7. Acoustic Properties
8. Thermal Properties
9. Diffusion
10. Defect Properties (Irradiation Defects See 11)
    10.1 Metals
    10.2 Non-Metals
11. Irradiation Effects (X-Ray Diffraction Investigations See 1 and 10)
12. Mechanical Properties (Plastic Deformations See 10)
    12.1 Metals
    12.2 Non-Metals
13. Electron States
    13.1 Band Structure
    13.2 Fermi Surfaces
    13.3 Surface and Interface States
    13.4 Impurity and Defect States
    13.5 Elementary Excitations (Phonons See 6)
        13.5.1 Excitons
        13.5.2 Plasmons
        13.5.3 Polarons
        13.5.4 Magnons
14. Electrical Properties. Transport Phenomena
    14.1 Metals. Semi-Metals
    14.2 Superconductivity. Superconducting Materials and Devices
    14.3 Semiconductors
        14.3.1 Films
        14.3.2 Surfaces and Interfaces
        14.3.3 Devices. Junctions (Contact Problems See 14.3.4)
        14.3.4 High-Field Phenomena, Space-Charge Effects, Inhomogeneities, Injected Carriers (Electroluminescence See 20.3; Junctions See 14.3.3)
    14.4 Dielectrics
        14.4.1 Ferroelectrics
15. Thermoelectric and Thermomagnetic Properties
16. Photoconductivity. Photovoltaic Effects
17. Emission of Electrons and Ions
    17.1 Field Emission Microscope Investigations

18. Magnetic Properties
    18.1 Paramagnetic Properties
    18.2 Ferromagnetic Properties
        18.2.1 Ferromagnetic Films
    18.3 Ferrimagnetic Properties
    18.4 Antiferromagnetic Properties

19. Magnetic Resonance

20. Optical Properties
    20.1 Spectra. Optical Constants (X-Ray Spectra See 20)
    20.2 Lasers
    20.3 Luminescence (X-Ray Spectra See 20)

21. Metals. Alloys
    21.1 Transition Metals (Group Sc to Zn) and Their Alloys
        21.1.1 Fe and Fe Alloys
    21.2 Alkali-Metals
    21.3 Alkali-Earth Metals
    21.4 Rare-Earth Metals
    21.5 Actinides
    21.6 Noble Metals
    21.7 Semi-Metals (Semiconducting Alloys See 22.7)

22. Semiconductors and Ionic Crystals
    22.1 Elements
        22.1.1 Germanium
        22.1.2 Silicon
        22.1.3 Group VI Elements
    22.2 III–V Compounds
        22.2.1 Arsenides
        22.2.2 Phospides
        22.2.3 Animonides
        22.2.4 Mixed Crystals
    22.3 IV–IV, II–V, and III–VI Compounds
        22.3.1 Mixed Crystals
    22.4 II–VI Compounds
        22.4.1 Sulphides
        22.4.2 Selenides
        22.4.3 Tellurides
        22.4.4 Mixed Crystals
    22.5 Halides
        22.5.1 Silver Halides
        22.5.2 Alkali Halides
        22.5.3 Alkali-Earth Halides
        22.5.4 Mixed Crystals
    22.6 Simple Oxides (CdO, ZnO See 22.4)
        22.6.1 Mixed Crystals
    22.7 Semiconducting Intermetallic Compounds
        22.7.1 Mixed Crystals
    22.8 Tertiary and Higher Compounds
        22.8.1 Salts of Oxy-Acids (Carbonates, Phosphates, Silicates, Including Mica etc.)
        22.8.2 Spinels, Garnets, Ferrites
    22.9 Organic Semiconductors

23. Solidified Gases

For comments on the use of the classification scheme see p. 6.

Some Comments on the Use of the Subject and Author Index

The classification system[1]) comprises 20 broad sections each devoted to a major field of solid state physics and 3 broad sections devoted to the main groups of substances. A section may contain a number of subsections dealing with aspects of this field or special substance, respectively, which tend to receive particular attention. A paper dealing with this field will either be classified under one or more of the subsections or, if none of these are relevant, it will just appear under the general section heading (i.e. sections and subsections are equivalent in this sense). For example, a paper dealing with problems of paramagnetism would appear under 18.1 but not 18, whereas a paper on diamagnetism would be quoted under 18; a paper dealing with GaP would appear under 22.2.2, not under 22 or 22.2, whereas a paper on AlN would be quoted under 22.2, and a paper on a II–IV or a IV–VI compound would be quoted under 22.

All papers will be quoted under each relevant section. There is made a distinction between one or two principal entries (refering to the main subject matter of the paper) and secondary entries (refering to subsidiary subject matter and substance, respectively). In the principal entries the title of the paper is given in its original language and in English. In addition, every principal entry is marked by * in front of the title of the paper. In the secondary entries the title is only given in English.

Each entry gives the volume, the number of the issue, the initial and final page numbers, and the year in which the paper appeared. A paper of physica status solidi (b) or physica status solidi (a) is marked by (b) or (a) in front of the line, respectively. The letter "K" before the page numbers indicates a short note published in the offset part of the journal. Each entry is followed by references of the other entries. These references give the subject classifications and the pages in the index where the other entries appear. The references to the principal entries are given in italics.

Review articles appear, at the head of the list of entries under a given subject and are given the subtitle "Review Articles". These articles are listed in the chronological order of their publication, irrespective of being a paper of physica status solidi (b) or physica status solidi (a). The subject classification for a review article consisting of several parts is based on the article as a whole rather than on the individual parts.

The entries for original papers, short notes, and errata are made together in a single list for each subject and this follows the entries for review articles. These are again ordered according to data of publication, irrespective of being a paper of physica status solidi (b) or physica status solidi (a). Original papers are sometimes published in several parts under the same principal title (in general with special subtitles). The various parts of papers of this type are treated separately and each part is classified according to its particular contents.

In quoting errata the following rule is obeyed. If the erratum refers to a paper quoted in the preceding index, it is listed in the chronological order of its publication. otherwise the erratum is quoted together with the paper to which it refers.

---

[1]) See cover two and three, and pages 4 and 5. For the case of losing the cover by bookbinding we repeat the classification scheme on pages 4 and 5.

In using the subject index attention should be given to the following details:

subject	quoted in	remarks
representations of space groups	1 and 13.1	
Alfen waves, helicon waves	13 and 20	
hot electrons	13 and 14.3.4	
transport phenomena	14 or subsection of 14	if not clearly to be assigned to 8 or 15
galvanomagnetic effects	14 or subsection of 14	not in 18
electro-optical effects	14 and 20 (or subsection of 14 and 20)	
piezoelectricity, electrostriction	14 or subsection of 14	not in 12 or subsection of 12; different electromechanical effects are quoted in 12 or subsection of 12
ionic conduction	14.4	not in 14.3
magneto-optical effects	18 and 20 (or subsection of 18 and 20)	possibly in additional sections, e.g. Faraday effect in 13
ferromagnetic films	18.2.1	not additionally in 1.5
X-ray spectra	20	but not X-ray studies of structure
MIS and MOS structures	the section assigned to the semiconductor	not in the section assigned to the oxide or the metal
MIM structures	the section assigned to the insulator	generally not in the section assigned to the metal
epitaxial layers	the section assigned to the layer	not in the section assigned to the substrate

The autor index gives the names of the authors in alphabetical order. Each name is followed by the bibliographic details of the paper and the numbers of the pages on which the paper appears in the subject index. A paper of physica status solidi (b) or physica status solidi (a) is again marked by (b) or (a) in front of the number of the volume, respectively. The principal entry is again given in italics. When several papers have been published by the same author they are arranged according to the dates of publication, irrespective of being a paper of physics status solidi (b) or physica status solidi (a).

Subject Index

1. Structure of Crystalline Solids

Original Papers and Short Notes

W. Herremans, M. David, and R. Gevers
*Crystal Potential near a Stacking Fault
(a) Vol. **70**, No. 1, p. 295–309 (1982).
See also *10*, p. *94*.

A. Andreeff, Th. Frauenheim, E. A. Goremychkin, H. Griessmann, B. Lippold, W. Matz, O. D. Chistyakov, and E. M. Savitskii
*Crystal Field in the Laves Phase Compound $PrNi_2$
(b) Vol. **111**, No. 2, p. 507–512 (1982).
See also 8, p. 87; *13.4*, p. *169*; 15, p. 246; 18.1, p. 266; 21.1, p. 349; 21.4, p. 370.

G. S. Dubey
*Calculation of the Static Structure Factor of the Classical Two-Dimensional Plasma
(b) Vol. **111**, No. 2, p. K99–K101 (1982).
See also *13*, p. *143*.

J. F. J. Jordan, A. Axmann, H. Egger, and J. Kalus
*Temperature Dependence of the Compressibility of d_8-Naphthalene
(a) Vol. **71**, No. 2, p. 457–462 (1982).
See also **6**, p. 75; 12.2, p. 135; 22.9, p. 505.

I. V. Stasyuk and R. A. Grigorchuk
*The Theory of the Strain Effects in the Crystals Described by the s(d)–f Model
(b) Vol. **112**, No. 1, p. 327–338 (1982).
See also 12, p. 129; *13*, p. *144*.

A. Castets, D. Gignoux, J. C. Gomes Sal, F. Rodrigues Gonsalves, and E. Roudaut
*Magnetic Properties and Structure of Tb_2Pt
(a) Vol. **73**, No. 2, p. 475–481 (1982).
See also *18.2*, p. *272*; 18.4, p. 281; 21.4, p. 371; 21.6, p. 378.

P. A. Alekseev, I. P. Sadikov, Yu. L. Shitikov, I. A. Markova, O. D. Chistyakov, E. M. Savitskii, and J. Kjems
*Crystalline Electric Field Effects in $Pr(La, Y)Al_3$
(b) Vol. **114**, No. 1, p. 161–167 (1982).
See also *13.4*, p. *176*; 21.4, p. 372.

1.1 Perfectly Periodic Structure

Original Papers and Short Notes

W. Bażela, I. Szott, and A. Szytuła
*Magnetic Properties of the $Co_xNi_{1-x}MnGe$ System in the Hexagonal Phase
(a) Vol. **69**, No. 1, p. K49 – K54 (1982).
Erratum
(a) Vol. **71**, No. 1, p. K131 (1982).
See also *18.4*, p. *279*; 22.8, p. 482.

H. S. Kim and S. S. Sheinin
*An Assessment of the High-Energy Approximation in the Dynamical Theory of Electron Diffraction
(b) Vol. **109**, No. 2, p. 807 – 816 (1982).
See also 21.6, p. 374.

V. I. Kushnir and E. V. Suvorov
*Investigation of the Dynamic Diffraction Focusing of X-Rays by a Homogeneously Bent Crystal in the Symmetric Laue Case
(a) Vol. **69**, No. 2, p. 483 – 490 (1982).
See also 22.1.2, p. 402.

K. Naukkarinen and T. Tuomi
*Explanation of Contrast Effects in Synchrotron X-Ray Topographs of Bent Crystals by Means of Multiple Diffraction
(a) Vol. **69**, No. 2, p. 687 – 697 (1982).
See also 12.2, p. 134; 22.1.2, p. 402.

L. M. Lityagina, Z. V. Malyushitskaya, T. A. Pashkina, and S. S. Kabalkina
*Isothermal Compression of Al to 10 GPa at 673 K
(a) Vol. **69**, No. 2, p. K147 – K150 (1982).
See also *12.1*, p. *130*; 21, p. 332.

P. Zaumseil and U. Winter
*Triple Crystal Diffractometer Investigations of Silicon Crystals with Different Collimators–Analyzer Arrangements
(a) Vol. **70**, No. 2, p. 497 – 505 (1982).
See also 10.2, p. 108; 22.1.2, p. 405.

E. R. de Gil
*Ternary Semiconducting Compounds with Chalcopyrite-Type Structure (I): Fundamental Equation and Its Physically Acceptable Solutions
(a) Vol. **70**, No. 2, p. 519 – 523 (1982).
See also *1.3*, p. *27*; 22.8, p. 484.

V. Holý
*The Coherence Description of the Dynamical X-Ray Diffraction from Randomly Disordered Crystals (I): General Formalism
(b) Vol. **111**, No. 1, p. 341 – 351 (1982).
See also 10.2, p. 109; 22, p. 386.

H. S. KIM, R. PEREZ, and S. S. SHEININ
*Experimental Evidence Showing that the Use of Symmetrical Laue Diffraction Conditions in the Dynamical Theory of Electron Diffraction Can Give Significant Errors in Diffraction Contrast Calculations
(a) Vol. **72**, No. 2, p. 561–568 (1982).
See also 22.1.2, p. 410.

R. S. NARAYANAN, J. SHANKAR, H. C. GUPTA, and B. B. TRIPATHI
*Theory of Interionic Forces in Alkali Halide Crystals
(b) Vol. **113**, No. 1, p. 339–346 (1982).
See also *4*, p. *64*; 12.2, p. 136; 22.5.2, p. 460.

R. GEVERS and M. DAVID
*Relativistic Theory of Electron and Positron Diffraction at High and Low Energies
(b) Vol. **113**, No. 2, p. 665–678 (1982).

V. T. BUBLIK, J. WILKE und A. T. PEREVERSEV
*Die Natur der festen Lösungen der überschüssigen Komponenten in unlegiertem Gallium-antimonid
(The Nature of Solid Solutions of the Excess Components in Non-Doped Gallium Antimonide)
(a) Vol. **73**, No. 2, p. K271–K273 (1982).
See also *1.4*, p. *37*; 22.2.3, p. 429.

SH. M. EFENDIEV, T. Z. KULIEVA, V. A. LOMONOV, M. I. CHIRAGOV, M. GRANDOLFO, and P. VECCHIA
*Crystal Structure of Bismuth Titanium Oxide $Bi_{12}TiO_{20}$
(a) Vol. **74**, No. 1, p. K17–K21 (1982).
See also 22.8, p. 490.

NGUYEN AN, G. WORM, and R. HERRMANN
*Symmetrical Grain Boundaries with Special Angles and Their Coincidence Site Lattices
(b) Vol. **114**, No. 2, p. 349–357 (1982).
See also 1.4, p. 37.

A. R. BADZIAN
*On the Geometrical Condition for X-Ray Scattering by Thermal Waves in Crystals
(a) Vol. **74**, No. 2, p. 637–649 (1982).
See also 6, p. 81; 22, p. 392; 22.1.1, p. 399; 22.1.2, p. 416; 22.5.2, p. 464.

1.2 Solid-State Phase Transformations

Review Articles

F. Reynaud
*Order–Disorder Transitions in Substitutional Solid Solutions
(a) Vol. **72**, No. 1, p. 11–59 (1982).
See also *1.3*, p. *25*; 21, p. 330; 21.1, p. 343; 21.1.1, p. 359; 21.6, p. 374.

P. P. Seregin, A. R. Regel, A. A. Andreev, and F. S. Nasredinov
A Study of Local Structure of Chalcogenide Vitreous Semiconductors and of States of Impurity Atoms in Them by Means of Mössbauer and Electron Spectroscopies
(a) Vol. **74**, No. 2, p. 373–394 (1982).
See also *2*, p. *45*; 5, p. 66; *13.4* p. *162*; 22, p. 382.

Original Papers and Short Notes

Yu. Ya. Boguslavskii
*Equation for the Melting Curve of Solids under High Pressure
(b) Vol. **109**, No. 1, p. 145–151 (1982).
See also *8*, p. *84*; 12.1, p. 130; 21.2, p. 365; 23, p. 509.

Yu. S. Stark and A. S. Sternberg
Application of BBGKY Equations to the Study of Ordering in Alloys
(b) Vol. **109**, No. 1, p. 403–409 (1982).
See also *1.3*, p. *25*; 21, p. 331.

P. Gluck
*Transitions in Fast Ionic Conductors
(b) Vol. **109**, No. 1, p. 437–439 (1982).
See also *14.4*, p. *234*; 22.5, p. 450.

M. Bartkowiak and S. Robaszkiewicz
*Charge Ordering in the Extended Hubbard Model. The Caron-Pratt Approximation Results
(b) Vol. **109**, No. 1, p. K11–K16 (1982).
See also *13*, p. *140*; 14, p. 190.

E. I. Terukov, N. Mattern, W. Reichelt, H. Oppermann, and M. Wolf
The Behaviour of Resistance and Thermopower and the Coexistence of Two Phases at Metal–Insulator Transitions
(a) Vol. **69**, No. 1, p. 97–101 (1982).
See also *14.1*, p. *193*; *14.3*, p. *202*; 15, p. 245; 22.6, p. 468.

A. Leblé, J. J. Rousseau, J. C. Fayet, J. Pannetier, J. L. Fourquet, and R. de Pape
Order–Disorder Transition of NH_4AlF_4 through EPR Investigations
(a) Vol. **69**, No. 1, p. 249–256 (1982).
See also *19*, p. *283*; 22.8, p. 481.

M. Drulis and B. Staliński
Low-Temperature Specific Heats of Non-Stoichiometric Neodynium Hydrides
(a) Vol. **69**, No. 1, p. 385–392 (1982).
See also *8*, p. *85*; 21.4, p. 369.

J. W. Brightwell, C. N. Buckley, and B. Ray
*Polymorphism in AgI and $Ag_{1-x}Cu_xI$
(a) Vol. **69**, No. 1, p. K11–K13 (1982).
See also *1.3*, p. *26*; 22.5.4, p. 467.

M. Onoda, T. Takahashi, and H. Nagasawa
*Nuclear Quadrupole Effect and Phase Transition in the Quasi-One-Dimensional Conductor β-$Na_{0.33}V_2O_5$
(b) Vol. **109**, No. 2, p. 793–797 (1982).
See also *19*, p. *283*; 22.8, p. 482.

A. Ghosh, L. K. Banerjee, and A. N. Basu
*Cohesion, Phase Transition, Elastic and Dielectric Properties of Group IV Semiconductors
(b) Vol. **109**, No. 2, p. 799–806 (1982).
See also *4*, p. *61*; 12.2, p. 134; 14.4, p. 234; 22.1, p. 393; 22.1.1, p. 395; 22.1.2, p. 401.

P. Trzaskoma, J. Zieliński, and M. Szopa
*On the Role of Phonons for the Pressure Induced Change of Valence
(b) Vol. **109**, No. 2, p. K129–K132 (1982).
See also *6*, p. *71*; 21.4, p. 369.

O. Simmich
*Description of Particle Growth by Means of a Difference Approximation
(a) Vol. **69**, No. 2, p. 657–662 (1982).
See also *10.1*, p. *97*; 21, p. 332.

T. Kanaizuka
Interpretation of Mössbauer Spectra of β-FeB and Its Low Temperature Modification, α-FeB
(a) Vol. **69**, No. 2, p. 739–744 (1982).
See also *5*, p. *66*; *18.2*, p. *268*; 21.1.1, p. 359.

H. Fischer, H.-D. Geiler, G. Götz, A. V. Dvurechenskii, B. P. Kashnikov, and L. S. Smirnov
Antiannealing Effects in Ion-Implanted and Electron Pulse Irradiated Quartz Crystals
(a) Vol. **69**, No. 2, p. K139–K142 (1982).
See also *11*, p. *119*; 22.6, p. 469.

A. Graja, R. Swietlik, J. Petzelt, and L. Dobiášová
*Phase Transition in $MTPP(TCNQ)_2$ Single Crystals by Infrared Reflection Measurements
(a) Vol. **69**, No. 2, p. K205–K208 (1982).
See also 6, p. 71; 20.1, p. 301; 22.9, p. 503.

A. I. Naskidashvili, I. A. Naskidashvili, L. S. Topchyan, V. N. Kozhanov, and E. P. Romanov
*Heat Capacity Anomalies in the Intermetallic Compound V_2Zr
(a) Vol. **69**, No. 2, p. K239–K242 (1982).
See also *8*, p. *85*; 21.1, p. 345.

R. Parrot and B. Canny
*ESR Study of the Monoclinic–Orthorhombic Phase Transition in Neodymium Pentaphosphates
(b) Vol. **110**, No. 1, p. 123–131 (1982).
See also *19*, p. *284*; 22.8.1, p. 493.

Z. Czapla, L. Sobczyk, and J. Mróz
*Isotopic Effect in Ferroelectric Rubidium Hydrogen Selenate Crystals
(a) Vol. **70**, No. 1, p. 73–78 (1982).
See also *14.4.1*, p. *240*; 22.8.1, p. 493.

V. I. Bogdanov, A. V. Ruban, and D. L. Fuks
Covalent Contribution to the Bond Energy of Phases with Ll_2 Structure
(b) Vol. **110**, No. 2, p. 431–435 (1982).
See also *1.3*, p. *27*; *4*, p. *62*; 21.1, p. 346; 21.6, p. 375.

F. Brehat, B. Wyncke, and M. El Sherif
*Lattice Mode Anharmonicity in Sodium Nitrite
(b) Vol. **110**, No. 2, p. 499–506 (1982).
See also *6*, p. *73*; 14.4.1, p. 241; *20.1*, p. *303*; 22.8.1, p. 494.

D. Lepski and P. Burck
*Precipitate Structure Analysis by Diffraction Contrast Image Matching for a Metastable Semicoherent Carbide Phase in Molybdenum (II): Image Matching
(a) Vol. **70**, No. 2, p. 571–581 (1982).
See also *1.4*, p. *33*; 10.1, p. 98; 12.1, p. 131; 21.1, p. 347.

P. Günter, R. Sanctuary, and F. Rohner
*Changes of Optical Properties at the Successive Phase Transitions in Rb_2ZnCl_4
(a) Vol. **70**, No. 2, p. 583–589 (1982).
See also *20.1*, p. *304*; 22.5.4, p. 467.

C. Boulesteix et B. Yangui
*Domains et interfaces engendrées au cours d'une transformation cristallographique coopérative — application au cas de la transformation hexagonale monoclinique des sequioxydes de terres rares (I): Dénombrement et position relative des domains équivalents (Domains and Interfaces Created during Cooperative Crystallographic Transformation — Application to Hexagonal-Monoclinic Transformation of Rare Earth Sesquioxides (I): Determination of the Number and Relative Position of Equivalent Domains)
(a) Vol. **70**, No. 2, p. 597–613 (1982).
Erratum
(a) Vol. **74**, No. 1, p. 361 (1982).
See also *1.4*, p. *33*; 22.6, p. 470.

J. Torres, C. Roucau, and R. Ayroles
*Investigation of the Interactions between Ferroelastic Domain Walls and of the Structural Transition in Lead Phosphate Observed by Electron Microscopy (I): Experimental Results
(a) Vol. **70**, No. 2, p. 659–669 (1982).
See also *1.4*, p. *34*; 22.8.1, p. 494.

Bai-Xin Liu and M-A. Nicolet
*Formation of Au–Ge Metastable Phases by Ion Mixing
(a) Vol. **70**, No. 2, p. 671–676 (1982).
See also *1.3*, p. *27*; 2, p. 48; 21.6, p. 376 ; 22.1.1, p. 396.

H. Ishida, R. Ikeda, and D. Nakamura
*Pre-Melting State of Methylammonium Iodide as Revealed by Proton Magnetic Resonance
(a) Vol. **70**, No. 2, p. K151 – K154 (1982).
See also *19*, p. *286*; 22.9, p. 504.

S. V. Misul and I. A. Belobrova
*Sequence of Phase Transitions in $(NH_4)_2ZnCl_4$
(a) Vol. **70**, No. 2, p. K167 – K170 (1982).
See also 14.4.1, p. 241; 19, p. 286; 22.8, p. 484.

T. Soma, H. Iwanami, and H.-Matsuo Kagaya
*Pressure–Volume Relation under High Pressure for Si–Ge Solid Solution
(b) Vol. **111**, No. 1, p. K23 – K26 (1982).
See also 4, p. 62; 12.2, p. 135; 22.1.1, p. 396; 22.1.2, p. 406.

K. Parliński
*Phase Transitions in KND_2
(b) Vol. **111**, No. 1, p. K55 – K59 (1982).
See also 22.9, p. 505.

L. G. Khvostantsev, A. I. Orlov, N. Kh. Abrikosov, T. E. Svechnikova, and S. N. Chizhevskaya
*Thermoelectric Properties and Phase Transitions in Bi_2Te_3 under Hydrostatic Pressure up to 9 GPa and Temperature up to 300 °C
(a) Vol. **71**, No. 1, p. 49 – 53 (1982).
See also 12.2, p. 135; 14.3, p. 206; *15*, p. *246*; 22.7, p. 479.

C. Kaito, N. Nakamura, K. Teranishi, S. Sekimoto, and M. Shiojiri
*High Resolution Electron Microscopic Studies of the Polymorphic Transformation and Crystal Structures of Low-Temperature Ag_2Se Phases
(a) Vol. **71**, No. 1, p. 109 – 119 (1982).
See also *1.1*, p. *12*; 22, p. 386.

J. Torrès, C. Roucau, and R. Ayroles
*Investigation of the Interactions between Ferroelastic Domain Walls and of the Structural Transition in Lead Phosphate Observed by Electron Mocrpscopy (II): Interpretation of the Interactions between Ferroelastic Domain Walls
(a) Vol. **71**, No. 1, p. 193 – 205 (1982).
See also *1.4*, p. *34*; 22.8.1, p. 494.

P. E. Tomaszewski and K. Łukaszewicz
*Orthorhombic Low-Temperature Phase of $LiKSO_4$
(a) Vol. **71**, No. 1, p. K53 – K55 (1982).
See also 22.8.1, p. 495.

G. T. Didrikhsons and H. Pfeiffer
*On Phase Diagrams of Binary Systems
(a) Vol. **71**, No. 1, p. K79 – K81 (1982).
See also *3*, p. *58*.

R. S. Singh, V. K. Jain, and J. Shanker
*Relative Stability and Structural Phase Transitions in Alkaline Earth Chalcogenide Crystals
(b) Vol. **111**, No. 2, p. 519 – 523 (1982).
See also 4, p. 63; 22, p. 387.

V. E. Schneider and E. E. Tornau
*On the Theory of Isostructural Phase Transitions in Crystals
(b) Vol. **111**, No. 2, p. 565–574 (1982).
See also 14.4.1, p. 242; 22.8, p. 484.

S. Robaszkiewicz and M. Bartkowiak
*Ground State Phase Diagram of the Extended Hubbard Model with Ising-Like Exchange Interactions
(b) Vol. **111**, No. 2, p. K107–K111 (1982).
See also *13*, p. *143*; 14, p. 191; 18, p. 261.

H. Arashi and M. Ishigame
*Raman Spectroscopic Studies of the Polymorphism in ZrO_2 at High Pressures
(a) Vol. **71**, No. 2, p. 313–321 (1982).
See also 6, p. 75; 20.1, p. 307; 22.6, p. 472.

V. B. Dudnikova and V. S. Urusov
*Solvus of $KCl{:}Sr^{2+}$ Solid Solution by Density Measurements of Single Crystals with the Flotation Method
(a) Vol. **71**, No. 2, p. 409–414 (1982).
See also *10.2*, p. *109*; 22.5.2, p. 458.

W. J. Kuhn, J. Petersson, and H. E. Müser
*Phase Transition Fronts in Single Crystals of the Antiferroelectric Squaric Acid
(a) Vol. **71**, No. 2, p. 483–492 (1982).
See also *14.4*, p. *236*; 22.9, p. 505.

J. K. Krüger, M. Pietralla, and H.-G. Unruh
*The Elastic Properties of Hexatriacontane Single Crystals at Their Various Phase Transitions
(a) Vol. **71**, No. 2, p. 493–504 (1982).
See also *6*, p. *75*; 12.2, p. 135; 22.9, p. 505.

O. Bertrand, Ph. Dufour, N. Floquet, and L. C. Dufour
*Preliminary Electron Microscopy Investigation of the Reduction into MoO_2 of the Higher Molybdenum Oxides. Mo_4O_{11} and CS Oxides $Mo_{18}O_{52}$, Mo_9O_{26}, Mo_8O_{23}
(a) Vol. **71**, No. 2, p. 511–521 (1982).
See also *1.3*, p. *28*; 22.6, p. 472.

M. T. Sebastian, D. Pandey, and P. Krishna
*X-Ray Diffraction Study of the 2H to 3C Solid State Transformation in Vapour Grown Single Crystals of ZnS
(a) Vol. **71**, No. 2, p. 633–640 (1982).
See also 10.2, p. 109; 22.4.1, p. 441.

G. Krabbes, E. I. Terukov und H. Oppermann
*Zeitabhängiges Verhalten des elektrischen Widerstandes beim α-Phasenübergang in stark dotiertem Cobalteisensulfid $Co_yFe_{1-y}S$
(Time Dependence of the Electrical Resistivity of α-Phase Transition in Highly Doped Cobalt Iron Sulfide ($Co_yFe_{1-y}S$))
(a) Vol. **71**, No. 2, p. K143–K146 (1982).
See also *14.3*, p. *207*; 22.8, p. 486.

A. V. Andreev, E. N. Tarasov, A. V. Deryagin, and S. M. Zadvorkin
*Influence of Symmetry of the Crystalline Structure on the Magnetic Anisotropy of Tb_2Co_7
(a) Vol. **71**, No. 2, p. K245–K247 (1982).
See also 1.3, p. 29; *18.2*, p. *271*; 21.1, p. 349; 21.4, p. 371.

M. D. Kaplan
*Metamagnetoelastic and Metamagnetoelectric Behaviour of the Jahn-Teller Antiferroelastics
(b) Vol. **112**, No. 1, p. 351–356 (1982).
See also 18.4, p. 280; 22.8, p. 487.

M. A. Korzhuev
On the Fundamental Energy Gap in GeTe
(b) Vol. **112**, No. 1, p. K39–K41 (1982).
See also *13.1*, p. *152*; 22.7, p. 479.

J. Echigoya and J. W. Edington
*A Transmission Electron Microscope Study of the Chalcocite–Djurleite Transformation in Topotactically Grown Thin Films of Cu_xS
(a) Vol. **72**, No. 1, p. 305–311 (1982).
See also *1.5*, p. *39*; 22, p. 388.

A. E. Moskalev, I. A. Belobrova, L. I. Zherebtsova, and I. P. Aleksandrova
*Ferroelectric Phase Transition in $(NH_4)_2ZnBr_4$
(a) Vol. **72**, No. 1, p. K19–K21 (1982).
See also *14.4.1*, p. *242*; 19, p. 288; 22.8, p. 487.

Y. A. Badr and S. Awad
*On the Low Temperature Phase Transition in Ammonium Sulfate
(a) Vol. **72**, No. 1, p. K27–K32 (1982).
See also 8, p. 87; *14.4.1*, p. *242*; 22.8.1, p. 495.

A. Heinrich, E. I. Terukov, W. Reichelt, H. Wagner, and H. Oppermann
*IR-Absorption in Doped and Undoped VO_2
(a) Vol. **72**, No. 1, p. K61–K64 (1982).
See also 6, p. 76; 20.1, p. 308; 22.6, p. 472.

E. Heiner
The Dielectric Response of a Charge Density Wave (CDW) Distorted Electron–Phonon System in the Precritical Region (I): The Self-Consistent Method
(b) Vol. **112**, No. 2, p. 489–500 (1982).
See also *13*, p. *144*; 14.4, p. 237; 22, p. 388.

H.-E. Ponath and S. V. Strizhevskii
Scattering of Surface Polaritons by the Order Parameter Fluctuations in the Vicinity of a Phase Transition Point
(b) Vol. **112**, No. 2, p. 533–542 (1982).
See also *13.3*, p. *160*; *13.5*, p. *179*.

T. D. Sands, J. Washburn, and R. Gronsky
*High Resolution Observations of Copper Vacancy Ordering in Chalcocite (Cu_2S) and the Transformation to Djurleite ($Cu_{1.97\ \text{to}\ 1.94}S$)
(a) Vol. **72**, No. 2, p. 551–559 (1982).
See also *1.5*, p. *40*; 22, p. 388.

O. Bostanjoglo and G. Hoffmann
Time-Resolved TEM of Transient Effects in Pulse Annealing of Ge and Ge–Te Films
(a) Vol. **73**, No. 1, p. 95 – 105 (1982).
See also *1.5*, p. *40*; 3, p. 59; 22.1.1, p. 398; 22.1.3, p. 417.

O. Simmich
*Theoretical Considerations on the Diffusion-Controlled Particle Growth in an Isotropic Medium
(a) Vol. **73**, No. 1, p. 107 – 115 (1982).
See also *10.1*, p. *100*; 21, p. 339.

G. Götz, H.-D. Geiler, and M. Wagner
Pulse Laser Induced High-Temperature Solid-Phase Annealing of Arsenic Implanted Silicon
(a) Vol. **73**, No. 1, p. 145 – 151 (1982).
See also *3*, p. *59*; 22.1.2, p. 412.

F. Watari, J. van Landuyt, P. Delavignette, S. Amelinckx, and N. Igata
X-Ray Peak Broadening as a Result of Twin Formation in Some Oxides Derived by Dehydration
(a) Vol. **73**, No. 1, p. 215 – 224 (1982).
See also *1.4*, p. *36*; 22.6, p. 474.

J. Singh and S. Ranganathan
*On the Mechanism of F.C.C. → H.C.P. Transformation
(a) Vol. **73**, No. 1, p. 243 – 248 (1982).
See also *1.4*, p. *36*; 21.1, p. 354; 21.1.1, p. 362.

G. Sorge, H. Beige, A. Almeida, and L. A. Shuvalov
*Elastic Nonlinearity of $KH_3(SeO_3)_2$ and $KD_3(SeO_3)_2$ Crystals
(a) Vol. **73**, No. 1, p. K63 – K66 (1982).
See also 12.2, p. 136; *14.4.1*, p. *243*; 22.8, p. 489.

A. V. Andreev, A. V. Deryagin, V. N. Moskalev, and N. V. Mushnikov
*On Crystalline Structure of Hydrides of $ErFe_2$ and $HoFe_2$
(a) Vol. **73**, No. 1, p. K69 – K71 (1982).
See also *1.3*, p. *30*; 18.2, p. 272; 21.1.1, p. 362; 21.4, p. 371.

R. N. Mukherjee, D. N. S. Srivastwa, and S. C. Sen
*Photoluminescence Spectrum of NH_4Br:Tl after First-Oder Crystallographic Phase Transformation
(b) Vol. **113**, No. 2, p. 731 – 736 (1982).
See also 20.1, p. 312; *20.3*, p. *326*; 22.5, p. 452.

B. Pura, J. Przedmojski, A. Jabłonka, and R. Dąbrowski
Anisotropy of X-Ray Critical Scattering in **4,4′**-Bi-Pentylobiphenyl Liquid Crystal
(b) Vol. **113**, No. 2, p. K119 – K122 (1982).
See also *2*, p. *53*; 22.9, p. 507.

T. Soma, Y. Iwayama, and H. Matsuo Kagaya
*Equation of State of Cu-Based Binary Alloys
(b) Vol. **113**, No. 2, p. K131 – K133 (1982).
See also *12.1*, p. *132*; 21.1, p. 355.

H. Hedler, W. Andrä, and G. Götz
*Activation of High As and Sb Concentrations in Silicon by Laser Irradiation
(a) Vol. **73**, No. 2, p. **333**–338 (1982).
See also 11, p. 126; 14.3, p. 210; 22.1.2, p. 413.

M. Paunov and A. Milchev
*2D Phase Transitions at High Densities
(a) Vol. **73**, No. 2, p. 339–**349** (1982).
See also 1.5, p. 41; 1.6, p. 44.

U. Straube, G. Sorge, and N. R. Ivanov
Temperature-Dependent Electrostriction Coefficient and Pseudoproper Phase Transition in TSCC
(a) Vol. **73**, No. 2, p. 467–474 (1982).
See also *12.2*, p. *137*; *14.4.1*, p. *243*; 22.8, p. 490.

B. Yangui et C. Boulesteix
*Domaines et interfaces engendrées au cours d'une transformation cristallographique coopérative — application au cas de la transformation hexagonale monoclinique des sesquioxydes de terres rares (II): Position relative et raccordement des variantes de domaines de rotation et de translation: raccordement en étoile
(Domains and Interfaces Created during Cooperative Crystallographic Transformation — Application to Hexagonal-Monoclinic Transformation of Rare Earth Sesquioxides (II): Relative Position and Connection of the Variants of Rotation and Translation Domains: Connection into Stars)
(a) Vol. **73**, No. 2, p. 515–527 (1982).
See also *1.4*, p. *37*; 22.6, p. 475.

F. Delannay
*On the Reduction of Orthorhombic MoO_3 to MoO_2
(a) Vol. **73**, No. 2, p. 529–537 (1982).
See also *1.4*, p. *37*; 22.6, p. 475.

M. Januszczyk, J. Pietrzak, and A. Stecki
*Electrical Conductivity of $Fe(PY)_3Cl_3 \cdot PY$ Complex Single Crystals near the Structural Phase Transition
(a) Vol. **73**, No. 2, p. K185–K188 (1982).
See also *14.3*, p. *210*; 22.9, p. 507.

A. Graja, E. Sopa, and L. Dobiášová
*Infrared Study of the Structural Phase Transition in a Ternary 1 D Salt $(TMA^+) \cdot \cdot (TCNQ)^{2/3-}(I_3^-)_{1/3}$
(a) Vol. **73**, No. 2, p. K237–K241 (1982).
See also *20.1*, p. *313*; 22.9, p. 507.

I. Onyszkiewicz, N. T. Malafaev, A. A. Murakhovskii, and J. Pietrzak
*Magnetic and FMR Investigation of the Structural Phase Transition in Cu-Ferrite with Cooperative Jahn-Teller Effect
(a) Vol. **73**, No. 2, p. K243–K247 (1982).
See also 18.3, p. 277; 19, p. 290; 22.8.2, p. 501.

D. BARB, E. BĂRBULESCU, and A. BĂRBULESCU
Diffuse Phase Transitions and Ferroelectric–Paraelectric Diagram for the $BaTiO_3$–$SrTiO_3$ System
(a) Vol. **74**, No. 1, p. 79–83 (1982).
See also *14.4.1*, p. *244*; 22.8.1, p. 497.

L. G. KHVOSTANTSEV, V. A. SIDOROV, L. E. SHELIMOVA, and N. KH. ABRIKOSOV
*Phase Transitions in GeTe at Hydrostatic Pressure up to 9.3 GPa
(a) Vol. **74**, No. 1, p. 185–192 (1982).
See also 12.2, p. 138; *14.3*, p. *211*; 22.7, p. 480.

V. P. KOVERDA, V. N. SKOKOV, and V. P. SKRIPOV
*Liquid–Solid Phase Transition in Discontinuous Metal Films
(a) Vol. **74**, No. 1, p. 343–351 (1982).
See also *1.5*, p. *41*; 21, p. 341.

K. PARLIŃSKI
*On the Phase Transition in Adamantane
(a) Vol. **74**, No. 1, p. K37–K40 (1982).
See also 22.9, p. 508.

V. I. ZINENKO
*The Cluster Investigation of the q-State Potts Model
(b) Vol. **114**, No. 2, p. 695–703 (1982).
See also 4, p. 65; 8, p. 90.

I. O. BASHKIN, V. F. DEGTYAREVA, YU. M. DERGACHEV, and E. G. PONYATOVSKII
*The Structural Transition from NaCl to CsCl Type in Cesium Hydride
(b) Vol. **114**, No. 2, p. 731–734 (1982).
See also 22.5, p. 452.

L. G. KHVOSTANTSEV and N. A. NIKOLAEV
*Investigation of the $\alpha \rightleftharpoons \alpha'$ Transformation of Cerium under Pressure
(b) Vol. **114**, No. 2, p. K135–K137 (1982).
See also 21.4, p. 372.

L. V. CHOUDARY, J. V. RAO, P. N. MURTY, and C. R. K. MURTY
*Phase Transition Studies of the Liquid Crystal H_pBBA
(a) Vol. **74**, No. 2, p. 431–435 (1982).
See also *2*, p. *55*; 22.9, p. 508.

L. KOMOROWSKI, A. CHYLA, and R. KOWAL
*Phase Transitions in n-Alkyltriphenylphosphonium Tetracyanoquinodimethanides
(a) Vol. **74**, No. 2, p. 453–457 (1982).
See also 8, p. 90; 14.3, p. 212; 19, p. 291; 22.9, p. 508.

J. ALBERS, A. KLÖPPERPIEPER, H. J. ROTHER, and K. H. EHSES
*Antiferroelectricity in Betaine Phosphate
(a) Vol. **74**, No. 2, p. 553–557 (1982).
See also *14.4.1*, p. *244*; 20.1, p. 316; 22.8.1, p. 498.

1.3 Alloys. Metallurgy

Review Article

F. REYNAUD
*Order–Disorder Transitions in Substitutional Solid Solutions
(a) Vol. **72**, No. 1, p. 11–59 (1982).
See also *1.2*, p. *14*; 21, p. 330; 21.1, p. 343; 21.1.1, p. 359; 21.6, p. 374.

Original Papers and Short Notes

H. BAKKER
*The Vibrational Entropy Change Associated with the Disordering of a Two-Dimensional Binary Alloy Using a Model of Nearest- and Next-Nearest-Neighbour Interactions
(b) Vol. **109**, No. 1, p. 211–216 (1982).
See also 4, p. 61; 8, p. 84; 21, p. 330.

YU. S. STARK and A. S. STERNBERG
*Application of BBGKY Equations to the Study of Ordering in Alloys
(b) Vol. **109**, No. 1, p. 403–409 (1982).
See also 1.2, p. 14; 21, p. 331.

G. VAN TENDELOO and S. AMELINCKX
*High Resolution Electron Microscopic and Electron Diffraction Study of the Au–Mg System (II): The Y-Phase and Some Observations on the $Au_{77}Mg_{23}$ Phase
(a) Vol. **69**, No. 1, p. 103–120 (1982).
See also *1.4*, p. *32*; 21.3, p. 367; 21.6, p. 374.

N. SAITO and T. YAMAGUCHI
Optical and Compositional Properties of Amorphous $Si_xC_yF_z$ Films Prepared by Reactive Sputtering
(a) Vol. **69**, No. 1, p. 133–138 (1982).
See also *2*, p. *45*; *20.1*, p. *299*; 22.1.2, p. 400.

L. B. DUBROVSKAYA, S. Z. NAZAROVA, and A. F. PREKUL
*Superconducting and Normal Properties of Solid Solutions $Nb_{1-x}Ta_xC_y$
(a) Vol. **69**, No. 1, p. 167–172 (1982).
See also 14.2, p. 199; 14.3, p. 203; 18, p. 259; 21, p. 331.

M. HIDA, H. MAEDA, N. KAMIJO, and H. TERAUCHI
*EXAFS Study on the Local Structure of Copper Crystals Deformed by Heavy Cold Working
(a) Vol. **69**, No. 1, p. 297–303 (1982).
See also 21.1, p. 344.

G. COCCO, L. SCHIFFINI, S. ENZO, and A. BENEDETTI
*A SAXS Study of Coarsening Mechanism of GP Zones in the Al–Li–Mg System
(a) Vol. **69**, No. 1, p. 343–348 (1982).
See also 10.1, p. 96; 21, p. 331; 21.2, p. 365; 21.3, p. 367.

J. TEUHO and K. MÄKINEN
*On the Out-of-Step Period in $Au_3Zn(H)$
(a) Vol. **70**, No. 1, p. K1–K3 (1982).
See also *1.4*, p. *33*; 21.1, p. 346; 21.6, p. 375.

H. KIDO, M. SHIMADA, and M. KOIZUMI
*Synthesis and Magnetic Properties of GdCoSi and GdMnSi
(a) Vol. **70**, No. 1, p. K23–K26 (1982).
See also *18.2*, p. *269*; 21.1, p. 346; 21.4, p. 370.

V. I. BOGDANOV, A. V. RUBAN, and D. L. FUKS
*Covalent Contribution to the Bond Energy of Phases with Ll_2 Structure
(b) Vol. **110**, No. 2, p. 431–435 (1982).
See also 1.2, p. 16; *4*, p. *62*; 21.1, p. 346; 21.6, p. 375.

V. PRABHAWALKAR and B. D. PADALIA
X-Ray Spectroscopic Study of Rare Earth–Gallium Intermetallics RGa_2
(b) Vol. **110**, No. 2, p. 659–668 (1982).
See also *20*, p. *294*; 21.4, p. 370.

V. I. CHERVINSKII, M. M. KANTOR, I. I. NOVIKOV, and R. M. SOFRONOVA
*The Nature of Planar Faults in a Dilute Molybdenum–Boron Alloy
(a) Vol. **70**, No. 2, p. 379–386 (1982).
See also 10.1, p. 98; 21, p. 334.

J. VAN LANDUYT, G. VAN TENDELOO, and S. AMELINCKX
*New Types of Periodic Antiphase Boundary Structures Derived from the DO_{22} Structure in the Gold–Manganese System
(a) Vol. **70**, No. 2, p. 407–414 (1982).
See also 1.4, p. 33; 21.1, p. 347; 21.6, p. 375.

A. I. LOTKOV, V. N. GRISHKOV, and V. V. FADIN
*The Influence of Palladium on the Martensitic Transformation of the Intermetallic Compound TiNi
(a) Vol. **70**, No. 2, p. 513–517 (1982).
See also *14.1*, p. *194*; 21.1, p. 347.

E. R. DE GIL
*Tenary Semiconducting Compounds with Chalcopyrite-Type Structure (I): Fundamental Equation and Its Physically Acceptable Solutions
(a) Vol. **70**, No. 2, p. 519–523 (1982).
See also *1.1*, p. *11*; 22.8, p. 484.

R. KUMAR, VISHWAMITTAR, K. CHANDRA, and S. K. DATE
*Mössbauer and Magnetic Studies of $Fe_{0.968-x}Mn_xSi_{0.032}$ Ternary Systems
(a) Vol. **70**, No. 2, p. 545–553 (1982).
See also 5, p. 67; 12.1, p. 131; 18, p. 260; 21.1.1, p. 360.

BAI-XIN LIU and M.-A. NICOLET
*Formation of Au–Ge Metastable Phases by Ion Mixing
(a) Vol. **70**, No. 2, p. 671–676 (1982).
See also *1.2*, p. *16*; 2, p. 48; 21.6, p. 376; 22.1.1, p. 396.

E. A. Vasilev and V. A. Virchenko
*Solid Solutions in the $(Fe_{1.22}Sb)_{1-x}(Fe_{1.68}Sn)_x$ System
(a) Vol. **70**, No. 2, p. K141–K143 (1982).
See also *18.2*, p. *270*; 21.1.1, p. 360.

I. Z. Rahman, D. Melville, and W. I. Khan
*Y–Co Alloys: Magnetic Identification of a Y_5Co_{19} Phase
(a) Vol. **70**, No. 2, p. K175–K179 (1982).
See also *18.2*, p. *270*; 21.1, p. 348; 21.4, p. 370.

P. Singh and K. N. Khanna
*Entropy of Mixing of Liquid Alloys
(b) Vol. **111**, No. 1, p. 111–116 (1982).
See also 2, p. 48; 8, p. 86; 21, p. 334; 21.1, p. 348.

I. G. Ratishvili
*The Low-Temperature Stable States of Some Transition Metal Hydrides
(a) Vol. **71**, No. 1, p. 223–232 (1982).
See also 4, p. 62; 21, p. 335.

T. Ikari, S. Shigetomi, and K. Hashimoto
Crystal Structure and Raman Spectra of InSe
(b) Vol. **111**, No. 2, p. 477–481 (1982).
See also *6*, p. *74*; *20.1*, p. *305*; 22.3, p. 433.

V. I. Bogdanov, A. V. Mikhailov, and D. L. Fuks
*Thermodynamics of Aluminium- and Magnesium-Base Interstitial Solutions (I): Analysis of Interaction Parameters
(b) Vol. **111**, No. 2, p. 483–489 (1982).
See also 21, p. 335; 21.3, p. 367.

T. Soma, Y. Kimura, and H.-Matsuo Kagaya
*Elastic-Stiffness Constants under Pressure of Cs–K, Cs–Rb, and Rb–K Solid Solutions
(b) Vol. **111**, No. 2, p. 679–683 (1982).
See also *12.1*, p. *131*; 21.2, p. 365.

B. Christ, G. Oelgart, and R. Stegmann
**ZAF* Correction for Electron Probe Microanalysis of $Bi_{1-x}Sb_x$ Alloys
(a) Vol. **71**, No. 2, p. 463–471 (1982).
See also 11, p. 122; 21.7, p. 380.

O. Bertrand, Ph. Dufour, N. Floquet, and L. C. Dufour
*Preliminary Electron Microscopy Investigation of the Reduction into MoO_2 of the Higher Molybdenum Oxides. Mo_4O_{11} and CS Oxides $Mo_{18}O_{52}$, Mo_9O_{26}, Mo_8O_{23}
(a) Vol. **71**, No. 2, p. 511–521 (1982).
See also *1.2*, p. *18*; 22.6, p. 472.

G. A. Scholz, R. F. Frindt, and A. E. Curzon
*Electron Diffraction Investigation of the Ag_xTaS_2 System (I): Superlattices in $Ag_{1/3}TaS_2$ and $Ag_{2/3}TaS_2$
(a) Vol. **71**, No. 2, p. 531–542 (1982).
See also 22.8, p. 485.

A. V. Andreev, E. N. Tarasov, A. V. Deryagin, and S. M. Zadvorkin
Influence of Symmetry of the Crystalline Structure on the Magnetic Anisotropy of Tb_2Co_7
(a) Vol. **71**, No. 2, p. K245–K247 (1982).
See also *1.2*, p. *19*; *18.2*, p. *271*; 21.1, p. 349; 21.4, p. 371.

V. I. Bogdanov, A. V. Mikhailov, and D. L. Fuks
*Thermodynamics of Aluminium- and Magnesium-Base Interstitial Solutions (II): Energy of Mixing of Binary and Ternary Alloys
(b) Vol. **112**, No. 1, p. 171–178 (1982).
See also 21, p. 336; 21.3, p. 367.

V. S. Urusov and V. B. Dudnikova
*Energy of Anion Polarisation in Layered Structures
(b) Vol. **112**, No. 1, p. 303–310 (1982).
See also *4*, p. *63*; 22.5.2, p. 458; 22.5.3, p. 465.

P. Colombet, M. Tremblet, and M. Danot
*Magnetic Properties of the Diluted B-Spinels $Cu_yCr_y^{III}M_{2-y}^{IV}S_4(M^{IV} = Sn, Ti)$
(a) Vol. **72**, No. 1, p. 105–116 (1982).
See also 5, p. 67; *18*, p. *261*; 19, p. 288; 22.8.2, p. 500.

H. van Swijgenhoven, L. M. Stals, and K. H. J. Buschow
*Phase Transformations in Amorphous Zr–Ni Alloys
(a) Vol. **72**, No. 1, p. 153–160 (1982).
See also *2*, p. *50*; 21.1, p. 350.

I. M. Reda, J. Hafner, P. Pongratz, A. Wagendristel, H. Bangert, and P. K. Bhat
*Amorphous Cu–Ag Films with High Stability
(a) Vol. **72**, No. 1, p. 313–324 (1982).
See also *2*, p. *50*; 14.1, p. 196; 21.1, p. 351; 21.6, p. 377.

Q. B. Yang
*Coincidence Coefficients of Two Space Lattices and Their Lattice Planes
(a) Vol. **72**, No. 1, p. 343–351 (1982).
See also 21.1, p. 351; 21.1.1, p. 361.

G. A. Scholz, R. F. Frindt, and A. E. Curzon
*Electron Diffraction Investigation of the Ag_xTaS_2 System (II): Superlattices, Structure and Charge Density Waves in Ag_xTaS_2
(a) Vol. **72**, No. 1, p. 375–390 (1982).
See also 22.8, p. 487.

S. Matthies
*Form Effects in the Description of the Orientation Distribution Function (ODF) of Texturized Materials by Model Components
(b) Vol. **112**, No. 2, p. 705–716 (1982).
See also *1.4*, p. *35*.

T. Soma, Y. Kimura, and H. M. Kagaya
*Equation of State of the Alloy System Composed of Alkali Metal Elements
(b) Vol. **112**, No. 2, p. K107–K109 (1982).
See also 12.1, p. 132; 21.2, p. 366.

S. Matthies and G. W. Vinel
*On the Reproduction of the Orientation Distribution Function of Texturized Samples from Reduced Pole Figures Using the Conception of a Conditional Ghost Correction
(b) Vol. **112**, No. 2, p. K111 – K114 (1982).
See also *1.4*, p. *35*; 21.1, p. 351.

S. Matthies and G. W. Vinel
*An Example Demonstrating a New Reproduction Method of the ODF of Texturized Samples from Reduced Pole Figures
(b) Vol. **112**, No. 2, p. K115 – K120 (1982).
See also *1.4*, p. *35*; 21.1, p. 351.

W. Hoyer, A. Müller, E. Thomas, and M. Wobst
X-Ray-Diffraction Investigation on Molten In–Te Alloys
(a) Vol. **72**, No. 2, p. 585 – 592 (1982).
See also *2*, p. *51*; 22.7, p. 479.

J. M. López and J. A. Alonso
*Semiempirical Study of Metastable Alloys Obtained by Ion Implantation in Metals and Semiconductors
(a) Vol. **72**, No. 2, p. 777 – 781 (1982).
See also *11*, p. *125*; 21, p. 338; 22, p. 389.

G. V. S. Sastry, C. Suryanarayana, and G. van Tendeloo
*A Structural Study of Vapour-Deposited Al–Pd Alloys
(a) Vol. **73**, No. 1, p. 267 – 278 (1982).
See also 1.4, p. 36; 21, p. 339; 21.1, p. 354.

A. V. Andreev, A. V. Deryagin, V. N. Moskalev, and N. V. Mushnikov
*On Crystalline Structure of Hydrides of $ErFe_2$ and $HoFe_2$
(a) Vol. **73**, No. 1, p. K69 – K71 (1982).
See also *1.2*, p. *21*; 18.2, p. 272; 21.1.1, p. 362; 21.4, p. 371.

V. I. Malyutin, V. E. Osukhovskii, Yu. D. Vorobiev, and A. S. Abramov
*The Structure Constant and Micromagnetic Structure of Etched $Ni_{80}Fe_{20}$ Films
(a) Vol. **73**, No. 1, p. K111 – K114 (1982).
See also *18.2.1*, p. *275*; 21.1, p. 354; 21.1.1, p. 362.

S. Matthies and K. Helming
*General Consideration of the Loss of Information on the Orientation Distribution Function of Texturized Samples in Pole Figure Measurements
(b) Vol. **113**, No. 2, p. 569 – 582 (1982).
See also *1.4*, p. *36*; 21.1, p. 354.

B. Lorenz and D. Ihle
*Band Effects on Charge Ordering in Fe_3O_4
(b) Vol. **113**, No. 2, p. 601 – 607 (1982).
See also *13*, p. *145*; 22.6, p. 474.

M. Eizenberg
*Silicide Formation Resulting from the Interfacial Reaction of Silicon and Thin Films of Ir–V Alloys and Bilayers
(a) Vol. **73**, No. 2, p. 483 – 495 (1982).
See also *14.3.4*, p. *232*; 21.1, p. 355; 22.1.2, p. 413.

R. A. Jago and P. L. Rossiter
*A Re-Examination of Long-Range Ordering at Fe–25 at% Ni
(a) Vol. **73**, No. 2, p. 497–502 (1982).
See also 1.4, p. 36; 21.1, p. 355; 21.1.1, p. 362; 22.8.2, p. 501.

D. E. G. Williams
*Spatial Resolution and Dense Random Packing
(a) Vol. **73**, No. 2, p. 539–544 (1982).
See also *2*, p. *53*; 21, p. 340.

B. Rönnefarth, R. Mattheis, and H. Dintner
*Deposition and Structure of Different Composite Ni–Cr Thin Films
(a) Vol. **73**, No. 2, p. K153–K156 (1982).
See also *1.5*, p. *41*; 2, p. 53; 21.1, p. 355.

R. V. Nandedkar and P. Delavignette
*On the Formation of a New Superstructure in the Zirconium–Aluminium System
(a) Vol. **73**, No. 2, p. K157–K160 (1982).
See also 21, p. 340.

R. A. Jago, P. E. Clark, and P. L. Rossiter
*The Santa Catharina Meteorite and the Equilibrium State of Fe–Ni Alloys
(a) Vol. **74**, No. 1, p. 247–254 (1982).
See also *5*, p. *68*; 18.2, p. 273; 21.1, p. 356; 21.1.1, p. 363.

R. Gerlach and H. Löffler
*Clustering Processes Proceeding in Two Al–2.0 at% Zn–1.3 at% Mg Alloys after Direct Quench to the Ageing Temperature
(a) Vol. **74**, No. 2, p. 517–528 (1982).
See also *10.1*, p. *102*; 21, p. 342; 21.1, p. 358; 21.3, p. 368.

S. G. Fries, J. A. H. da Jornada, and A. Maciel
*Experimental Study of the Electric Field Gradient in In_5Bi_3
(a) Vol. **74**, No. 2, p. 661–666 (1982).
See also *13.4*, p. *178*; 22.7, p. 480.

A. Baldan and D. R. F. West
*The Strength of Certain Ni–Al–Ta and Ni–Al–Hf Alloys Containing γ' and β-Phases
(a) Vol. **74**, No. 2, p. 705–713 (1982).
See also *12.1*, p. *133*; 21, p. 342; 21.1, p. 358.

V. K. Singh, M. Singh, and S. Bhan
*Metastable Phases in $Fe_{75}Sn_{25}$–$Co_{75}Sn_{25}$ Alloys
(a) Vol. **74**, No. 2, p. K115–K117 (1982).
See also 21.1.1, p. 364.

1.4 Microstructure
(Magnetic Domains See 18; Ferroelectric Domains See 14.4.1)

Original Papers and Short Notes

G. van Tendeloo and S. Amelinckx
*High Resolution Electron Microscopic and Electron Diffraction Study of the Au–Mg System (II): The Y-Phase and Some Observations on the $Au_{77}Mg_{23}$ Phase
(a) Vol. **69**, No. 1, p. 103–120 (1982).
See also *1.3*, p. *25*; 21.3, p. 367; 21.6, p. 374.

C. Manolikas
*Electron Microscopic Study of the Ordered and Disordered Phases in Ga_2Se_3
(a) Vol. **69**, No. 1, p. 393–405 (1982).
See also *1.3*, p. *26*; 22, p. 383.

S. K. Maksimov and E. N. Nagdaev
*Electron Microscopy Images of Composition-Modulated Crystals (II): Dilatational Contrast Images
(a) Vol. **69**, No. 2, p. 505–512 (1982).
See also 10.2, p. 105; 22.2.4, p. 430.

P. K. Madden and S. B. Fisher
*Void Swelling Studies of Stainless Steel Containing Silicon and Titanium
(a) Vol. **69**, No. 2, p. 569–588 (1982).
See also 11, p. 119; 21.1.1, p. 359.

G. van Tendeloo and S. Amelinckx
Electron Diffraction and Electron Microscopic Study of the Alloy Au_2Mn and Its Formation Mechanism
(a) Vol. **69**, No. 2, p. 589–602 (1982).
See also *1.3*, p. *26*; 21.1, p. 344; 21.6, p. 375.

L. Castellani, P. Gondi, C. Patuelli, and R. Berti
*On the EBIC Contrast of Dislocations in Si
(a) Vol. **69**, No. 2, p. 677–685 (1982).
See also *10.2*, p. *105*; 22.1.2, p. 402.

H. K. D. H. Bhadeshia
Diffusional Transformations: A Theory for the Formation of Superledges
(a) Vol. **69**, No. 2, p. 745–750 (1982).
See also *1.3*, p. *26*; 22.8.2, p. 499.

A. Hauet, P. Duval et P. Matletras
*Ordre à courte distance et précipitation cohérente dans la phase α d'un cupro–aluminium (Short-Range Order and Coherent Precipitation in the α-Phase of Copper–Aluminium Alloy)
(a) Vol. **69**, No. 2, p. K135–K138 (1982).
See also *1.3*, p. *26*; 21, p. 332; 21.1, p. 345.

R. Hamar et J. P. Riquet
*Estimation de l'énergie d'un joint de phases métalliques
(Estimate of the Boundary Energy of Metallic Phases)
(a) Vol. **70**, No. 1, p. 127–138 (1982).
See also *10.1*, p. *97*; 21, p. 333; 21.1, p. 346.

G. VAN TENDELOO, J. VAN LANDUYT, and S. AMELINCKX
High Resolution Electron Microscopic and Electron Diffraction Study of Gold-Manganese Alloys with a Composition in the Vicinity of Au_3Mn
(a) Vol. **70**, No. 1, p. 145–158 (1982).
See also *1.3*, p. *26*; 21.1, p. 346; 21.6, p. 375.

L. PASEMANN, H. BLUMTRITT, and R. GLEICHMANN
*Interpretation of the EBIC Contrast of Dislocations in Silicon
(a) Vol. **70**, No. 1, p. 197–209 (1982).
See also *10.2*, p. *106*; 14.3.3, p. 220; 22.1.2, p. 403.

J. TEUHO and K. MÄKINEN
*On the Out-of-Step Period in $Au_3Zn(H)$
(a) Vol. **70**, No. 1, p. K1–K3 (1982).
See also *1.3*, p. *27*; 21.1, p. 346; 21.6, p. 375.

E. F. KRIMMEL
Slip Line Free Silicon in Large-Area Multiple-Scan Annealing with a Line-Focused Electron Beam
(a) Vol. **70**, No. 1, p. K63–K65 (1982).
See also *11*, p. *120*; 22.1.2, p. 404.

H. KUWAMOTO
*Domain Effects Accompanying the Metal-Insulator Transition in $(V_{0.99}Cr_{0.01})_2O_3$
(b) Vol. **110**, No. 2, p. K175–K178 (1982).
See also *14*, p. *191*; 22.6.1, p. 477.

R. I. KUZNETSOVA, N. N. ZHUKOV, O. A. KAIBYSHEV, and R. Z. VALIEV
*Mechanism of Superplastic Deformation of Coarse-Grained Materials
(a) Vol. **70**, No. 2, p. 371–378 (1982).
See also *10.1*, p. *98*; 21, p. 334.

J. VAN LANDUYT, G. VAN TENDELOO, and S. AMELINCKX
New Types of Periodic Antiphase Boundary Structures Derived from the DO_{22} Structure in the Gold-Manganese System
(a) Vol. **70**, No. 2, p. 407–414 (1982).
See also *1.3*, p. *27*; 21.1, p. 347; 21.6, p. 375.

D. LEPSKI and P. BURCK
*Precipitate Structure Analysis by Diffraction Contrast Image Matching for a Metastable Semicoherent Carbide Phase in Molybdenum (II): Image Matching
(a) Vol. **70**, No. 2, p. 571–581 (1982).
See also *1.2*, p. *16*; 10.1, p. 98; 12.1, p. 131; 21.1, p. 347.

C. BOULESTEIX et B. YANGUI
*Domains et interfaces engendrées au cours d'une transformation cristallographique coopérative – application au cas de la transformation hexagonale monoclinique des sesquioxydes de terres rares (I): Dénombrement et position relative des domains équivalents (Domains and Interfaces Created during Cooperative Crystallographic Transformation – Application to Hexagonal-Monoclinic Transformation of Rare Earth Sesquioxides (I): Determination of the Number and Relative Position of Equivalent Domains)
(a) Vol. **70**, No. 2, p. 597–613 (1982).
Erratum
(a) Vol. **74**, No. 1, p. 361 (1982).
See also *1.2*, p. *16*; 22.6, p. 470.

S. K. Maksimov and E. N. Nagdaev
*Electron Microscopy Images of Composition-Modulated Crystals (III): Moiré Fringe Images
(a) Vol. **72**, No. 1, p. 135–145 (1982).
See also 10.2, p. 110; 22.2.4, p. 430.

S. Matthies
*From Effects in the Description of the Orientation Distribution Function (ODF) of Texturized Materials by Model Components
(b) Vol. **112**, No. 2, p. 705–716 (1982).
See also *1.3*, p. *29*.

S. Matthies and G. W. Vinel
*On the Reproduction of the Orientation Distribution Function of Texturized Samples from Reduced Pole Figures Using the Conception of a Conditional Ghost Correction
(b) Vol. **112**, No. 2, p. K111–K114 (1982).
See also *1.3*, p. *30*; 21.1, p. 351.

S. Matthies and G. W. Vinel
*An Example Demonstrating a New Reproduction Method of the ODF of Texturized Samples from Reduced Pole Figures
(b) Vol. **112**, No. 2, p. K115–K120 (1982).
See also *1.3*, p. *30*; 21.1, p. 351.

L. U. Ogbuji
*Origin of Long-Period Polytypism in Polycrystalline SiC
(a) Vol. 72, No. 2, p. 455–461 (1982).
See also *1.1*, p. *12*; 22.3, p. 434.

L. N. Bezmaternykh, V. G. Mashchenko, N. A. Sokolova, and V. L. Temerov
*Interaction of Magnetostatic and Magnetoelastic Waves with Growth Inhomogeneities in Iron Garnet Single Crystals
(a) Vol. **72**, No. 2, p. 469–474 (1982).
See also *18*, p. *262*; 22.8.2, p. 501.

E. Ziegler, W. Siegel, H. Blumtritt, and O. Breitenstein
Electrical and EBIC Investigations of GaP Grain Boundaries
(a) Vol. **72**, No. 2, p. 593–605 (1982).
See also *14.3.3*, p. *223*; 22.2.2, p. 425.

B. Yangui
*Mise en évidence de nouvelles macles par microscopie électronique conventionelle dans les sesquioxydes de terres rares monocliniques (B–Ln_2O_3)
(Electron Microscopic Evidence of New Twins in Monoclinic Rare-Earth Sesquioxides (B–Ln_2O_3))
(a) Vol. **72**, No. 2, p. 631–643 (1982).
See also *1.2*, p. *20*; 22.6, p. 473.

A. N. Anagnostopoulos, C. Manolikas, D. Papadopoulos, and J. Spyridelis
Composition Faults in $ZnIn_2S_4$(III) Layered Crystals and Their Influence on the Anisotropic Conductivity of This Compound
(a) Vol. **72**, No. 2, p. 731–736 (1982).
See also *14.3*, p. *209*; 22.8, p. 488.

M. S. Bapna
*An Estimate of the Precipitation Hardening by Coherent Particles
(a) Vol. **72**, No. 2, p. 799–807 (1982).
See also *10.1*, p. *100*; 21, p. 338; 21.1, p. 352; 21.6, p. 377.

P. G. Self, M. P. Shaw, and W. M. Stobbs
*Anomalous Contrast in Weak Beam Images of Stacking Faults. Observation and Calculation
(a) Vol. **73**, No. 1, p. 37–48 (1982).
See also 21, p. 339; 21.1, p. 353; 21.1.1, p. 362.

F. Watari, J. van Landuyt, P. Delavignette, S. Amelinckx, and N. Igata
*X-Ray Peak Broadening as a Result of Twin Formation in Some Oxides Derived by Dehydration
(a) Vol. **73**, No. 1, p. 215–224 (1982).
See also 1.2, p. 21; 22.6, p. 474.

J. Singh and S. Ranganathan
*On the Mechanism of F.C.C. → H.C.P. Transformation
(a) Vol. **73**, No. 1, p. 243–248 (1982).
See also *1.2*, p. *21*; 21.1, p. 354; 21.1.1, p. 362.

G. V. S. Sastry, C. Suryanarayana, and G. van Tendeloo
A Structural Study of Vapour-Deposited Al–Pd Alloys
(a) Vol. **73**, No. 1, p. 267–278 (1982).
See also *1.3*, p. *30*; 21, p. 339; 21.1, p. 354.

I. G. Salisbury
*On the Effect of Lattice Dilatation upon Weak-Beam Contrast from Stacking Faults in Transmission Electron Microscopy
(a) Vol. **73**, No. 1, p. K11–K13 (1982).
See also 21.1, p. 354; 22.1.2, p. 412.

M. Wilkens, H. Föll, and C. B. Carter
*A Further Comment on the Weak-Beam Contrast of Stacking Faults in Silicon
(a) Vol. **73**, No. 1, p. K15–K19 (1982).
See also 22.1.2, p. 412.

S. Matthies and K. Helming
*General Consideration of the Loss of Information on the Orientation Distribution Function of Texturized Samples in Pole Figure Measurements
(b) Vol. **113**, No. 2, p. 569–582 (1982).
See also *1.3*, p. *30*; 21.1, p. 354.

G. S. Kandaurova, L. G. Onoprienko, and N. I. Sokolovskaya
*The Effect of Twin Microstructure on the Magnetic Properties of Alloys with CuAuI Superlattice Ordering
(a) Vol. **73**, No. 2, p. 351–360 (1982).
See also *18.2*, p. *272*; 21.1, p. 355; 21.1.1, p. 362.

M. J. Marcinkowski
Unified Theory of Grain Boundaries. An Update
(a) Vol. **73**, No. 2, p. 409–419 (1982).
See also *10*, p. *95*.

R. A. Jago and P. L. Rossiter
A Re-Examination of Long-Range Ordering at Fe–25 at% Ni
(a) Vol. **73**, No. 2, p. 497–502 (1982).
See also *1.3*, p. *31*; 21.1, p. 355; 21.1.1, p. 362; 22.8.2, p. 501.

B. Yangui et C. Boulesteix
*Domaines et interfaces engendrées au cours d'une transformation cristallographique coopérative — application au cas de la transformation hexagonale monoclinique des sesquioxydes de terres rares (II): Position relative et raccordement des variantes de domaines de rotation et de translation: raccordement en étoile
(Domains and Interfaces Created during Cooperative Crystallographic Transformation — Application to Hexagonal-Monoclinic Transformation of Rare Earth Sesquioxides (II): Relative Position and Connection of the Variants of Rotation and Translation Domains: Connection into Stars)
(a) Vol. **73**, No. 2, p. 515–527 (1982).
See also *1.2*, p. *22*; 22.6, p. 475.

F. Delannay
*On the Reduction of Orthorhombic MoO_3 to MoO_2
(a) Vol. **73**, No. 2, p. 529–537 (1982).
See also *1.2*, p. *22*; 22.6, p. 475.

V. T. Bublik, J. Wilke und A. T. Pereversev
*Die Natur der festen Lösungen der überschüssigen Komponenten in unlegiertem Galliumantimonid
(The Natur of Solid Solutions of the Excess Components in Non-Doped Gallium Antinonide)
(a) Vol. **73**, No. 2, p. K271–K273 (1982).
See also *1.1*, p. *13*; 22.2.3, p. 429.

H. Endert and W. Melle
*Laser-Induced Damage in KDP Crystals. The Influence of Growth Ghosts and Growth Bands
(a) Vol. **74**, No. 1, p. 141–148 (1982).
See also *11*, p. *127*; 20.2, p. 317; 22.8.1, p. 498.

I. G. Salisbury
*Weak-Beam Contrast from Stacking Faults in Silicon
(a) Vol. **74**, No. 1, p. 353–359 (1982).
See also 22.1.2, p. 415.

Nguyen An, G. Worm, and R. Herrmann
*Symmetrical Grain Boundaries with Special Angles and Their Coincidence Site Lattices
(b) Vol. **114**, No. 2, p. 349–357 (1982).
See also *1.1*, p. *13*.

M. Čerňanský
*Broadening and Shift of Diffraction Lines Due to Partial Coherence of X-Rays
(b) Vol. **114**, No. 2, p. 365–372 (1982).
See also *10.1*, p. *101*; 21.1, p. 357.

V. A. Melik-Shakhnazarov, I. A. Naskidashvili, V. I. Serdobintsev, and V. I. Savin
*Anelastic Effects in the β-Phase of Niobium Hydride
(a) Vol. **74**, No. 2, p. 593–602 (1982).
See also *7*, p. *83*; *12.1*, p. *133*; 21, p. 342.

J. Mahy, J. van Landuyt, and S. Amelinckx
*Diffraction Evidence for Superstructure in Vanadium Diselenide
(a) Vol. **74**, No. 2, p. K89–K91 (1982).
See also 22, p. 392.

P. Anuradha, M. Ramakrishna Murthy and I. V. K. Bhagavan Raju
*Study of Growth and Cross Defects of $NaNO_3$ Single Crystals
(a) Vol. **74**, No. 2, p. K93–K96 (1982).
See also *3*, p. *60*; 22.8.1, p. 498.

1.5 Films

Review Article

Z. M. Jarzębski
Preparation and Physical Properties of Transparent Conducting Oxide Films
(a) Vol. **71**, No. 1, p. 13–41 (1982).
See also 3, p. 57; 13.1, p. 148; *14.3.1*, p. *213*; *20.1*, p. *298*; 22.6, p. 468; 22.6.1, p. 477.

Original Papers and Short Notes

I. E. Bolotov and V. Yu. Kolosov
*Electron Microscope Investigation of Crystals Based on Bend-Contour Arrangement (I): Relationship between Bend-Contour Arrangement and Bend Geometry
(a) Vol. **69**, No. 1, p. 85–96 (1982).
See also 10.2, p. 103; 22.1.3, p. 417.

M. P. Shearer, S. K. Sen, and C. L. Bauer
*Investigation of Interfacial Reactions in Bimetallic Thin-Film Couples of Al/Cu and Pb/Cu by Measurement of Contact Resistance
(a) Vol. **69**, No. 1, p. 139–149 (1982).
See also 9, p. 91; 14.1, p. 193; 21, p. 331; 21.1, p. 343.

G. Eichhorn and U. Rensch
*MOCVD of AlN on Silicon
(a) Vol. **69**, No. 1, p. K3–K6 (1982).
See also *3*, p. *57*; 22.2, p. 418.

E. Harnik, S. Kovnovich, and T. Chernobelskaya
Secondary Acoustoelectric Current in Ultrathin Metal Films Deposited on a Piezoelectric Substrate
(a) Vol. **69**, No. 2, p. K179–K181 (1982).
See also *7*, p. *82*; 21.6, p. 375.

A. V. Dvurechenskii, B. P. Kashnikov, L. S. Smirnov, R. Grötzschel, and R. Klabes
Synthesis of Solid Solutions at Electron Pulse Processing of Ge–Si Structures
(a) Vol. **69**, No. 2, p. K183–K187 (1982).
See also *3*, p. *57*; *11*, p. *119*; 22.1.1, p. 395; 22.1.2, p. 402.

V. V. Nemoshkalenko, A. G. Lesnik, Yu. V. Kudryavtsev, I. V. Lezhnenko, L. I. Nikolaev, and L. S. Voskrekasenko
*Electronic Structure of Fe–45 Al Films
(b) Vol. **110**, No. 1, p. 239–243 (1982).
See also *13*, p. *141*; 20, p. 294; 21.1.1, p. 360.

W. Kleinn
*Restwiderstandsbestimmung aus Strom–Spannungs-Kurven kurzer freitragender Elektrotransportproben epitaktischer Goldschichten
(Determination of the Residual Resistivity from Current–Voltage Characteristics of Short Support-Free Samples of Epitaxial Gold Films)
(a) Vol. **70**, No. 1, p. 219–232 (1982).
See also *14.1*, p. *194*; 21.6, p. 375.

H. Müller
*Simulation of Oriented Crystallization in Evaporated Antimony Thin Films
(a) Vol. **70**, No. 1, p. 249–255 (1982).
See also *3*, p. *57*; 21.7, p. 379.

K. Ishida, Y. Matsumoto, and K. Taguchi
*Lattice Defects in LPE InP–InGaAsP–InGaAs Structure Epitaxial Layers on InP Substrates
(a) Vol. **70**, No. 1, p. 277–286 (1982).
See also *10.2*, p. *106*; 22.2.4, p. 430.

O. Bostanjoglo
Time-Resolved TEM of Pulsed Crystallization of Amorphous Si and Ge Films
(a) Vol. **70**, No. 2, p. 473–481 (1982).
See also *3*, p. *57*; 22.1.1, p. 396; 22.1.2, p. 405.

W. Andrä and H. Danan
*Interface Energy as an Origin of Intrinsic Stress in Thin Films
(a) Vol. **70**, No. 2, p. K145–K149 (1982).
See also *12*, p. *129*.

H.-D. Geiler, F. Thrum, and G. Götz
*Pb-Silicide Reactions Induced by a Millisecond Laser Pulse
(a) Vol. **70**, No. 2, p. K159–K162 (1982).
See also *11*, p. *121*; 22.1.2, p. 406.

C. Ammer
*LEED Intensity Analysis of Epitaxial Ag/Au Systems (I): The Influence of the Single Ion–Core Scattering on the LEED Spectra
(a) Vol. **71**, No. 1, p. 135–144 (1982).
See also *1.6*, p. *42*; 21.6, p. 376.

A. D. Pogrebnyak, V. A. Kuzminikh, and K. P. Arefev
*Positron Profiles and Positron Annihilation in Thin Layers
(a) Vol. **71**, No. 1, p. 145–151 (1982).
See also *1.6*, p. *43*; 21, p. 335; 22.2.1, p. 421.

K. Bewilogua, D. Dietrich, G. Holzhüter, and Ch. Weissmantel
*Structure of Amorphous Carbon Films
(a) Vol. **71**, No. 1, p. K57–K59 (1982).
See also *2*, p. *49*; 22.1, p. 393.

C. Ammer, M. Klaua, and H. Bethge
*LEED Intensity Analysis of Epitaxial Ag/Au Systems (II): LEED Study of the Initial Stages of Epitaxy
(a) Vol. **71**, No. 2, p. 415–428 (1982).
See also *1.6*, p. *43*; 21.6, p. 376.

J. Echigoya and J. W. Edington
*A Transmission Electron Microscope Study of the Chalcocite–Djurleite Transformation in Topotactically Grown Thin Films of Cu_xS
(a) Vol. **72**, No. 1, p. 305–311 (1982).
See also *1.2*, p. *19*; 22, p. 388.

BAI-XIN LIU, M.-A. NICOLET, and S. S. LAU
*Ti–Au Metallic Glasses Formed by Ion Mixing
(a) Vol. **73**, No. 1, p. 183–188 (1982).
See also 2, p. 52; *11*, p. *126*; 21.1, p. 353; 21.6, p. 377.

K. S. MAKSIMOV, L. A. BONDARENKO, and G. N. GAIDUKOV
*Stability of Composition Automodulated Complex Epitaxial Films
(a) Vol. **73**, No. 1, p. K129–K132 (1982).
See also 22.2.4, p. 431.

M. PAUNOV and A. MILCHEV
2D Phase Transitions at High Densities
(a) Vol. **73**, No. 2, p. 339–349 (1982).
See also *1.2*, p. *22*; **1.6**, p. **44**.

M. M. EL OKER, E. A. ABOU-SAIF, S. A. EL SAHHAR, and A. A. MOHAMED
*Optical Properties and Structure of Thin Indium Films
(a) Vol. **73**, No. 2, p. 389–394 (1982).
See also *20.1*, p. *313*; 21, p. 340.

B. RÖNNEFARTH, R. MATTHEIS, and H. DINTNER
*Deposition and Structure of Different Composite Ni–Cr Thin Films
(a) Vol. **73**, No. 2, p. K153–K156 (1982).
See also *1.3*, p. *31*; 2, p. 53; 21.1, p. 355.

P. GÖRNERT, M. NEVŘIVA, J. ŠIMŠOVÁ, W. ANDRÄ, W. SCHÜPPEL, P. ŠUMŠÁI, and R. BUBÁKOVÁ
*Co Containing Garnet Films with Low Magnetization
(a) Vol. **74**, No. 1, p. 107–112 (1982).
See also *18.2.1*, p. *275*; 22.8.2, p. 502.

H. HENNIG, K.-H. HECKNER, D. HIRSCH, and H. LADWIG
*The Influence of the Preparation on Layer Properties of Transparent Indium–Tin Oxide Electrodes
(a) Vol. **74**, No. 1, p. 133–139 (1982).
See also *14.3.1*, p. *216*; 22.6.1, p. 477.

V. P. KOVERDA, V. N. SKOKOV, and V. P. SKRIPOV
*Liquid–Solid Phase Transition in Discontinuous Metal Films
(a) Vol. **74**, No. 1, p. 343–351 (1982).
See also *1.2*, p. *23*; 21, p. 341.

B. I. BELEVTSEV, YU. F. KOMNIK, V. I. ODNOKOZOV, and A. V. FOMIN
*Discovery of Low-Temperature Metastable State in Amorphous Bismuth–Hydrogen Systems
(b) Vol. **114**, No. 2, p. 467–474 (1982).
See also 2, p. 55; *14.2*, p. *201*; 21.7, p. 381.

G. ANDRÄ, H.-D. GEILER, G. GÖTZ, K. H. HEINIG, and H. WOITTENNEK
*Explosive Liquid-Phase Crystallization of Thin Silicon Films during Pulse Heating
(a) Vol. **74**, No. 2, p. 511–515 (1982).
See also *3*, p. *59*; 22.1.2, p. 416.

L. V. RAO, B. S. NAIDU, and P. J. REDDY
*Structure and Optical Energy Gap Measurements in As–Te–Cd Chalcogenide Thin Films
(a) Vol. **74**, No. 2, p. K137–K140 (1982).
See also 2, p. 56; *20.1*, p. *316*; 22.8, p. 491.

1.6 Surfaces

Original Papers and Short Notes

A. D. Pogrebnyak, V. A. Kuzminikh, and K. P. Arefev
*Positron Profiles and Positron Annihilation in Thin Layers
(a) Vol. 71, No. 1, p. 145–151 (1982).
See also *1.5*, p. *39*; 21, p. 335; 22.2.1, p. 421.

J.-M. Derochette
*Adsorption of H_2 and N_2 on Ir(100) Studied by Field Emission
(a) Vol. 71, No. 1, p. K99–K102 (1982).
See also *17.1*, p. *258*; 21, p. 335.

C. Ammer, M. Klaua, and H. Bethge
*LEED Intensity Analysis of Epitaxial Ag/Au Systems (II): LEED Study of the Initial Stages of Epitaxy
(a) Vol. 71, No. 2, p. 415–428 (1982).
See also *1.5*, p. *39*; 21.6, p. 376.

V. G. Kohn, M. V. Kovalchuk, R. M. Imamov, B. G. Zakharov, and E. F. Lobanovich
*The Investigation of Disturbed Surface Layer Structure by the External Photoeffect under X-Ray Diffraction Conditions
(a) Vol. 71, No. 2, p. 603–610 (1982).
See also *17*, p. *256*; 22.1.1, p. 397; 22.1.2, p. 408; 22.2.1, p. 421.

A. Nimmrichter, H. Pfeiffer, and W. Haubenreisser
*Step Growth of Many-Component Systems
(a) Vol. 71, No. 2, p. K239–K243 (1982).
See also *3*, p. *58*; 22.8.2, p. 500.

J. Kojnok, A. Szász, L. Kertész, and A. S. Sulakov
*Surface Corrosion of Mg-Containing Al-Based Alloys
(a) Vol. 72, No. 1, p. 131–134 (1982).
See also 21, p. 337.

H. Solbrig
*On the Ewald Summation Technique for 2D Lattices
(a) Vol. 72, No. 1, p. 199–205 (1982).

H. Pfeiffer
*Master Equation Approach to Step Growth
(a) Vol. 72, No. 1, p. 239–249 (1982).
See also *3*, p. *58*.

Th. Wirth, V. Atzrodt, and H. Lange
*Composition Changes of NiSi and PtSi Due to Ar^+ Ion Bombardment Determined from AES Measurements
(a) Vol. 72, No. 1, p. K89–K93 (1982).
See also *11*, p. *124*; 21, p. 337.

S. P. Mhaskar and P. K. Naik
*Effect of Surface Relaxation on Thermal Desorption of Helium from Polycrystalline Molybdenum
(a) Vol. 72, No. 2, p. 515–520 (1982).
See also *11*, p. *124*; 21, p. 338.

H. C. Eaton, A. C. Chou, and J. C. Carter
*Topographic Analysis of Grain Boundary Dislocations in Cold-Worked Tungsten Using Field Ion Microscopy
(a) Vol. 72, No. 2, p. 817–824 (1982).
See also *10.1*, p. *100*; 17.1, p. 258; 21, p. 338.

2. Non-Crystalline State

Review Articles

H. Böttger and V. V. Bryksin
Hopping Conductivity in Ordered and Disordered Systems (III)
(b) Vol. **113**, No. 1, p. 9–49 (1982).
See also 6, p. 70; 13, p. 139; 13.5.3, p. 188; *14.3*, p. *202*; 22, p. 382.

P. P. Seregin, A. R. Regel, A. A. Andreev, and F. S. Nasredinov
A Study of Local Structure of Chalcogenide Vitreous Semiconductors and of States of Impurity Atoms in Them by Means of Mössbauer and Electron Spectroscopies
(a) Vol. **74**, No. 2, p. 373–394 (1982).
See also 1.2, p. 14; 5, p. 66; *13.4*, p. *162*; 22, p. 382.

Original Papers and Short Notes

A. Bobák
*Effect of Fluctuations of the Coordination Number on Some Properties of Amorphous Ferromagnets
(b) Vol. **109**, No. 1, p. 161–166 (1982).
See also *18.2*, p. *267*.

O. Uemura, M. Itoh, and T. Satow
*Electronic Properties of the Tl–$AgTlTe_2$ System in the Liquid State
(b) Vol. **109**, No. 1, p. 361–368 (1982).
See also *14.3*, p. *202*; 15, p. 245; 18, p. 259; 22.8, p. 481.

K. S. Dubey
Lattice Thermal Conductivity of Noncrystalline Material at Low Temperatures. Application to Borosilicate Glass
(a) Vol. **69**, No. 1, p. 121–126 (1982).
See also *6*, p. *71*; *8*, p. *84*; 22.8.1, p. 492.

N. Saito and T. Yamaguchi
*Optical and Compositional Properties of Amorphous $Si_xC_yF_z$ Films Prepared by Reactive Sputtering
(a) Vol. **69**, No. 1, p. 133–138 (1982).
See also 1.3, p. 25; *20.1*, p. *299*; 22.1.2, p. 400.

S. Ahmed and D. Sigurd
*Electrical Properties of Glow Discharge a-Si:H Thin Films Prepared with a RF Plasma Etch Unit
(a) Vol. **69**, No. 1, p. 305–314 (1982).
See also *14.3.1*, p. *213*; 22.1.2, p. 400.

A. H. Abou El Ela, N. Abdelmohsen, and H. H. A. Labib
Dielectric Relaxation Study of $CuTlSe_2$ in the Solid and Liquid Phases
(a) Vol. **69**, No. 1, p. 377–384 (1982).
See also 13.4, p. 163; *14.4*, p. *234*; 22.8, p. 481.

J. L. McLaughlin
*The WCA Method and Pseudopotentials for Simple Liquid Metals
(b) Vol. **110**, No. 1, p. K45–K50 (1982).
See also *4*, p. *62*; 21, p. 333.

L. Schumann, A. Lehmann, H. Sobotta, V. Riede, U. Teschner, and K. Hübner
Infrared Studies of Reactively Sputtered SiO_x Films in the Composition Range $0.2 \leqq x \leqq 1.9$
(b) Vol. **110**, No. 1, p. K69–K73 (1982).
See also *6*, p. *72*; *20.1*, p. *302*; 22.6, p. 469.

H. Bińczycka, O. Gzowski, L. Murawski, and J. Sawicki
Mössbauer Effect and Electrical Conductivity in TeO_2–Fe_2O_3 Glasses
(a) Vol. **70**, No. 1, p. 51–55 (1982).
See also *5*, p. *66*; *14.3*, p. *204*; 22.6.1, p. 477.

J. Kočka, J. Stuchlík, O. Štika, M. Láznička, O. Renner, and E. Krouský
*Physical Properties of a-Si:H Prepared at about 1 Torr
(a) Vol. **70**, No. 1, p. 93–99 (1982).
See also *14.3.1*, p. *214*; 20.1, p. 302; 22.1.2, p. 403.

W. Gräfe
*Description of Internal Friction Background of Glasses by Means of Migration Processes
(a) Vol. **70**, No. 1, p. 101–108 (1982).
See also *12.2*, p. *134*; 22.6, p. 469.

P. Koštial
*Surface Wave Investigations of Se Amorphous Thin Films Deposited on $LiNbO_3$
(a) Vol. **70**, No. 1, p. K55–K58 (1982).
See also *7*, p. *82*; 22.1.3, p. 417.

H. Overhof
*The Temperature Dependence of the Spin Resonance in Boron Doped a-Si:H
(b) Vol. **110**, No. 2, p. 521–530 (1982).
See also *13.4*, p. *166*; *19*, p. *285*; 22.1.2, p. 404.

M. van der Meer, R. Schuchardt, and R. Keiper
*Strong-Field Hopping in Disordered Semiconductors. A Problem of Directed Percolation
(b) Vol. **110**, No. 2, p. 571–580 (1982).
See also *14.3*, p. *205*; 22, p. 384.

B. Oleš, A. M. Oleś, K. A. Chao, and Yi-Chen Cheng
*A Model for n-Methyl-Phenazinium Tetracyanoquinodimethan
(b) Vol. **110**, No. 2, p. K111–K116 (1982).
See also *13*, p. *142*; 14, p. 190; 18, p. 260; 22.9, p. 504.

G. A. Petrakovskii, K. A. Sablina, V. P. Ikonnikov, I. A. Volkov, and A. G. Klimenko
*The Effect of Amorphization on the Magnetic Properties of $KFeS_2$
(a) Vol. **70**, No. 2, p. 507–511 (1982).
See also *18*, p. *260*; 22.8, p. 483.

J. D. Livingston
Magnetomechanical Properties of Amorphous Metals
(a) Vol. **70**, No. 2, p. 591–596 (1982).
See also *12.1*, p. *131*; *18.2*, p. *269*; 21, p. 334.

K. Bewilogua, D. Dietrich, G. Holzhüter, and Ch. Weissmantel
*Structure of Amorphous Carbon Films
(a) Vol. **71**, No. 1, p. K57–K59 (1982).
See also *1.5*, p. *39*; 22.1, p. 393.

C. Viger and G. Fleury
*Spectral Response of Amorphous Se at 233 K
(a) Vol. **71**, No. 1, p. K103–K105 (1982).
See also *16*, p. *251*; 22.1.3, p. 417.

L. P. Kazakova, L. Tóth, and M. A. Taguyrdzhanov
*Reversible Photoinduced Change of Hole Drift Mobility in Amorphous AsSe Thin Films
(a) Vol. **71**, No. 1, p. K107–K109 (1982).
See also *16*, p. *251*; 22, p. 386.

J. P. Schaffer, J. H. Lind, P. L. Jones, and F. H. Cocks
*Photoconductivity of Ion-Plated Amorphous Hydrogenated Silicon
(a) Vol. **71**, No. 1, p. K111–K115 (1982).
See also *16*, p. *252*; 22.1.2, p. 407.

A. Lehmann, L. Schumann, H. Sobotta, V. Riede, U. Teschner, and K. Hübner
LO–TO Mode Splittings and Dynamic Effective Charges of Oxygen in Reactively Sputtered SiO_x Determined by IR Spectroscopy
(b) Vol. **111**, No. 2, p. K103–K106 (1982).
See also *6*, p. *75*; *20.1*, p. *306*; 22.6, p. 472.

F. Schauer, M. Jedlička, and J. Macků
Amorphous–Crystalline Heterojunction with Built-in Interface Charge (II): Non-Steady State Transport
(a) Vol. **71**, No. 2, p. 335–343 (1982).
See also 13.3, p. 159; *14.3.3*, p. *222*; 22, p. 387; 22.4.2, p. 443.

X.-Z. Dong, B. Gröger, T. Jendrysik, and H. Kronmüller
Annealing Effect of Domain Patterns of the Nearly Non-Magnetostrictive Amorphous Alloy $Co_{58}Ni_{10}Fe_5Si_{11}B_{16}$
(a) Vol. **71**, No. 2, p. 441–449 (1982).
See also *18*, p. *261*; *20*, p. *295*; 22.8, p. 485.

T. Satow, O. Uemura, and Y. Sagara
*The Neutron Diffraction Study of Liquid Antimony and Bismuth Chalcogenides
(a) Vol. **71**, No. 2, p. 555–561 (1982).
See also 14.3, p. 207; 22.7, p. 479.

A. V. Dvurechenskii, I. A. Ryazantsev, and V. A. Dravin
Doping of Amorphous Silicon by Manganese Ion Implantation
(a) Vol. **71**, No. 2, p. K133–K136 (1982).
See also *11*, p. *123*; 13.4, p. 170; *14.3.1*, p. *215*; 22.1.2, p. 408.

A. H. Abou El Ela, N. Abdelmohsen, and H. H. A. Labib
Electrical Conductivity, Thermoelectric Power, and Thermal Conductivity of $AgTlSe_2$ in the Solid and Liquid State
(a) Vol. **71**, No. 2, p. K219–K224 (1982).
See also *8*, p. *87*; 14.3, p. 208; 15, p. 247; 22.8, p. 486.

J. Goebbels, H. R. Khan, L. Lüders, and Ch. J. Raub
NMR on Metastable Pt–P and Ni–Pt–P
(b) Vol. **112**, No. 1, p. 115–118 (1982).
See also *19*, p. *288*; 21.1, p. 350.

S. SANTHANAM, B. K. SAMANTARAY, and A. K. CHAUDHURI
X-Ray Studies on the Microstructure of Vacuum Evaporated SnTe Thin Films
(a) Vol. **72**, No. 2, p. 521–527 (1982).
See also *1.5*, p. *40*; 16, p. 252; 22.7, p. 479.

W. HOYER, A. MÜLLER, E. THOMAS, and M. WOBST
*X-Ray Diffraction Investigation on Molten In–Te Alloys
(a) Vol. **72**, No. 2, p. 585–592 (1982).
See also 1.3, p. 30; 22.7, p. 479.

R. GÖRING, K. KNEIPP, and H. NASS
*Investigation of Glasses in the BaO–B_2O_3–GeO_3 System with Low BaO Content by NMR and Raman Spectroscopy
(a) Vol. **72**, No. 2, p. 623–630 (1982).
See also 6, p. 78; 19, p. 289; 20.1, p. 309; 22.6, p. 473.

B. RAUSCHENBACH und K. HOHMUTH
*Bildung amorpher Metall–Metalloid-Verbindungen durch Ionenimplantation
(Formation of Amorphous Metal–Metalloid Compounds by Ion Implantation)
(a) Vol. **72**, No. 2, p. 667–678 (1982).
See also *11*, p. *125*; 21, p. 338; 21.1, p. 352; 21.1.1, p. 361.

A. G. REVESZ and S. H. WEMPLE
The Optical Properties of Noncrystalline Silicon and $Si_{1-x}H_x$ Films
(a) Vol. **72**, No. 2, p. 721–729 (1982).
See also *1.5*, p. *40*; *4*, p. *63*; 20, p. 296; 22.1.2, p. 411.

I. ČERVEŇ and J. WEIS
*Influence of Extrapolation of the Interference Function of Amorphous $Pd_{83}Si_{17}$ on the Radial Distribution Function
(a) Vol. **72**, No. 2, p. K181–K184 (1982).
See also 21.1, p. 352.

P. SWARUP, R. K. JAIN, S. N. VERMA, S. CHARAN, and D. M. TANDLE
*LPE Growth of GaInAlAs on GaAs
(a) Vol. **72**, No. 2, p. K189–K192 (1982).
See also *13.1*, p. *153*; 20.1, p. 310; 20.3, p. 325; 22.2.4, p. 431.

C. ANCE, J. P. FERRATON, J. M. BERGER, and F. DE CHELLE
*Calculation Model for Valence Electrons and Hydrogen Concentration in a-Si:H
(b) Vol. **113**, No. 1, p. 105–111 (1982).
Erratum
(b) Vol. **114**, No. 2, p. 742 (1982).
See also *13.1*, p. *153*; 14.4, p. 237; 22.1.2, p. 412.

R. LENK
*Integral Representation of the Incoherent Field in Disordered Materials (III): Diffusion, Einstein Relation, and Landauer Formula
(b) Vol. **113**, No. 1, p. 179–183 (1982).
See also 9, p. 92; 14, p. 191.

R. DUTTA, P. K. BANERJEE, and S. S. MITRA
Optical and Electrical Properties of Hydrogenated Amorphous Silicon Carbide
(b) Vol. **113**, No. 1, p. 277–284 (1982).
See also *14.3.1*, p. *215*; *20.1*, p. *310*; 22.3, p. 434.

B. PURA, J. PRZEDMOJSKI, A. JABŁONKA, and R. DĄBROWSKI
*Anisotropy of X-Ray Critical Scattering in 4,4'-Bi-Pentylobiphenyl Liquid Crystal
(b) Vol. 113, No. 2, p. K119 – K122 (1982).
See also 1.2, p. 21; 22.9, p. 507.

B. V. KOBRIN, R. M. KUPRIYANOVA, V. S. MINAEV, E. P. PROKOPIEV, and V. P. SHANTAROVICH
*Study of Imperfections in Ge–Te Semiconductors by the Positron Annihilation Technique
(a) Vol. 73, No. 2, p. 321 – 324 (1982).
See also *10.2*, p. *114*; 22.7, p. 479.

D. E. G. WILLIAMS
*Spatial Resolution and Dense Random Packing
(a) Vol. 73, No. 2, p. 539 – 544 (1982).
See also *1.3*, p. *31*; 21, p. 340.

V. YU. BALANDIN, A. V. DVURECHENSKII, and L. N. ALLEKSANDROV
The Polycenter Crystallization of Amorphous Silicon Layers at Pulse Annealing
(a) Vol 73, No. 2, p. 587 – 592 (1982).
See also *3*, p. *59*; 22.1.2, p. 413.

H. RATAJCZAK and I. GOŚCIAŃSKA
Thermal Hall Hysteresis Loops in Thin Amorphous HoCoFilms
(a) Vol. 73, No. 2, p. 633 – 639 (1982).
See also *14.1*, p. *197*; *18.3*, p. *277*; 21.1, p. 355; 21.4, p. 371.

B. RÖNNEFARTH, R. MATTHEIS, and H. DINTNER
Deposition and Structure of Different Composite Ni–Cr Thin Films
(a) Vol. 73, No. 2, p. K153 – K156 (1982).
See also *1.3*, p. *31*; *1.5*, p. *41*; 21.1, p. 355.

F. SCHAUER, V. ŠMÍD, O. ZEMŠKAL, and L. ŠTOURAČ
*Evaluation of Space-Charge-Limited Currents in a-Si:H/c-Si Heterojunction
(a) Vol. 73, No. 2, p. K199 – K203 (1982).
See also *14.3.4*, p. *232*; 22.1.2, p. 414.

E. G. SHAROYAN and H. A. SAMUELYAN
ESR and Dielectric Anomalies in Amorphous Cu-Phthalocyanine–Iodine Compounds
(a) Vol. 73, No. 2, p. K213 – K216 (1982).
See also *14.4*, p. *238*; *19*, p. *290*; 22.9, p. 507.

V. P. PAN, I. V. ZOLOTUKHIN, S. E. LITVIN, B. G. NIKITIN, and A. S. SHPIGEL
*Electrical Properties of Lanthanum–Aluminium Glass
(a) Vol. 73, No. 2, p. K287 – K291 (1982).
See also 8, p. 89; *14.1*, p. *197*; 14.2, p. 201; 21, p. 341.

V. L. BONCH-BRUEVICH and TO THI BICH NGUYET
Low Temperature Conductivity and Optical Transitions of a Two-Dimensional Disordered Semiconductor under Weak Localization Conditions
(b) Vol. 114, No. 1, p. 67 – 71 (1982).
See also *14.3*, p. *211*; *20.1*, p. *314*; 22, p. 391.

R. LENK
*Integral Representation of the Incoherent Field in Disordered Materials (IV): Relation to Multiple Scattering Theory and Inclusion of Pair Correlations
(b) Vol. 114, No. 1, p. 113 – 120 (1982).
See also 13, p. 146; 14, p. 191.

B. I. Belevtsev, Yu. F. Komnik, V. I. Odnokozov, and A. V. Fomin
Discovery of Low-Temperature Metastable State in Amorphous Bismuth–Hydrogen Systems
(b) Vol. **114**, No. 2, p. 467–474 (1982).
See also *1.5*, p. *41*; *14.2*, p. *201*; 21.7, p. 381.

D. J. González, J. A. Alonso, and M. P. Iñiguez
*Cellular Density Functional Theory of the Heat of Formation of Disordered Simple Alloys. Influence of the Cell Partitioning
(b) Vol. **114**, No. 2, p. 495–501 (1982).
See also *13*, p. *147*; 21.3, p. 368.

B. Hüttner and W. Pompe
*A New Theoretical Approach to the Low-Temperature Specific Heat of Disordered Solids
(b) Vol. **114**, No. 2, p. 503–513 (1982).
See also *8*, p. *90*; 22.1, p. 394; 22.6, p. 476.

R. Lenk
*Integral Representation of the Incoherent Field in Disordered Materials (V): Wave Diffusion in a Correlated Medium
(b) Vol. **114**, No. 2, p. 515–522 (1982).
See also **14**, p. **192**.

J. Lange and H. Bässler
Temperature Dependence of the Hopping Mobility in Amorphous Tetracene
(b) Vol. **114**, No. 2, p. 561–569 (1982).
See also *14.3.1*, p. *216*; *16*, p. *254*; 22.9, p. 508.

N. Ishii, S. Oozora, M. Kumeda, and T. Shimizu
*Calculations of g-Value and Linewidth of the ESR Signal in Amorphous Silicon Nitride
(b) Vol. **114**, No. 2, p. K111–K114 (1982).
See also *19*, p. *291*; 22, p. 392.

H. A. Brown
*On the Curie Temperatures of Amorphous Ferromagnets
(b) Vol. **114**, No. 2, p. K189–K194 (1982).
See also *18.2*, p. *274*.

L. V. Choudary, J. V. Rao, P. N. Murty, and C. R. K. Murty
*Phase Transition Studies of the Liquid Crystal H_pBBA
(a) Vol. **74**, No. 2, p. 431–435 (1982).
See also *1.2*, p. *23*; 22.9, p. 508.

U. Henniger and R. Keiper
*Diffusion Theory for Continous Metal–Amorphous Semiconductor Contacts
(a) Vol. **74**, No. 2, p. 497–509 (1982).
See also *14.3.4*, p. *233*.

W. Scharff, J.-W. Erben, A. Wolf, K. Breuer, Chr. Weissmantel, R. Klabes, H. Woitennek, K.-H. Heinig, M. Voelskov, J. Matthäi, and A. Schmidt
*Flash-Lamp-Induced Crystal Growth of Silicon on Amorphous Substrates Using Artificial Surface-Relief Structures
(a) Vol. **74**, No. 2, p. 545–552 (1982).
See also *3*, p. *60*; 11, p. 128; 22.1.2, p. 416.

3. Crystal Growth

Review Article

Z. M. Jarzębski
Preparation and Physical Propertics of Transparent Conducting Oxide Films
(a) Vol. **71**, No. 1, p. 13–41 (1982).
See also 1.5, p. 38; 13.1, p. 148; *14.3.1*, p. *213*; *20.1*, p. *298*; 22.6, p. 468; 22.6.1, p. 477.

Original Papers and Short Notes

G. Eichhorn and U. Rensch
*MOCVD of AlN on Silicon
(a) Vol. **69**, No. 1, p. K3–K6 (1982).
See also *1.5*, p. *38*; 22.2, p. 418.

Th. Klupsch
Towards a Molecular Theory of the Solid–Liquid Phase Boundary. Thermodynamic and Structural Properties
(b) Vol. **109**, No. 2, p. 535–542 (1982).
See also *1.6*, p. *42*.

A. V. Dvurechenskii, B. P. Kashnikov, L. S. Smirnov, R. Grötzschel, and R. Klabes
*Synthesis of Solid Solutions at Electron Pulse Processing of Ge–Si Structures
(a) Vol. **69**, No. 2, p. K183–K187 (1982).
See also 1.5, p. 38; *11*, p. *119*; 22.1.1, p. 395; 22.1.2, p. 402.

H. Müller
*Simulation of Oriented Crystallization in Evaporated Antimony Thin Films
(a) Vol. **70**, No. 1, p. 249–255 (1982).
See also *1.5*, p. *39*; 21.7, p. 379.

E. Wolf
The Nature and Formation Mechanism of Anomalous Defects in Dislocation Free FZ-Silicon Crystals Caused by Hydrogen Doping
(a) Vol. **70**, No. 1, p. K59–K61 (1982).
See also *10.2*, p. *106*; 22.1.2, p. 404.

K. Tempelhoff and N. van Sung
*Formation of Self-Disorder Agglomerates in Dislocation-Free Silicon during Crystal Growth
(a) Vol. **70**, No. 2, p. 441–449 (1982).
See also 10.2, p. 107; 22.1.2, p. 404.

O. Bostanjoglo
*Time-Resolved TEM of Pulsed Crystallization of Amorphous Si and Ge Films
(a) Vol. **70**, No. 2, p. 473–481 (1982).
See also *1.5*, p. *39*; 22.1.1, p. 396; 22.1.2, p. 405.

E. Ziegler, A. Heinrich, H. Oppermann, and G. Stöver
Growth and Electrical Properties of Non-Stoichiometric ZnO Single Crystals Doped with Co
(a) Vol. **70**, No. 2, p. 563–570 (1982).
See also 13.4, p. 167; *14.3*, p. *205*; 22.4, p. 437.

J. Strzeszewski and K. Świderczak
Electron Microscopic Studies of Crystallization and Magnetic Domain Structure of Mn–Ga Thin Films
(a) Vol. **73**, No. 1, p. 31–36 (1982).
See also *1.2*, p. *20*; *18.2.1*, p. *275*; 21.1, p. 353.

O. Bostanjoglo and G. Hoffmann
Time-Resolved TEM of Transient Effects in Pulse Annealing of Ge and Ge–Te Films
(a) Vol. **73**, No. 1, p. 95–105 (1982).
See also *1.2*, p. *21*; *1.5*, p. *40*; 22.1.1, p. 398; 22.1.3, p. 417.

G. Götz, H.-D. Geiler, and M. Wagner
*Pulse Laser Induced High-Temperature Solid-Phase Annealing of Arsenic Implanted Silicon
(a) Vol. **73**, No. 1, p. 145–151 (1982).
See also *1.2*, p. *21*; 22.1.2, p. 412.

J. Narayan and O. W. Holland
*Solid-Phase-Epitaxial Growth in Ion-Implanted Silicon
(a) Vol. **73**, No. 1, p. 225–236 (1982).
See also 2, p. 52; *11*, p. *126*; 22.1.2, p. 412.

V. Yu. Balandin, A. V. Dvurechenskii, and L. N. Aleksandrov
*The Polycenter Crystallization of Amorphous Silicon Layers at Pulse Annealing
(a) Vol. **73**, No. 2, p. 587–592 (1982).
See also 2, p. 53; 22.1.2, p. 413.

H.-D. Geiler, E. Glaser, G. Götz, and M. Wagner
*Explosive Liquid-Phase Crystallization of Ion-Implanted Silicon
(a) Vol. **73**, No. 2, p. K161–K163 (1982).
See also 22.1.2, p. 414.

R. Hergt and H. Pfeiffer
*Jackson Factor and Growth Kinetics of Garnet LPE Films of Singular Orientation
(a) Vol. **74**, No. 1, p. 97–105 (1982).
See also 22.8.2, p. 502.

G. Grange, R. Landers, and B. Mutaftschiev
*The Bubble Method: A Tool for Growth Morphology Studies of the Crystal–Melt Interface
(a) Vol. **74**, No. 1, p. 233–245 (1982).
See also *1.6*, p. *44*; 22.5.4, p. 467.

A. I. Aleksandrovskii and P. Posmykiewicz
*Impurity Striations in Melt-Grown Crystals
(a) Vol. **74**, No. 1, p. K23–K26 (1982).
See also *10*, p. *95*.

G. Andrä, H.-D. Geiler, G. Götz, K. H. Heinig, and H. Woittennek
*Explosive Liquid-Phase Crystallization of Thin Silicon Films during Pulse Heating
(a) Vol. **74**, No. 2, p. 511–515 (1982).
See also *1.5*, p. *41*; 22.1.2, p. 416.

4. Bonding Properties

Review Article

F. Bechstedt
Electronic Relaxation Effects in the Core Level Spectra of Solids
(b) Vol. **112**, No. 1, p. 9–49 (1982).
See also *13.1*, p. *148*; 17, p. 255; 20, p. 293; 21, p. 330; 22, p. 382; 23, p. 509.

Original Papers and Short Notes

R. N. Joarder and R. V. Gopala Rao
*Effective Pair Potential and Thermodynamic Properties of Liquid Noble Metals
(b) Vol. **109**, No. 1, p. 137–144 (1982).
See also 6, p. 70; 21.6, p. 374.

H. Bakker
The Vibrational Entropy Change Associated with the Disordering of a Two-Dimensional Binary Alloy Using a Model of Nearest- and Next-Nearest-Neighbour Interactions
(b) Vol. **109**, No. 1, p. 211–216 (1982).
See also *1.3*, p. *25*; 8, p. 84; 21, p. 330.

J. Shanker, D. P. Agrawal, and A. K. Rajauria
*Analysis of the Relative Stability of the NaCl, CsCl, and Cubic ZnS Structures in Alkali Halide Crystals
(b) Vol. **109**, No. 1, p. K23–K26 (1982).
See also *1*, p. *9*; 22.5.2, p. 454.

J. Shanker and A. K. Rajauria
*Theory of Interatomic Forces in Rare Gas Crystals
(b) Vol. **109**, No. 2, p. 711–716 (1982).
See also 12.2, p. 134; 23, p. 509.

L. Skála
*The CNDO/BW Study of Lithium Clusters (II): B.C.C. Clusters
(b) Vol. **109**, No. 2, p. 733–741 (1982).
See also *13*, p. *140*; 21.2, p. 365.

A. Ghosh, L. K. Banerjee, and A. N. Basu
*Cohesion, Phase Transition, Elastic and Dielectric Properties of Group IV Semiconductors
(b) Vol. **109**, No. 2, p. 799–806 (1982).
See also *1.2*, p. *15*; 12.2, p. 134; 14.4, p. 234; 22.1, p. 393; 22.1.1, p. 395; 22.1.2, p. 401.

J. Shanker and G. D. Jain
*Cohesive Energies and Bulk Modulus for Alkali Halide Mixed Crystals
(b) Vol. **110**, No. 1, p. 257–262 (1982).
See also *12.2*, p. *134*; 22.5.4, p. 467.

L. Skála
*The CNDO/BW Study of Lithium Clusters (III): Different Crystallographic Structures
(b) Vol. **110**, No. 1, p. 299–306 (1982).
See also *13*, p. *141*; 21.2, p. 365.

R. S. SINGH, V. K. JAIN, and J. SHANKER
Relative Stability and Structural Phase Transitions in Alkaline Earth Chalcogenide Crystals
(b) Vol. **111**, No. 2, p. 519–523 (1982).
See also *1.2*, p. *17*; 22, p. 387.

R. FELLENBERG, P. STREUBEL, and A. MEISEL
*Chemical Shifts of Auger Lines in Solids on the Example of the $KL_{23}L_{23}$ Transition in Silicon and Its Compounds
(b) Vol. **112**, No. 1, p. 55–60 (1982).
See also 17, p. 256; 22.1.2, p. 409.

R. H. RAUTIOAHO
*An Interatomic Pair Potential for Aluminium. Calculation of Stacking Fault Energy
(b) Vol. **112**, No. 1, p. 83–89 (1982).
See also 6, p. 76; 21, p. 336.

P. DEUS, V. FREI, O. BILEK, and W. CORDTS
Estimation of Force Constants and Phonon Frequencies of Ternary Semiconducting Compounds with Chalcopyrite Structure from Specific Heat Data
(b) Vol. **112**, No. 1, p. 269–277 (1982).
See also *6*, p. *76*; 22.8, p. 486.

V. S. URUSOV and V. B. DUDNIKOVA
*Energy of Anion Polarisation in Layered Structures
(b) Vol. **112**, No. 1, p. 303–310 (1982).
See also *1.3*, p. *29*; 22.5.2, p. 458; 22.5.3, p. 465.

N. A. KULAGIN and D. T. SVIRIDOV
*Change of the Effective Occupation Numbers of 3d-Shells of Mn^{2+} Ions in Manganese Halide Crystals
(b) Vol. **112**, No. 1, p. K61–K66 (1982).
See also *13.4*, p. *171*; 22.5, p. 451.

V. RAMAMURTHY
Nonequivalence of De Launay and Clark, Gazis, and Wallis Angular Force Models
(b) Vol. **112**, No. 2, p. 463–471 (1982).
See also *6*, p. *77*.

J. SHANKER and K. SINGH
Analysis of the Grüneisen Parameter and Its Volume Dependence in Alkali Halide Crystals
(b) Vol. **112**, No. 2, p. 615–620 (1982).
See also *6*, p. *77*; 12.2, p. 136; 22.5.2, p. 459.

R. K. SINGH and D. K. NEB
*Role of Zero-Point Motion and Three-Body Effects in the Study of Crystal Properties of Rare Gas Solids
(b) Vol. **112**, No. 2, p. 735–746 (1982).
See also 6, p. 77; 8, p. 88; 12.2, p. 136; 23, p. 509.

A. G. REVESZ and S. H. WEMPLE
*The Optical Properties of Noncrystalline Silicon and $Si_{1-x}H_x$ Films
(a) Vol. **72**, No. 2, p. 721–729 (1982).
See also *1.5*, p. *40*; 2, p. 51; 20, p. 296; 22.1.2, p. 411.

V. P. GUPTA and C. P. AGRAWAL
*Rydberg Potential in Diatomic Ionic Crystals
(b) Vol. **114**, No. 1, p. 265–270 (1982).
See also *6*, p. *80*; 8, p. 89; 12.2, p. 137; 22.5.2, p. 462; 22.6, p. 475.

V. K. JAIN and J. SHANKER
*Analysis of Interionic Potentials in Alkali Chalcogenide Crystals
(b) Vol. **114**, No. 1, p. 271–275 (1982).
See also 12.2, p. 137; 22, p. 391.

B. B. TRIPATHI, H. C. GUPTA, and M. V. N. AMBIKA PRASAD
Effect of Polarizability in the Phonon Spectrum of Zinc-Blende Compounds
(b) Vol. **114**, No. 1, p. K11–K14 (1982).
See also *6*, p. *80*; 22.2.2, p. 426; 22.2.3, p. 429.

T. S. SHI, S. N. SAHU, G. S. OEHRLEIN, A. HIRAKI, and J. W. CORBETT
*Models for the Hydrogen-Related Defect–Impurity Complexes and Si–H Infrared Bands in Crystalline Silicon
(a) Vol. **74**, No. 1, p. 329–341 (1982).
See also *10.2*, p. *116*; 11, p. 128; 20.1, p. 315; 22.1.2, p. 414.

J. Z. DAMM and Z. CHVOJ
*Optimization of Tosi-Fumi Ionic Radii for F.C.C. Alkali Halide Crystals
(b) Vol. **114**, No. 2, p. 413–418 (1982).
See also 22.5.2, p. 463.

L. SKÁLA, L. KÜNNE, H.-G. FRITSCHE, and H. MÜLLER
*Cluster Model for Interstitial Hydrogen Bonding in Nickel
(b) Vol. **114**, No. 2, p. 439–448 (1982).
See also 10.1, p. 102; 13.4, p. 177; 21.1, p. 357.

M. KUCHARCZYK and S. OLSZEWSKI
*Repulsive Energy and the Grüneisen Parameter of Alkali Halides Calculated on the Basis of a Quantum-Statistical ab Initio Theory
(b) Vol. **114**, No. 2, p. 589–598 (1982).
See also 6, p. 81; 12.2, p. 138; 22.5.2, p. 463.

R. K. SINGH and S. P. SANYAL
*Debye-Waller Factors for Alkaline Earth Oxides
(b) Vol. **114**, No. 2, p. 637–643 (1982).
See also *6*, p. *81*; 22.6, p. 476.

J. SHANKER and R. SUNDARAJ
*Dielectric and Anharmonic Behaviour of Alkali Halide Crystals
(b) Vol. **114**, No. 2, p. 659–665 (1982).
See also 12.2, p. 138; 14.4, p. 239; 22.5.2, p. 463.

V. I. ZINENKO
The Cluster Investigation of the q-State Potts Model
(b) Vol. **114**, No. 2, p. 695–703 (1982).
See also *1.2*, p. *23*; 8, p. 90.

5. Mössbauer Spectroscopy

Review Article

P. P. Seregin, A. R. Regel, A. A. Andreev, and F. S. Nasredinov
A Study of Local Structure of Chalcogenide Vitreous Semiconductors and of States of Impurity Atoms in Them by Means of Mössbauer and Electron Spectroscopies
(a) Vol. **74**, No. 2, p. 373 – 394 (1982).
See also 1.2, p. 14; *2*, p. *45*; *13.4*, p. *162*; 22, p. 382.

Original Papers and Short Notes

G. N. Nadjaryan, R. G. Gabrielyan, A. R. Mkrtchyan, and R. M. Khanamiryan
*Inverse Problem in Mössbauer Spectroscopy in Case of Low-Frequency Modulation
(b) Vol. **109**, No. 1, p. 131 – 136 (1982).

C. Bansal and K. N. Shrivastava
*Mössbauer Isomer Shifts of $^{40}K^+$ in KF, KCl and KBr Lattices
(b) Vol. **109**, No. 2, p. 743 – 748 (1982).
See also 22.5.2, p. 455.

T. Kanaizuka
*Interpretation of Mössbauer Spectra of β-FeB and Its Low Temperature Modification, α-FeB
(a) Vol. **69**, No. 2, p. 739 – 744 (1982).
See also 1.2, p. 15; *18.2*, p. *268*; 21.1.1, p. 359.

P. Paul, S. P. Puri, D. Raj, and R. K. Puri
*Force Constant Change in Sn–Impurity Systems: ^{107}Pd:^{119}Sn, ^{197}Au:^{119}Sn, and $^{63.5}$Cu:^{119}Sn
(b) Vol. **110**, No. 1, p. 81 – 87 (1982).
See also 6, p. 72; 10.1, p. 97; 21.1, p. 345; 21.6, p. 375.

H. Bińczycka, O. Gzowski, L. Murawski, and J. Sawicki
*Mössbauer Effect and Electrical Conductivity in TeO_2–Fe_2O_3 Glasses
(a) Vol. **70**, No. 1, p. 51 – 55 (1982).
See also 2, p. 47; *14.3*, p. *204*; 22.6.1, p. 477.

O. Aikala
*On the Use of Annular γ-Ray Sources to Measure Directional Compton Profiles
(b) Vol. **110**, No. 2, p. K107 – K110 (1982).
See also 22.6, p. 470.

T. P. Arsenio, Z. Arguello, P. H. Domingues, N. C. Furtado, and C. A. Taft
*Temperature Dependence of Hyperfine Field and Resonance Fraction in Single Crystal $RbFeS_2$
(b) Vol. **110**, No. 2, p. K129 – K131 (1982).
See also *18.4*, p. *280*; 22.8.2, p. 500.

E. N. Ovchinnikova and R. N. Kuzmin
*Group-Theoretical Approach to the Investigation of Mössbauer Diffraction Patterns
(b) Vol. **110**, No. 2, p. K137 – K139 (1982).

R. Kumar, Vishwamittar, K. Chandra, and S. K. Date
Mössbauer and Magnetic Studies of $Fe_{0.958-x}Mn_xSi_{0.032}$ Ternary Systems
(a) Vol. **70**, No. 2, p. 545–553 (1982).
See also *1.3*, p. *27*; 12.1, p. 131; 18, p. 260; 21.1.1, p. 360.

P. P. Seregin and F. S. Nasredinov
Study of Gold and Platinum Impurity Atom State in Vitreous Arsenic Selenide
(a) Vol. **70**, No. 2, p. 769–775 (1982).
See also *2*, p. *48*; *10.2*, p. *108*; 14.3, p. 206; 17, p. 256; 22, p. 385.

E. Antoncik and Gu Bing-Lin
*Calibration of the Isomer Shift for the 67.03 and 13.3 keV Transitions of ^{73}Ge
(b) Vol. **111**, No. 1, p. 261–270 (1982).
See also 13, p. 142; 22.1.1, p. 396.

G. L. Whittle, S. J. Campbell, and A. M. Stewart
*A Mössbauer Study of the Magnetic Hyperfine Field Distribution in the Amorphous Alloy $Fe_{32}Ni_{36}Cr_{14}P_{12}B_6$
(a) Vol. **71**, No. 1, p. 245–251 (1982).
See also 2, p. 48; 18.2, p. 270; 21.1, p. 349; 21.1.1, p. 361.

E. Z. Liverts and A. K. Zhetbaev
*Sternheimer Quadrupole Factors in Ferric Compounds
(b) Vol. **111**, No. 2, p. 469–475 (1982).
See also 22.8.2, p. 500.

A. A. Bahgat and M. K. Fayek
*Absolute Evaluation of Ferrous and Ferric Concentration in Ca Amphibole
(a) Vol. **71**, No. 2, p. 575–581 (1982).
See also 22.8, p. 486.

Ph. Bauer and J. M. Genin
*Dipolar Contribution and Quadrupole Effect in $GdFe_2$
(a) Vol. **71**, No. 2, p. K149–K152 (1982).
See also 18.3, p. 276; 21.1.1, p. 361.

J. Janicki, J. Pietrzak, A. Porębska, and J. Suwalski
*Mössbauer Study of Copper Ferrites
(a) Vol. **72**, No. 1, p. 95–98 (1982).
See also *18.3*, p. *276*; 22.8.2, p. 500.

P. Colombet, M. Trimblet, and M. Danot
Magnetic Properties of the Diluted B-Spinels $Cu_yCr_y^{III}M_{2-y}^{IV}S_4$ (M^{IV} = Sn, Ti)
(a) Vol. **72**, No. 1, p. 105–116 (1982).
See also *1.3*, p. *29*; *18*, p. *261*; 19, p. 288; 22.8.2, p. 500.

V. I. Nikolaev, V. S. Rusakov, and N. I. Chistyakova
*On the Pressure Effect on the Hyperfine Magnetic Field in Spinel-Type Ferrites
(a) Vol. **72**, No. 1, p. K45–K47 (1982).
See also 22.8.2, p. 500.

P. Orlandi, A. Magistris, and A. Rigamonti
*Mössbauer and Conductivity Study of Ferroelastic Ferric Molybdate
(a) Vol. **72**, No. 2, p. 543–549 (1982).
See also *14*, p. *191*; 22.8.1, p. 496.

E. Wieser, G. Krabbes, and E. I. Terukov
Detection of the Metastable Simultaneous Occurrence of High- and Low-Temperature States in $Co_yFe_{1-y}S$ by Mössbauer Spectroscopy
(a) Vol. **72**, No. 2, p. 695–699 (1982).
See also *1.2*, p. *20*; 14, p. 191; 18, p. 262; 22.8, p. 488.

K. K. P. Srivastava and S. N. Mishra
*Magnetic Relaxation Effects in the Mössbauer Spectrum of Fe^{2+} in Gillespite ($BaFeSi_4O_{10}$)
(b) Vol. **113**, No. 1, p. 93–104 (1982).
See also 22.8, p. 488.

A. V. Mitin
*Ultrasound Modulation of γ-Radiation in a Medium with Inhomogeneous Distribution of Mössbauer Nuclei
(b) Vol. **113**, No. 1, p. 113–124 (1982).
See also *7*, p. *83*.

T. Tomov, D. Klissurski, and I. Mitov
*Mössbauer Study of the Formation of Solid Solutions in the α-Fe_2O_3–Al_2O_3 System
(a) Vol. **73**, No. 1, p. 249–254 (1982).
See also 18, p. 263; 22.6, p. 474.

K. Surendranath and C. Bansal
*Mössbauer Effect Study of Polymetallic Nodules from Indian Ocean Bed
(a) Vol. **73**, No. 1, p. K133–K136 (1982).
See also 22.8, p. 489.

B. A. S. de Barros, Jr., C. A. dos Santos, and I. J. R. Baumvol
Search for Intermetallic Phases in Radiation-Enhanced Diffusion of Tin into Steels
(a) Vol. **73**, No. 1, p. K137–K140 (1982).
See also *9*, p. *92*; *11*, p. *126*; 21.1.1, p. 362.

F. J. Litterst and V. D. Gorobchenko
*Mössbauer Relaxation Spectra for Nuclear Motion in an Octahedral Cage with Reorienting Electric Field Gradient
(b) Vol. **113**, No. 2, p. K135–K138 (1982).

G. Albanese, M. Carbucicchio, L. Pareti, and S. Rinaldi
*Cation Distribution and Magnetic Properties of $BaZn_2Fe_{16-x}In_xO_{27}$ and $BaZn_2Fe_{16-x}Sc_xO_{27}$ Hexagonal Ferrites
(a) Vol. **73**, No. 2, p. K193–K197 (1982).
See also *18*, p. *263*; 22.8.2, p. 501.

A. E. Gelyasin and E. A. Vasilev
Influence of Temperature on Structure and Properties of $Mn_{1+x}Sb_{1-x}Sn_x$ Alloys
(a) Vol. **73**, No. 2, p. K221–K224 (1982).
See also *18.2*, p. *272*; *18.3*, p. *277*; 21, p. 341.

Prem Paul and S. P. Puri
Impurity–Host to Host–Host Coupling Force Constant Ratio for ^{57}Fe in Rare Gas Solids
(b) Vol. **114**, No. 1, p. 213–220 (1982).
See also *4*, p. *64*; *6*, p. *80*; 10.2, p. 116; 23, p. 509.

R. A. Jago, P. E. Clark, and P. L. Rossiter
*The Santa Catharina Meteorite and the Equilibrium State of Fe–Ni Alloys
(a) Vol. **74**, No. 1, p. 247–254 (1982).
See also *1.3*, p. *31*; 18.2, p. 273; 21.1, p. 356; 21.1.1, p. 363.

M. Shimada, S. Shimono, and M. Koizumi
*Mössbauer Effect Studies on $(Sr_{1-x}Sm_x)FeO_3$ Synthesized under High Oxygen Pressure Conditions
(a) Vol. **74**, No. 1, p. K13–K16 (1982).
See also 22.8.1, p. 498.

K. Choudhury, B. K. Das, C. K. Majumdar, and M. Adhikari
*Study of Corrosion of Steel in Acid Medium by Mössbauer Spectroscopy
(a) Vol. **74**, No. 1, p. K27–K30 (1982).
See also 21.1.1, p. 363.

L. Takács, A. Vértes, and H. Leidheiser, Jr.
*Effect of Stresses on the Mössbauer Line-Intensities in Nickel Electrodeposits of [100] Texture
(a) Vol. **74**, No. 1, p. K45–K48 (1982).
See also *12.1*, p. *133*; 18.2, p. 274; 21.1, p. 357.

A. Antonini, V. Fano, I. Ortalli, and A. Vera
*Mössbauer Study of the $Pb_xSn_{1-x}{}^{125}Te$ System
(b) Vol. **114**, No. 2, p. 585–588 (1982).
See also 22.8, p. 491.

P. H. Domingues, T. P. Arsenio, N. C. Furtado, and C. A. Taft
*Quadrupolar and Magnetic Hyperfine Anisotropy in Cesium Dithioferrate
(b) Vol. **114**, No. 2, p. K161–K163 (1982).
See also 18.4, p. 281; 22.8.2, p. 502.

C. Gohy, A. Gérard, and F. Grandjean
*Mössbauer Study of Wustite and Mangano-Wustite
(a) Vol. **74**, No. 2, p. 583–591 (1982).
See also 22.6, p. 476; 22.6.1, p. 478.

F. S. Nasredinov and A. V. Ermolaev
Study of Compensation of Excess Charge of Tin and Iron Impurity Atoms in Cobalt Monoxide by Mössbauer Spectroscopy
(a) Vol. **74**, No. 2, p. 631–636 (1982).
See also *10.2*, p. *117*; 22.6, p. 476.

6. Lattice Dynamics. Phonons

Review Article

Original Papers and Short Notes

K. S. Dubey
*Lattice Thermal Conductivity of Noncrystalline Material at Low Temperatures. Application to Borosilicate Glass
(a) Vol. **69**, No. 1, p. 121–126 (1982).
See also 2, p. 45; *8*, p. *84*; 22.8.1, p. 492.

H.-J. Fitting and J.-U. Friemann
Monte-Carlo Studies of the Electron Mobility in SiO_2
(a) Vol. **69**, No. 1, p. 349–358 (1982).
See also *13*, p. *140*; *14.3.4*, p. *227*; 22.6, p. 468.

L. Wojtczak, T. Siklós, B. Mrygon, and J. Mielnicki
*Lattice Constant Fluctuations in Anharmonic Crystals
(b) Vol. **109**, No. 2, p. 483–493 (1982).

V. Shamrai, K. Bohmhammel and G. Wolf
The Heat Capacity of Nb_3SnH_x in Dependence on the Hydrogen Content
(b) Vol. **109**, No. 2, p. 511–517 (1982).
See also *8*, p. *85*; 14.2, p. 199; 21, p. 331.

E. Gutsche
*Non-Condon Approximations and the Static Approach in the Theory of Non-Radiative Multiphonon Transitions
(b) Vol. **109**, No. 2, p. 583–597 (1982).
See also *13.4*, p. *164*; 20.3, p. 319; 22, p. 383.

K. Peuker, R. Enderlein, A. Schenk, and E. Gutsche
*Theory of Non-Radiative Multiphonon Capture Processes. Solution of Old Controversies
(b) Vol. **109**, No. 2, p. 599–606 (1982).
See also *13.4*, p. *164*; 20.3, p. 319; 22, p. 383.

F. Sukegawa, M. Kobayashi, and I. Yokota
*Collective Motion in Silver Chalcogenides
(b) Vol. **109**, No. 2, p. 717–725 (1982).
See also *13.5.2*, p. *187*; 14, p. 190; 22, p. 383.

W. Kraak, R. Herrmann, and H. Haupt
*Investigation of the Charge Carrier Scattering in Bi under High Hydrostatic Pressure
(b) Vol. **109**, No. 2, p. 785–792 (1982).
See also *14*, p. *190*; 21.7, p. 379.

C. Trallero Giner
*Long-Wave Secondary Radiation in Polar Semiconductors as a Function of Temperature
(b) Vol. **109**, No. 2, p. 817–824 (1982).
See also *20*, p. *294*; 22, p. 384.

P. Trzaskoma, J. Zieliński, and M. Szopa
*On the Role of Phonons for the Pressure Induced Change of Valence
(b) Vol. **109**, No. 2, p. K129–K132 (1982).
See also *1.2*, p. *15*; 21.4, p. 369.

A. Graja, R. Swietlik, J. Petzelt, and L. Dobiášová
Phase Transition in $MTPP(TCNQ)_2$ Single Crystals by Infrared Reflection Measurements
(a) Vol. **69**, No. 2, p. K205–K208 (1982).
See also *1.2*, p. *15*; 20.1, p. 301; 22.9, p. 503.

F. Brehat, B. Wyncke, and M. El Sherif
*Lattice Mode Anharmonicity in Sodium Nitrite
(b) Vol. **110**, No. 2, p. 499–506 (1982).
See also 1.2, p. 16; 14.4.1, p. 241; *20.1*, p. *303*; 22.8.1, p. 494.

A. Beckmann, E. Kolley, and W. Kolley
*On Superconductivity in Narrow-Band Alloys
(b) Vol. **110**, No. 2, p. 557–564 (1982).
See also *14.2*, p. *199*; 21.1, p. 347.

S. Bednarek and U. Rössler
*Method of Invariants Applied to Indirect Gap Absorption
(b) Vol. **110**, No. 2, p. 565–570 (1982).
See also *20.1*, p. *303*; 22.1.2, p. 404; 22.2.2, p. 424.

R. M. Agrawal and R. P. S. Rathore
*Interactions in Non-Simple F.C.C. Metals
(b) Vol. **110**, No. 2, p. 677–686 (1982).
See also *4*, p. *62*; 12.1, p. 131; 21, p. 334; 21.1, p. 347.

M. Sekita, M. Ishii, and Y. Miyazawa
*Raman Scattering in $Bi_2Ge_3O_9$
(b) Vol. **110**, No. 2, p. K141–K144 (1982).
See also *20.1*, p. *304*; 22.8, p. 483.

I. Z. Kostadinov and Vu Thi Binh
*Temperature Dependence of the Phonon Spectra of β-AgI
(b) Vol. **111**, No. 1, p. 53–58 (1982).
See also 20.1, p. 304; 22.5.1, p. 453.

K. N. Shrivastava
*Second- and Third-Order Processes in the Theory of Spin–Lattice Relaxation
(b) Vol. **111**, No. 1, p. 77–82 (1982).
See also *13*, p. *142*; 18.1, p. 265; 19, p. 286.

J. L. Miane, F. Carmona, and P. Delhaes
Microwave Measurements of the Thermal Expansion of Organic Metals
(b) Vol. **111**, No. 1, p. 235–245 (1982).
See also *8*, p. *86*; 14.1, p. 195; 22.9, p. 505.

M. S. Kushwaha
*The Bond-Bending Force Model (BBFM) for Phonons in β-SiC
(b) Vol. **111**, No. 1, p. 337–340 (1982).
See also 22.3, p. 433.

K. R. Zhdanov and F. S. Rakhmenkulov
*Influence of Electron–Phonon Interaction on Electronic Heat Capacity of Nb
(b) Vol. **111**, No. 1, p. 361–364 (1982).
See also *8*, p. *86*; 21, p. 334.

T. M. Gasymov, A. A. Katanov, and F. K. Kocharli
*Propagation of Strong Electromagnetic Waves in Kane Semiconductors under Conditions of Electron and Phonon Heating
(b) Vol. **111**, No. 1, p. 383–395 (1982).
See also *13*, p. *143*; 14.3.4, p. 228; 22, p. 386.

T. SOMA and H. MATSUO
*Debye-Waller Factor of Si and Ge
(b) Vol. **111**, No. 2, p. K93–K97 (1982).
See also 22.1.1, p. 397; 22.1.2, p. 407.

A. LEHMANN, L. SCHUMANN, H. SOBOTTA, V. RIEDE, U. TESCHNER, and K. HÜBNER
*LO–TO Mode Splittings and Dynamic Effective Charges of Oxygen in Reactively Sputtered SiO_x Determined by IR Spectroscopy
(b) Vol. **111**, No. 2, p. K103–K106 (1982).
See also 2, p. 49; *20.1*, p. *306*; 22.6, p. 472.

A. N. ABBASOV, K. R. ALLAKHVERDIEV, F. G. ALIEV, and R. M. ZAMANOVA
*Far-Infrared Absorption Spectra of $CdInAlS_4$ Single Crystals
(b) Vol. **111**, No. 2, p. K129–K131 (1982).
See also *20.1*, p. *306*; 22.8, p. 485.

V. P. IZVEKOV, A. WATTERICH, T. GYÖRGYI, and E. PUNGOR
*Infrared Properties of Doped and Irradiated NaCl Crystals
(b) Vol. **111**, No. 2, p. K153–K156 (1982).
See also *20.1*, p. *307*; 22.5.2, p. 458.

H. ARASHI and M. ISHIGAME
*Raman Spectroscopic Studies of the Polymorphism in ZrO_2 at High Pressures
(a) Vol. **71**, No. 2, p. 313–321 (1982).
See also *1.2*, p. *18*; 20.1, p. 307; 22.6, p. 472.

J. F. J. JORDAN, A. AXMANN, H. EGGER, and J. KALUS
Temperature Dependence of the Compressibility of d_8-Naphthalene
(a) Vol. **71**, No. 2, p. 457–462 (1982).
See also *1*, p. *10*; 12.2, p. 135; 22.9, p. 505.

J. K. KRÜGER, M. PIETRALLA, and H.-G. UNRUH
*The Elastic Properties of Hexatriacontane Single Crystals at Their Various Phase Transitions
(a) Vol. **71**, No. 2, p. 493–504 (1982).
See also *1.2* , p. *18*; 12.2, p. 135; 22.9, p. 505.

K. BOHMHAMMEL, P. DEUS, G. KÜHN, and W. MÖLLER
*Specific Heat, Debye Temperature, and Related Properties of Chalcopyrite Semiconducting Compounds $CuGaSe_2$, $CuGaTe_2$, and $CuInTe_2$
(a) Vol. **71**, No. 2, p. 505–510 (1982).
See also *8*, p. *87*; 22.8, p. 485.

M. G. KAPLUNOV, K. I. POKHODNIYA, and YU. G. BORODKO
Temperature Dependence of Reflection Spectra of Conductive TCNQ Salts
(a) Vol. **71**, No. 2, p. 649–656 (1982).
See also 14.3.4, p. 229; *20.1*, p. *307*; 22.9, p. 505.

S. R. SINGH and S. SINGH
*Phonon Scattering in Plastically Deformed Ge
(b) Vol. **112**, No. 1, p. 51–54 (1982).
See also 10.2, p. 110; 22.1.1, p. 397.

B. S. UMAROV, V. S. GORELIK, M. M. SUSHINSKII, and J. F. VETELINO
*Raman Scattering from Nonfundamental States in Ammonium Chloride Crystals at Low Temperatures
(b) Vol. **112**, No. 1, p. 69–73 (1982).
See also *20.1*, p. *307*; 22.5, p. 451.

R. H. RAUTIOAHO
An Interatomic Pair Potential for Aluminium. Calculation of Stacking Fault Energy
(b) Vol. **112**, No. 1, p. 83–89 (1982).
See also *4*, p. *63*; 21, p. 336.

S. GHOSH and V. K. AGRAWAL
Effect of Magnetostatic Field on the Stimulated Raman Scattering of Transversely Polarized Electromagnetic Waves and Raman Instability in Semiconductor Plasmas
(b) Vol. **112**, No. 1, p. 119–131 (1982).
See also *13*, p. *144*; *20*, p. *295*; 22.2.3, p. 428.

T. SOMA, Y. KIMURA, and H.-M. KAGAYA
*Compression Effect on the Lattice Dynamics of Alkali Metals
(b) Vol. **112**, No. 1, p. 151–154 (1982).
See also 12.1, p. 131; 21.2, p. 366.

P. DEUS, V. FREI, O. BILEK, and W. CORDTS
*Estimation of Force Constants and Phonons Frequencies of Ternary Semiconducting Compounds with Chalcopyrite Structure from Specific Heat Data
(b) Vol. **112**, No. 1, p. 269–277 (1982).
See also 4, p. 63; 22.8, p. 486.

V. M. YARTSEV
*Electron–Molecular Vibration Coupling in Trimerized Organic Ion-Radical Semiconductors
(b) Vol. **112**, No. 1, p. 279–287 (1982).
See also 20.1, p. 308; 22.9, p. 506.

G. MADORE, J. D. N. CHEEKE, D. HUET, and J.-P. MANEVAL
*Photo-Excited Acoustic Phonons in InSb and Ge
(b) Vol. **112**, No. 1, p. K1–K4 (1982).
See also 22.1.1, p. 397; 22.2.3, p. 428.

F. KH. ABDULABEV, A. A. ABDUMALIKOV, and A. S. RAKHMATOV
*The Dynamics of Solitons in One-Dimensional Molecular Crystals
(b) Vol. **112**, No. 1, p. K5–K10 (1982).
See also *13.5*, p. *179*; 22, p. 387.

H. NEUMANN, W. KISSINGER, R. D. TOMLINSON, and N. AVGERINOS
*Two-Phonon Absorption Spectra in $CuInTe_2$
(b) Vol. **112**, No. 1, p. K19–K23 (1982).
See also *20.1*, p. *308*; 22.8, p. 487.

A. HEINRICH, E. I. TERUKOV, W. REICHELT, H. WAGNER, and H. OPPERMANN
IR-Absorption in Doped and Undoped VO_2
(a) Vol. **72**, No. 1, p. K61–K64 (1982).
See also *1.2*, p. *19*; 20.1, p. 308; 22.6, p. 472.

I. SH. AKHMADULLIN and R. V. SABUROVA
*Phonon Scattering by Paraelectrical Centers in Crystals (I): General
(b) Vol. **112**, No. 2, p. 367–378 (1982).
See also *13.4*, p. *171*.

R. H. RAUTIOAHO
An Interatomic Pair Potential for Copper. Calculation of Stacking Fault Energy
(b) Vol. **112**, No. 2, p. 391–398 (1982).
See also *10.1*, p. *99*; 21.1, p. 351

V. Ramamurthy
*Nonequivalence of De Launay and Clark, Gazis and Wallis Angular Force Models
(b) Vol. **112**, No. 2, p. 463–471 (1982).
See also 4, p. 63.

H. D. Lutz and G. Kliche
Far-Infrared Reflection Spectra, Optical and Dielectric Constants, Effective Charges, and Lattice Dynamics of the Skutterudites CoP_3,, $CoAs_3$, and $CoSb_3$
(b) Vol. **112**, No. 2, p. 549–557 (1982).
See also 13.5.2, p. 187; *20.1*, p. *309*; 22, p. 388.

J. Shanker and K. Singh
*Analysis of the Grüneisen Parameter and its Volumne Dependence in Alkali Halide Crystals
(b) Vol. **112**, No. 2, p. 615–620 (1982).
See also 4, p. 63; 12.2, p. 136; 22.5.2, p. 459.

R. Khanna
*Force Model Dependence and Asymptotic Behaviour of the Static Lattice Green Function for H.C.P. Crystals
(b) Vol. **112**, No. 2, p. 667–674 (1982).
See also 10.1, p. 99; 21.1, p. 351; 21.3, p. 368.

O. P. Burmistrova and V. A. Kosobukin
Radiative Decay of Surface Collective Excitations in Polarizable Cylinders
(b) Vol. **112**, No. 2, p. 675–683 (1982).
See also *13.3*, p. *160*; *13.5.2*, p. *187*; 20, p. 296.

R. K. Singh and S. L. Chaplot
*Lattice Mechanics of Molecular Solid Sodium Nitrate
(b) Vol. **112**, No. 2, p. 717–724 (1982).
See also 22.8.1, p. 496.

R. K. Singh and D. K. Neb
Role of Zero-Point Motion and Three-Body Effects in the Study of Crystal Properties of Rare Gas Solids
(b) Vol. **112**, No. 2, p. 735–746 (1982).
See also *4*, p. *63*; 8, p. 88; 12.2, p. 136; 23, p. 509.

K. R. Allakhverdiev, M. M. Godzhaev, A. I. Nadzhafov, and R. M. Sardarly
*Light Scattering Spectra of the Solid Solutions $TlGa_xIn_{1-x}Se_{2x}S_{2(1-x)}$
(b) Vol. **112**, No. 2, p. K93–K97 (1982).
See also *20.1*, p. *309*; 22.8, p. 488.

T. Soma, Y. Katabira, and H. M. Kagaya
*Mode Grüneisen Parameter for Long-Wave Phonons of Si–Ge Solid Solution
(b) Vol. **112**, No. 2, p. K99–K102 (1982).
See also *8*, p. *88*; 12.2, p. 136; 22.3, p. 434.

Yu. E. Perlin, A. A. Kaminskii, M. G. Blazha, and V. N. Enakii
Multiphonon Nonradiative Transitions of TR^{3+} Ions in $Y_3Al_5O_{12}$ and $Lu_3Al_5O_{12}$ Garnets
(b) Vol. **112**, No. 2, p. K125–K130 (1982).
See also *13.4*, p. *172*; *20*, p. *296*; 22.8.2, p. 501.

O. A. C. Nunes
*Parametric Instability of Phonons in Free-Carrier Semiconductors
(b) Vol. **112**, No. 2, p. K131–K135 (1982).
See also 22.2.3, p. 428.

L. A. Tarkhnishvili, K. A. Kvavadze, and M. M. Nadareishvili
Investigation of Concentration Dependence of the Low Temperature Heat Capacity of KCl with OH^- Molecular Impurity
(b) Vol. **113**, No. 1, p. K27 – K30 (1982).
See also *8*, p. *88*; 22.5.2, p. 460.

V. P. Iordatii and L. Z. Kon
Influence of the Phonon Frequency Dispersion on the Critical Temperature of Superconductors
(b) Vol. **113**, No. 1, p. K67 – K71 (1982).
See also *14.2*, p. *200*.

K. I. Wysokiński and A. L. Kuzemsky
Electron-Phonon Interaction in Disordered Transition Metal Alloys
(b) Vol. **113**, No. 2, p. 409 – 420 (1982).
See also *2*, p. *52*; 8, p. 89; 21.1, p. 354.

S. Yamazaki, M. Kishi, and T. Katoda
*Calculation of Excess Free Energy of Mixing and Analysis of Phonon Spectra with Clustering Parameter in III–V Pseudobinary Alloys
(b) Vol. **113**, No. 2, p. 421 – 430 (1982).
See also 22.2.4, p. 431.

A. I. Bobrysheva, I. I. Jeru, and S. A. Moskalenko
The Isotopic Shift of the Excitonic Absorption Lines in Cu_2O Crystals
(b) Vol. **113**, No. 2, p. 439 – 445 (1982).
See also *13.5.1*, p. *184*; *20.1*, p. *311*; 22.6, p. 474.

M. Ya. Valakh, M. P. Lisitsa, G. S. Pekar, G. N. Polysskii, V. I. Sidorenko, and A. M. Yaremko
*Anharmonic Coupling of Phonon Modes in Mixed $Zn_xCd_{1-x}Se$ Crystals
(b) Vol. **113**, No. 2, p. 635 – 645 (1982).
See also *20.1*, p. *312*; 22.4.4, p. 449.

R. D. Bijalwan and P. N. Ram
*Low-Temperature Lattice Specific Heat of CuAl
(b) Vol. **113**, No. 2, p. 691 – 696 (1982).
See also *8*, p. *89*; 21, p. 340; 21.1, p. 354.

G. Gómez-Santos and F. Yndurain
New Theoretical Method for Studying Elementary Excitations in Crystalline Disordered Alloys. Application to Phonons in Ni_xPt_{1-x}
(b) Vol. **113**, No. 2, p. 697 – 707 (1982).
See also *13.5*, p. *179*; 21.1, p. 354; 21.6, p. 378.

J. Shanker and K. Singh
*Analysis of Crystal Binding and Grüneisen Parameter in Cesium Halides, Silver Halides, and Alkaline Earth Oxides
(b) Vol. **113**, No. 2, p. 737 – 745 (1982).
See also *4*, p. *64*; 22.5.1, p. 453; 22.5.2, p. 461; 22.6, p. 474.

T. G. Kerimova, E. Yu. Salaev, A. Sh. Khidirov, N. G. Dervishov, and Sh. M. Efendiev
*Effect of Cationic Ordering on Vibrational Spectra of $CdIn_2S_4$
(b) Vol. **113**, No. 2, p. K107 – K109 (1982).
See also *20.1*, p. *312*; 22.8, p. 489.

P. Kleinert
*Coherent-Potential Approximation for Vibrations in $Ga_{1-x}In_xSb$ and $Ga_{1-x}In_xAs$ Mixed Crystals
(b) Vol. **114**, No. 2, p. 459–466 (1982).
See also 20.1, p. 315; 22.2.4, p. 431.

M. Kucharczyk and S. Olszewski
Repulsive Energy and the Grüneisen Parameter of Alkali Halides Calculated on the Basis of a Quantum-Statistical ab Initio Theory
(b) Vol. **114**, No. 2, p. 589–598 (1982).
See also *4*, p. *65*; 12.2, p. 138; 22.5.2, p. 463.

V. M. Fomin and E. P. Pokatilov
Conductivity of Semiconductors Influenced by Strong Electromagnetic Radiation
(b) Vol. **114**, No. 2, p. 615–625 (1982).
See also 13, p. 147; *14.3.4*, p. *232*; 22, p. 391.

R. K. Singh and S. P. Sanyal
*Debye-Waller Factors for Alkaline Earth Oxides
(b) Vol. **114**, No. 2, p. 637–643 (1982).
See also *4*, p. *65*; 22.6, p. 476.

A. M. Kuznetsov and J. Ulstrup
On the Theory of Long-Range Electron Hopping in Polar Media
(b) Vol. **114**, No. 2, p. 673–683 (1982).
See also *13.4*, p. *177*; 22, p. 392.

E. Ya. Glushko
Nonhomogeneous Kinetics of Quasiparticles in Steady-State Packet Representation
(b) Vol. **114**, No. 2, p. 685–694 (1982).
See also 13, p. 147; *13.5.1*, p. *186*.

V. V. Hizhnyakov
*Vibrational Relaxation in the Excited Electronic State
(b) Vol. **114**, No. 2, p. 721–730 (1982).
See also 20.3, p. 328; 22.5.2, p. 463.

K. Bohmhammel, P. Deus, and H. A. Schneider
*Investigation of Acoustic Phonon Modes of GaP and Compound Semiconductors $A^{II}B^{IV}C_2^{V}$ by Specific Heat Measurements
(a) Vol. **74**, No. 2, p. 485–492 (1982).
See also *8*, p. *90*; 22.2.2, p. 426; 22.8, p. 491.

A. R. Badzian
On the Geometrical Condition for X-Ray Scattering by Thermal Waves in Crystals
(a) Vol. **74**, No. 2, p. 637–649 (1982).
See also *1.1*, p. *13*; 22, p. 392; 22.1.1, p. 399; 22.1.2, p. 416; 22.5.2, p. 464.

7. Acoustic Properties

Original Papers and Short Notes

P. Palanichamy and S. P. Singh
*Change in Electron Mass and Collision Frequency Due to Acoustic Wave Propagation in Piezoelectric Semiconductors
(b) Vol. **109**, No. 1, p. 119–123 (1982).
See also *13*, p. *139*; 22.4.1, p. 439.

A. Witek
*The Effect of Illumination on Anomalous Ultrasonic Attenuation in Doped Single Crystals of $Bi_{12}GeO_{20}$
(a) Vol. **69**, No. 1, p. 67–70 (1982).
See also 16, p. 249; 22.6.1, p. 477.

N. A. Tyapunina and V. V. Blagoveshchenskii
*Dislocation Behaviour and Multiplication under Ultrasound
(a) Vol. **69**, No. 1, p. 77–83 (1982).
See also *10*, p. *94*.

E. Harnik, S. Kovnovich, and T. Chernobelskaya
*Secondary Acoustoelectric Current in Ultrathin Metal Films Deposited on a Piezoelectric Substrate
(a) Vol. **69**, No. 2, p. K179–K181 (1982).
See also 1.5, p. 38; 21.6, p. 375.

S. Ghosh and V. K. Agarwal
*Excitation of Acoustohelicon Waves Due to Modulation of a Laser Beam in Longitudinally Magnetised Semiconductors
(b) Vol. **110**, No. 1, p. K37–K40 (1982).
See also 13, p. 141; 18, p. 260; *20*, p. *294*; 22, p. 384.

P. Koštial
*Surface Wave Investigations of Se Amorphous Thin Films Deposited on $LiNbO_3$
(a) Vol. **70**, No. 1, p. K55–K58 (1982).
See also *2*, p. *47*; 22.1.3, p. 417.

S. V. Tarasenko and G. K. Chepurnykh
Study of Relaxation Processes in Antiferromagnets Using Sound Absorption Measurements
(b) Vol. **110**, No. 2, p. 507–511 (1982).
See also *13.5.4*, p. *189*; *18.4*, p. *279*.

O. N. Awasthi and B. K. Basu
*On Magnetoelastic Dispersion in a Nickel Single Crystal
(a) Vol. **70**, No. 2, p. 713–717 (1982).
See also 18.2, p. 269; 21.1, p. 347.

J. Braciník and J. Štelina
*Electric Field in the Acoustoelectric Stationary Domains of the Second Mode. Simple Model
(a) Vol. **70**, No. 2, p. K135–K139 (1982).
See also *14.3.4*, p. *228*; 22.2.3, p. 427.

H. Zeile, U. Harten, and K. Lassmann
Ultrasonic Investigation of the Acceptor Ground State of Si(B) (I): The Longitudinal and Transverse Relaxation Rates of the Strain-Split Two-Level System
(b) Vol. **111**, No. 1, p. 213–220 (1982).
See also *13.4*, p. *168*; 22.1.2, p. 406.

H. Zeile and K. Lassmann
Ultrasonic Investigation of the Acceptor Ground State of Si(B) (II): The Distribution of Splittings
(b) Vol. **111**, No. 2, p. 555–564 (1982).
See also *13.4*, p. *169*; 22.1.2, p. 407.

H. Białas and G. Schauer
*Circular Acoustic Birefringence in Quartz Observed by Time-of-Flight Measurement
(a) Vol. **72**, No. 2, p. 679–686 (1982).
See also 22.6, p. 473.

A. I. Mitsek, D. I. Sirota, and N. P. Kolmakova
*On the Theory of Acoustic Properties of Many-Domain Ferromagnets
(b) Vol. **113**, No. 1, p. 79–86 (1982).
See also 18.2, p. 271; 18.4, p. 280.

A. V. Mitin
*Ultrasound Modulation of γ-Radiation in a Medium with Inhomogeneous Distribution of Mössbauer Nuclei
(b) Vol. **113**, No. 1, p. 113–124 (1982).
See also *5*, p. *68*.

J. P. Jernot and J. L. Chermant
Model to Predict the Tensile Strength of Sintered Materials
(a) Vol. **73**, No. 2, p. 579–586 (1982).
See also *12.1*, p. *133*; 14.1, p. 197; 21.1.1, p. 363.

V. R. Velasco
*Influence of Surface Stress on the Surface Elastic Waves in Anisotropic Cubic Crystals
(b) Vol. **114**, No. 1, p. 35–38 (1982).

J. Ulner
*Temperatur- und Magnetfeldabhängigkeit einer transversalen Schallgeschwindigkeit im tetragonalen Paramagneten mit dem Gesamtdrehimpuls $J = 1$
(Temperature and Magnetic-Field Dependence of a Transversal Sound Velocity in the Tetragonal Paramagnet with Total Angular Momentum $J = 1$)
(b) Vol. **114**, No. 2, p. 735–739 (1982).
See also 18.1, p. 266.

V. M. Polyanovskii
*The Influence of Finite Conduction Bandwith on Gigantic Acoustomagnetic Oscillations in Semiconductors
(b) Vol. **114**, No. 2, p. K115–K117 (1982).
See also *19*, p. *291*.

V. A. Melik-Shakhazarov, I. A. Naskidashvili, V. I. Serdobintsev, and V. I. Savin
*Anelastic Effects in the β-Phase of Niobium Hydride
(a) Vol. **74**, No. 2, p. 593–602 (1982).
See also 1.4, p. 37; *12.1*, p. *133*; 21, p. 342.

8. Thermal Properties

Review Articles

Original Papers and Short Notes

K. S. Arntson and M. B. Langley
A Phenomenological Theory for Hysteresis in the Ionic Conductivity of Solid Electrolytes
(a) Vol. **69**, No. 1, p. 185–192 (1982).
See also *10.2*, p. *103*; *14.4*, p. *234*; 22.5, p. 450.

H. Gruber and E. Krautz
Magnetoresistance and Conductivity in the Binary System Titanium–Oxygen (I): Titanium Oxides with Metallic Conductivity
(a) Vol. **69**, No. 1, p. 287–295 (1982).
See also *14.1*, p. *193*; *14.3*, p. *203*; 15, p. 245; 18, p. 259; 22.6, p. 468.

M. Drulis and B. Staliński
*Low-Temperature Specific Heats of Non-Stoichiometric Neodymium Hydrides
(a) Vol. **69**, No. 1, p. 385–392 (1982).
See also 1.2, p. 15; 21.4, p. 369.

S. C. Mishra and P. L. Srivastava
*Thermodynamic Properties of Liquid Alkali Metals and Their Equiatomic Alloys
(b) Vol. **109**, No. 2, p. 457–461 (1982).
See also 2, p. 46; 21.2, p. 365.

V. Shamrai, K. Bohmhammel, and G. Wolf
*The Heat Capacity of Nb_3SnH_x in Dependence on the Hydrogen Content
(b) Vol. **109**, No. 2, p. 511–517 (1982).
See also 6, p. 71; 14.2, p. 199; 21, p. 331.

A. N. Chakravarti, K. P. Ghatak, S. Ghosh, and A. K. Chowdhury
*Effect of Heavy Doping on the Electronic Heat Capacity in Semiconductors
(b) Vol. **109**, No. 2, p. 705–710 (1982).
See also 13.4, p. 164; 22.2.1, p. 420

A. Kumar and G. C. Srivastava
*Thermal Conductivity Analysis of Germanium at Low Temperatures
(b) Vol. **109**, No. 2, p. K123–K127 (1982).
See also 22.1.1, p. 395.

I. Ansara, A. Pasturel, and K. H. J. Buschow
*Enthalpy Effects in Amorphous Alloys and Intermetallic Compounds in the System Zr–Cu
(a) Vol. **69**, No. 2, p. 447–453 (1982).
See also 2, p. 46; 21.1, p. 344.

K. N. Kalaidjiev, M. P. Mikhailov, G. I. Bozhanov, and St. R. Stoyanov
*Influence of a Constant Electric Field on the Heat-Conductivity Coefficient of TGS near the Phase Transition Temperature
(a) Vol. **69**, No. 2, p. K163–K166 (1982).
See also *14.4.1*, p. *240*; 22.8.1, p. 493.

A. I. Naskidashvili, I. A. Naskidashvili, L. S. Topchyan, V. N. Kozhanov, and E. P. Romanov
*Heat Capacity Anomalies in the Intermetallic Compound V_2Zr
(a) Vol. **69**, No. 2, p. K239–K242 (1982).
See also *1.2*, p. *15*; 21.1, p. 345.

K. P. Ghatak, K. K. Ghosh, H. M. Mukherjee, and A. N. Chakravarti
*Effect of Size Quantization on the Electronic Heat Capacity in Ultrathin Films of Non-Parabolic Semiconductors
(b) Vol. **110**, No. 1, p. 323–330 (1982).
See also 22.2.3, p. 427.

A. ANDREEFF, TH. FRAUENHEIM, E. A. GOREMYCHKIN, H. GRIESSMANN, B. LIPPOLD, W. MATZ, O. D. CHISTYAKOV, and E. M. SAVITSKII
*Crystal Field in the Laves Phase Compound $PrNi_2$
(b) Vol. **111**, No. 2, p. 507–512 (1982).
See also *1*, p. *10*; *13.4*, p. *169*; 15, p. 246; 18.1, p. 266; 21.1, p. 349; 21.4, p. 370.

P. VAROTSOS, K. ALEXOPOULOS, and K. NOMICOS
*Comments on the Pressure Variation of the Gibbs Energy for Bound and Unbound Defects
(b) Vol. **111**, No. 2, p. 581–590 (1982).
See also 9, p. 92; *10*, p. *94*; 14.4, p. 236; 21, p. 335; 21.1, p. 349; 21.6, p. 376; 22.5.2, p. 458.

I. C. DA CUNHA LIMA, M. FABBRI, and A. FERREIRA DA SILVA
*Diagonal and Off-Diagonal Disorder in Doped Semiconductors
(b) Vol. **111**, No. 2, p. K69–K74 (1982).
See also *14.3*, p. *207*; 22.1.2, p. 407.

K. RAMANI and A. M. GHODGAONKAR
Lattice Parameters of Alkali Halides by the Simons-Parr-Finlan Technique
(a) Vol. **71**, No. 2, p. 345–350 (1982).
See also *12.2*, p. *135*; 22.5.2, p. 458.

K. BOHMHAMMEL, P. DEUS, G. KÜHN, and W. MÖLLER
*Specific Heat, Debye Temperature, and Related Properties of Chalcopyrite Semiconducting Compounds $CuGaSe_2$, $CuGaTe_2$, and $CuInTe_2$
(a) Vol. **71**, No. 2, p. 505–510 (1982).
See also *6*, p. *75*; 22.8, p. 485.

A. H. ABOU EL ELA, N. ABDELMOHSEN, and H. H. A. LABIB
*Electrical Conductivity, Thermoelectric Power, and Thermal Conductivity of AgTlSe in the Solid and Liquid State
(a) Vol. **71**, No. 2, p. K219–K224 (1982).
See also 2, p. 49; 14.3, p. 208; 15, p. 247; 22.8, p. 486.

P. KISTAIAH, K. SATYANARAYANA MURTHY, and K. V. KRISHNA RAO
*X-Ray Determination of Lattice Parameter and Thermal Expansion of the Compound $CdIn_2Se_4$
(a) Vol. **71**, No. 2, p. K225–K229 (1982).
See also *1.1*, p. *12*; 22.8, p. 486.

A. SAHLING, P. FRACH, and E. HEGENBARTH
*The Heat Capacity of $LaNi_2$, $LaNi_5$, $PrNi_2$ and $PrNi_5$ in the Temperature Range between 0.3 and 7 K
(b) Vol. **112**, No. 1, p. 243–250 (1982).
See also 21.1, p. 350; 21.4, p. 371.

J. DEL CERRO, S. RAMOS, and J. ISIDORO
*Thermal Study of α-$LiIO_3$ Single Crystals below Room Temperature
(a) Vol. **72**, No. 1, p. K23–K26 (1982).
See also 14.4, p. 237; 22.8.1, p. 495.

Y. A. BADR and S. AWAD
On the Low Temperature Phase Transition in Ammonium Sulfate
(a) Vol. **72**, No. 1, p. K27–K32 (1982).
See also *1.2*, p. *19*; *14.4.1*, p. *242*; 22.8.1, p. 495.

H. GRASSI, A. BRAU, and J. P. FARGES
Heat Pulse Diffusion in TEA $(TCNQ)_2$
(b) Vol. **112**, No. 2, p. 633–640 (1982).
See also *14.3*, p. *208*; 22.9, p. 506.

R. K. SINGH and D. K. NEB
Role of Zero-Point Motion and Three-Body Effects in the Study of Crystal Properties of Rare Gas Solids
(b) Vol. **112**, No. 2, p. 735–746 (1982).
See also *4*, p. *63*; 6, p. 77; 12.2, p. 136; 23, p. 509.

T. SOMA, Y. KATABIRA, and H. M. KAGAYA
*Mode Grüneisen Parameter for Long-Wave Phonons of Si–Ge Solid Solution
(b) Vol. **112**, No. 2, p. K99–K102 (1982).
See also *6*, p. *77*; 12.2, p. 136; 22.3, p. 434.

M. HIDAKA, K. INOUE, B. J. GARRARD, and B. M. WANKLYN
Study of Structural Phase Transitions in the Layer Compound $RbVF_4$
(a) Vol. **72**, No. 2, p. 809–816 (1982).
See also *1.2*, p. *20*; 6, p. 78; 22.8, p. 488.

S. PARVIAINEN
*Magnetic Contribution to the Thermal Expansion of $TiFe_{0.5}Co_{0.5}$
(a) Vol. **72**, No. 2, p. K141–K144 (1982).
See also *18.2*, p. *271*; 21.1, p. 352; 21.1.1, p. 361.

E. FISCHER, W. HÄSSLER, and E. HEGENBARTH
*Glass-Like Behaviour in the Thermal Conductivity of $Sr_{1-x}Ba_x(Nb_2O_6)$ Single Crystal
(a) Vol. **72**, No. 2, p. K169–K171 (1982).
See also *14.4.1*, p. *242*; 22.8.1, p. 496.

H.-J. KROKOSZINSKI, C. SANTANDREA, E. GMELIN, and K. BÄRNER
*Specific Heat Anomaly Connected with a High-Spin–Low-Spin Transition in Metallic $MnAs_{1-x}P_x$
(b) Vol. **113**, No. 1, p. 185–195 (1982).
See also 18.2, p. 272; 21.1, p. 352.

L. A. TARKHNISHVILI, K. A. KVAVADZE, and M. M. NADAREISHVILI
*Investigation of Concentration Dependence of the Low Temperature Heat Capacity of KCl with OH^- Molecular Impurity
(b) Vol. **113**, No. 1, p. K27–K30 (1982).
See also 6, p. 79; 22.5.2, p. 460.

C. R. BROOKS, P. J. MESCHTER, and T. G. KOLLIE
*The Magnetic Heat Capacity of the Configurationally Disordered Ni–25 at% Fe Alloy
(a) Vol. **73**, No. 1, p. 189–198 (1982).
See also 12.1, p. 132; 14.1, p. 197; 18.2, p. 272; 21.1, p. 353; 21.1.1, p. 362.

J. GARBARCZYK, M. WASIUCIONEK, W. JAKUBOWSKI, and A. GRODZIŃSKI
*TGA and DTA Studies on Me-Stabilized β″-Alumina (Me = Co, Ni, Cu, Zn) Exposed to Moist Air
(a) Vol. **73**, No. 1, p. K97–K100 (1982).
See also *10.2*, p. *114*; 22.6, p. 474.

I. Henning, M. Mertig, R. Plath, G. Pompe, E. Hegenbarth, and R. Schalge
*Glass-Like Behaviour of $Sr_{1-x}Ba_xNb_2O_6$ (SBN) Single Crystals Demonstrated by Heat Capacity Measurements
(a) Vol. **73**, No. 1, p. K105 – K109 (1982).
See also *14.4.1*, p. *243*; 22.8, p. 489.

K. I. Wysokiński and A. L. Kuzemsky
Electron–Phonon Interaction in Disordered Transition Metal Alloys
(b) Vol. **113**, No. 2, p. 409 – 420 (1982).
See also *2*, p. *52*; 6, p. 79; 21.1, p. 354.

R. D. Bijalwan and P. N. Ram
*Low-Temperature Lattice Specific Heat of CuAl
(b) Vol. **113**, No. 2, p. 691 – 696 (1982).
See also *6*, p. *79*; 21, p. 340; 21.1, p. 354.

J. Grammatikakis
Measurements of the Ionic Conductivity of NaI and Calculation of the Bulk Modulus and Specific Heat of Vacancies
(b) Vol. **113**, No. 2, p. 721 – 726 (1982).
See also *10.2*, p. *114*; 12.2, p. 137; *14.4*, p. *238*; 22.5.2, p. 461.

V. P. Pan, I. V. Zolotukhin, S. E. Litvin, B. G. Nikitin, and A. S. Shpigel
Electrical Properties of Lanthanum-Aluminium Glass
(a) Vol. **73**, No. 2, p. K287 – K291 (1982).
See also *2*, p. *53*; *14.1*, p. *197*; 14.2, p. 201; 21, p. 341.

V. P. Gupta and C. P. Agrawal
Rydberg Potential in Diatomic Ionic Crystals
(b) Vol. **114**, No. 1, p. 265 – 270 (1982).
See also *4*, p. *65*; *6*, p. *80*; 12.2, p. 137; 22.5.2, p. 462; 22.6, p. 475.

B. N. Tewari, B. S. Bhargava, and K. N. Khanna
*Surface Entropy of Liquid Metals
(b) Vol. **114**, No. 1, p. K31 – K34 (1982).
See also *2*, p. *54*; 21, p. 341; 21.1, p. 356; 21.2, p. 366; 21.3, p. 368; 21.7, p. 381.

P. Gluck
On the Extended Hubbard Model in One Dimension
(b) Vol. **114**, No. 1, p. K47 – K50 (1982).
See also *13.5*, p. *180*.

Y. Nishino and T. Imura
Viscoelastic Behaviour of Oxide Films on Silicon Crystals
(a) Vol. **74**, No. 1, p. 193 – 200 (1982).
See also *12.2*, p. *138*; 22.1.2, p. 414.

V. I. Pecherskaya and D. N. Bolshutkin
*Electrical Resistance and Thermal Expansion of Fe–18Cr–(10 to 25) Ni Alloys in the Paramagnetic Temperature Range
(a) Vol. **74**, No. 1, p. 285 – 290 (1982).
See also *14.1*, p. *198*; 21.1, p. 357; 21.1.1, p. 363.

G. A. Petrakovskii, A. G. Klimenko, F. S. Rakhmenkulov, and A. V. Baranov
*Specific Heat and Low Field Magnetic Properties of $Co(S_{1-x}Se_x)_2$
(a) Vol. **74**, No. 1, p. K31 – K35 (1982).
See also *18*, p. *263*; 21.1, p. 357.

9. Diffusion

Review Articles

P. VAROTSOS and K. ALEXOPOULOS
Current Methods of Lattice Defect Analysis Using Dilatometry and Self-Diffusion. Critical Review and Proposals
(b) Vol. **110**, No. 1, p. 9–31 (1982).
See also 8, p. 84; *10.1*, p. *96*; 21, p. 330.

A. B. DOKTOROV and E. A. KOTOMIN
*Theory of Tunneling Recombination of Defects Stimulated by Their Motion (I): General Formalism
(b) Vol. **114**, No. 1, p. 9–34 (1982).
See also *10.2*, p. *103*.

E. A. KOTOMIN and A. B. DOKTOROV
*Theory of Tunneling Recombination of Defects Stimulated by Their Motion (II): Three Recombination Mechanisms
(b) Vol. **114**, No. 2, p. 287–318 (1982).
See also *10.2*, p. *103*; 22.5.2, p. 454.

Original Papers and Short Notes

M. P. SHEARER, S. K. SEN, and C. L. BAUER
Investigation of Interfacial Reactions in Bimetallic Thin-Film Couples of Al/Cu and Pb/Cu by Measurement of Contact Resistance
(a) Vol. **69**, No. 1, p. 139–149 (1982).
See also *1.5*, p. *38*; 14.1, p. 193; 21, p. 331; 21.1, p. 343.

E. NEBAUER, E. BRINK, P. JANIETZ and T. LIECKE
*Are Au Profiles Well Measurable in n-GaAs by Electroetching?
(a) Vol. **69**, No. 1, p. K65–K68 (1982).
See also 14.3.2, p. 217; 22.2.1, p. 420.

S. LOUALICHE, C. LUCAS, P. BARUCH, J. P. GAILLIARD, and J. C. PFISTER
*Theoretical Model for Radiation Enhanced Diffusion and Redistribution of Impurities. Comparison with Experiments
(a) Vol. **69**, No. 2, p. 663–676 (1982).
See also 11, p. 119; 22.1.2, p. 402.

M. L. DALAL, S. SIVARAMAN, and Y. V. G. S. MURTI
Thermal Dissociation Energy of F-Centers in Alkali Halides
(b) Vol. **110**, No. 1, p. 349–356 (1982).
See also *10.2*, p. *105*; 22.5.2, p. 456.

Y. SAKAGAMI
*Random Walk of a Di-Vacancy in an F.C.C. Lattice
(b) Vol. **111**, No. 1, p. 321–330 (1982).
See also 10, p. 94.

E. MANN
*On the Diffusion Theory of Kink-Pair Formation on Dislocations
(b) Vol. **111**, No. 2, p. 541–553 (1982).
See also *10*, p. *94*; 12, p. 129.

P. Varotsos, K. Alexopoulos, and K. Nomicos
Comments on the Pressure Variation of the Gibbs Energy for Bound and Unbound Defects
(b) Vol. **111**, No. 2, p. 581–590 (1982).
See also *8*, p. *87*; *10*, p. *94*; 14.4, p. 236; 21, p. 335; 21.1, p. 349; 21.6, p. 376; 22.5.2, p. 458.

E. V. Kučys, O. E. Panchuk, J. V. Kameneckas, and D. F. Junevičius
On the Kinetics of the Process of Decay of Supersaturated Solid Copper or Silver Solution in Cadmium Telluride
(a) Vol. **71**, No. 2, p. K157–K160 (1982).
See also *10.2*, p. *110*; 14.3, p. 207; 22.4.3, p. 446.

A. H. van Ommen
Vacancies in NiGa
(a) Vol. **72**, No. 1, p. 273–280 (1982).
See also *10.1*, p. *99*; 21.1, p. 350.

L. S. Wieluński, B. M. Paine, B. X. Liu, C.-D. Lien, and M.-A. Nicolet
Ion Mixing of Ni and Pt Layers on Si
(a) Vol. **72**, No. 1, p. 399–404 (1982).
See also *11*, p. *123*; 22.1.2, p. 409.

M. Ogino, Y. Oana, and M. Watanabe
*The Diffusion Coefficient of Germanium in Silicon
(a) Vol. **72**, No. 2, p. 535–541 (1982).
See also 22.1.2, p. 410.

O. Hildebrand
*Anomalous Impurity Diffusion in III–V Compounds: The Consequence of Self-Induced Field Effects
(a) Vol. **72**, No. 2, p. 575–584 (1982).
See also 14.3.3, p. 223; 22.2, p. 418.

R. Lenk
Integral Representation of the Incoherent Field in Disordered Materials (III): Diffusion Einstein Relation, and Landauer Formula
(b) Vol. **113**, No. 1, p. 179–183 (1982).
See also *2*, p. *51*; 14, p. 191.

R. Damköhler und Th. Heumann
*Experimentelle Untersuchungen zur Klärung des Diffusionsverhaltens kupferreicher Kupfer–Nickel-Legierungen
(Experimental Investigation on the Diffusion Behaviour of Copper-Rich Copper–Nickel Alloys)
(a) Vol. **73**, No. 1, p. 117–127 (1982).
See also 21.1, p. 353.

N. P. Tishchenko
*Self-Diffusion in Rare-Gas Solids
(a) Vol. **73**, No. 1, p. 279–286 (1982).
See also 23, p. 509.

B. A. S. de Barros, Jr., C. A. dos Santos, and I. J. R. Baumvol
*Search for Intermetallic Phases in Radiation-Enhanced Diffusion of Tin into Steels
(a) Vol. **73**, No. 1, p. K137–K140 (1982).
See also 5, p. 68; *11*, p. *126*; 21.1.1, p. 362.

J. Maier and G. Schwitzgebel
*Theoretical Treatment of the Diffusion Coupled with Reaction, Applied to the Example of a Binary Solid Compound MX
(b) Vol. **113**, No. 2, p. 535–547 (1982).
See also 22, p. 390.

T. Łukaszewicz
*Studies on the Diffusion of Impurities in II–VI Compounds Based on the Two-Sample Luminescence Method
(a) Vol. **73**, No. 2, p. 611–616 (1982).
See also 20.3, p. 327; 22.4.1, p. 441; 22.4.2, p. 444.

T. Łukaszewicz
*The Diffusion of Atoms and Luminescence Centres in II–VI Compounds
(a) Vol. **74**, No. 1, p. 307–312 (1982).
See also 20.3, p. 328; 22.4.2, p. 444.

R. Klabes, J. Matthäi, M. Voelskow, and S. Mutze
Pulsed Incoherent Light Annealing of Arsenic and Phosphorus Implanted Polycrystalline Silicon
(a) Vol. **74**, No. 1, p. K5–K7 (1982).
See also *11*, p. *128*; *14.3.1*, p. *216*; 22.1.2, p. 415.

I. T. Suprun
*Forced Pipe Diffusion of Mobile Pinning Points
(a) Vol. **74**, No. 2, p. 411–419 (1982).
See also 10, p. 95; 12, p. 129.

10. Defect Properties (Irradiation Defects See 11)

Original Papers and Short Notes

J. Kubát
*A Cooperative Mechanism Limiting the Spectra Width of Stress Relaxation Processes in Solids
(b) Vol. **111**, No. 2, p. 599–606 (1982).
See also *12*, p. *129*.

J. Lothe, V. L. Indenbom, and V. A. Chamrov
*Elastic Field and Self-Force of Dislocations Emerging at the Free Surfaces of an Anisotropic Halfspace
(b) Vol. **111**, No. 2, p. 671–677 (1982).
See also *12*, p. *129*.

R. C. Ecob
The Geometrical Simulation of the Structure of F.C.C./B.C.C. Interfaces
(a) Vol. **71**, No. 2, p. 399–407 (1982).
See also *1.4*, p. *34*.

J. Perez, F. Fouquet, and Y. He
*Homogeneous Flow in Metallic Glasses
(a) Vol. **72**, No. 1, p. 289–300 (1982).
See also *2*, p. *50*.

E. K. Naimi
*The Amplitude Dependence of Internal Friction and the Young's Modulus Defect in Solids Due to Anharmonism of Dislocation Motion at Low Vibration Amplitudes
(a) Vol. **72**, No. 2, p. 825–832 (1982).
See also *12*, p. *129*; 21, p. 338; 22.5, p. 452.

A. M. Dulgin
*On the Correlation between the Non-Monotonous Character of Flow Stress and the Mobile Dislocation Density
(a) Vol. **73**, No. 2, p. 315–320 (1982).

M. J. Marcinkowski
*Unified Theory of Grain Boundaries. An Update
(a) Vol. **73**, No. 2, p. 409–419 (1982).
See also 1.4, p. 36.

W. Püschl, R. Frydman, and G. Schoeck
*The Strength of the Dislocation Forest for 30° and 60° Dislocations
(a) Vol. **74**, No. 1, p. 211–216 (1982).

A. I. Aleksandrovskii and P. Posmykiewicz
*Impurity Striations in Melt-Grown Crystals
(a) Vol. **74**, No. 1, p. K23–K26 (1982).
See also *3*, p. *59*.

I. T. Suprun
Forced Pipe Diffusion of Mobile Pinning Points
(a) Vol. **74**, No. 2, p. 411–419 (1982).
See also *9*, p. *93*; 12, p. 129.

K. N. Subramanian
*Development of Deformation Bands
(a) Vol. **74**, No. 2, p. 683–695 (1982).

10.1 Metals

Review Articles

P. VAROTSOS and K. ALEXOPOULOS
*Current Methods of Lattice Defect Analysis Using Dilatometry and Self-Diffusion. Critical Review and Proposals
(b) Vol. **110**, No. 1, p. 9–31 (1982).
See also 8, p. 84; 9, p. 91; 21, p. 330.

T. A. PARKHOMENKO and V. V. PUSTOVALOV
*The Low-Temperature Yield Stress Anomaly in Metals and Alloys
(a) Vol. **74**, No. 1, p. 11–42 (1982).
See also 12.1, p. 130; 21, p. 330.

Original Papers and Short Notes

H.-J. KAUFMANN and P. P. PAL-VAL
*Peculiarities of High-Amplitude Dislocation Internal Friction in Molybdenum Single Crystals of High Purity in the Temperature Range 5.9 to 300 K
(a) Vol. **69**, No. 1, p. 127–132 (1982).
See also *12.1*, p. *130*; 21, p. 331.

H. J. BLYTHE, F. WALZ, and H. KRONMÜLLER
*Investigation of the Magnetic Aftereffect in Dilute Fe–Ni Alloys after Low-Temperature Neutron Irradiation
(a) Vol. **69**, No. 1, p. 237–240 (1982).
See also 11, p. 118; *18.2*, p. *267*; 21.1, p. 344; 21.1.1, p. 359.

S. B. FISHER and P. K. MADDEN
*The Calculation of the Effect of Point Defect Trapping on Void Growth
(a) Vol. **69**, No. 1, p. 257–268 (1982).
See also 11, p. 118; 21.1.1, p. 359.

A. LOISEAU and A. LASALMONIE
*Contrast of the Dislocations for $\boldsymbol{g} \cdot \boldsymbol{b} = 0, 1, 2$. Application to α-Titanium
(a) Vol. **69**, No. 1, p. 315–324 (1982).
See also 21, p. 331.

G. COCCO, L. SCHIFFINI, S. ENZO, and A. BENEDETTI
A SAXS Study of Coarsening Mechanism of GP Zones in the Al–Li–Mg System
(a) Vol. **69**, No. 1, p. 343–348 (1982).
See also *1.3*, p. *25*; 21, p. 331; 21.2, p. 365; 21.3, p. 367.

H. J. BLYTHE, F. DWORSCHAK, I. M. RICHARDSON, and F. WALZ
*Magnetic Relaxation Spectra of Some Dilute Binary Iron-Based Alloys after Low-Temperature Electron Irradiation
(a) Vol. **69**, No. 1, p. K97–K101 (1982).
See also 11, p. 118; *18.2*, p. *267*; 21.1.1, p. 359.

H. MÜHLBACH
*Dislocation Relaxation Peaks in High-Purity Molybdenum Single Crystals
(a) Vol. **69**, No. 2, p. 615–624 (1982).
See also *12.1*, p. *130*; 21, p. 331.

H. E. HANSEN and K. PETERSEN
*Surface Damage from Diamond-Saw Cutting of High Purity Mo Studied by Positrons
(a) Vol. 69, No. 2, p. 625–629 (1982).
See also *1.6*, p. *42*; 21, p. 332.

O. SIMMICH
*Description of Particle Growth by Means of a Difference Approximation
(a) Vol. 69, No. 2, p. 657–662 (1982).
See also *1.2*, p. *15*; 21, p. 332.

P. PAUL, S. P. PURI, D. RAJ, and R. K. PURI
Force Constant Change in Sn–Impurity Systems: ^{107}Pd:^{119}Sn, ^{197}Au: ^{119}Sn,and $^{63.5}$Cu:^{119}Sn
(b) Vol. 110, No. 1, p. 81–87 (1982).
See also *5*, p. *66*; 6, p. 72; 21.1, p. 345; 21.6, p. 375.

F. M. MANSY, N. K. GOBRAN, and G. SAID
*High-Temperature Creep Mechanism and Dissolution Process in Al–9.5 at% Zn Alloy
(a) Vol. 70, No. 1, p. 57–61 (1982).
See also 14.1, p. 194; 21, p. 333; 21.1, p. 345.

G. S. MURTHY and D. H. SASTRY
*Impression Creep of Zinc and the Rate-Controlling Dislocation Mechanism of Plastic Flow at High Temperatures
(a) Vol. 70, No. 1, p. 63–71 (1982).
See also 12.1, p. 130; 21.1, p. 345.

T. TROEV, A. KRUSTANOV, V. P. SHANTAROVICH, L. G. ARAVIN, and B. P. MOLIN
*Doppler Broadening and Positron Lifetime Measurements in Al–49 at% Zn at Different Temperatures
(a) Vol. 70, No. 1, p. 87–92 (1982).
See also 21, p. 333; 21.1, p. 346.

L. M. CASPERS, A. VAN VEEN, M. R. YPMA, and G. J. VAN DER KOLK
*Helium Precipitation in α-Fe
(a) Vol. 70, No. 1, p. 109–125 (1982).
See also 21.1.1, p. 360

R. HAMAR et J. P. RIQUET
*Estimation de l'énergie d'un joint de phases métalliques
(Estimate of the Boundary Energy of Metallic Phases)
(a) Vol. 70, No. 1, p. 127–138 (1982).
See also *1.4*, p. *32*; 21, p. 333; 21.1, p. 346.

J. X. ZHANG, T. S. KÊ, G. Y. LI, G. FANTOZZI, P. F. GOBIN, and M. WELLER
*Low-Temperature Cold Work Internal Friction Peaks in Al–0.5 wt% (0.21 at%) Cu
(a) Vol. 70, No. 1, p. 159–165 (1982).
See also *12.1*, p. *130*; 21, p. 333; 21.1, p. 346.

H. MIZUBAYASHI and S. OKUDA
*Low Temperature Internal Friction in Deformed Ta
(a) Vol. 70, No. 1, p. 257–268 (1982).
See also *12.1*, p. *130*; 21, p. 333.

S. A. MAHMOUD, M. A. SEMARY, and Z. FARID
*The Effect of Plastic Deformation on the Coherency of the Θ′-Phase
(a) Vol. 70, No. 1, p. K77–K79 (1982).
See also *12.1*, p. *130*; 21, p. 333; 21.1, p. 346.

H. Löffler and R. W. Gould
*On the Influence of Small Magnesium Additions on the Decomposition Behaviour of an Al–Zn (6 at%) Mother Alloy
(a) Vol. **72**, No. 1, p. 147–152 (1982).
See also 14.1, p. 195; 21, p. 337; 21.1, p. 350.

H. J. Kolkman
*Deviations from Schmid's Law in Thin Copper Single Crystals
(a) Vol. **72**, No. 1, p. 189–198 (1982).
See also 21.1, p. 350.

A. Luft and Ch. Ritschel
*Deformation and Slip Behaviour of Cold-Worked Molybdenum Single Crystals at Elevated Temperature (I): The Effect of Crystals Orientation
(a) Vol. **72**, No. 1, p. 225–237 (1982).
See also 21, p. 337.

A. H. van Ommen
*Vacancies in NiGa
(a) Vol. **72**, No. 1, p. 273–280 (1982).
See also 9, p. 92; 21.1, p. 350.

T. A. Ryumshina and V. I. Zaitsev
*Influence of Compressibility Anisotropy on the Dislocation Structure of Zinc, Bismuth, and Antimon Bicrystals
(a) Vol. **72**, No. 1, p. K11–K13 (1982).
See also 21.1, p. 351; 21.7, p. 380.

R. H. Rautioaho
*An Interatomic Pair Potential for Copper. Calculation of Stacking Fault Energy
(b) Vol. **112**, No. 2, p. 391–398 (1982).
See also 6, p. 76; 21.1, p. 351.

K. Masuda
*Calculation of Binding Energies of Small Vacancy Clusters in B.C.C. Transition Metals
(b) Vol. **112**, No. 2, p. 609–614 (1982).
See also *13.4*, p. *172*; 21, p. 337; 21.1.1, p. 361.

R. Khanna
Force Model Dependence and Asymptotic Behaviour of the Static Lattice Green Function for H.C.P. Crystals
(b) Vol. **112**, No. 2, p. 667–674 (1982).
See also *6*, p. *77*; 21.1, p. 351; 21.3, p. 368.

M. Pstrokoński and W. Chomka
*Measurements of the Internal Friction and the Shear Modulus in Cold-Worked Ordered and Disordered Ag_3Mg Alloys
(a) Vol. **72**, No. 2, p. 475–482 (1982).
See also *12.1*, p. *132*; 21.3, p. 368; 21.6, p. 377.

G. A. Malygin
*Impurity Atmospheres and Effective Activation Parameters of Dislocation Motion with Special Reference to Al–Mg and Zr–O Alloys
(a) Vol. **72**, No. 2, p. 493–501 (1982).
See also 21, p. 337; 21.1, p. 352; 21.3, p. 368.

M. S. Bapna
*An Estimate of the Precipitation Hardening by Coherent Particles
(a) Vol. **72**, No. 2, p. 799–807 (1982).
See also *1.4*, p. *35*; 21, p. 338; 21.1, p. 352; 21.6, p. 377.

H. C. Eaton, A. C. Chou, and J. C. Carter
*Topographic Analysis of Grain Boundary Dislocations in Cold-Worked Tungsten Using Field Ion Microscopy
(a) Vol. **72**, No. 2, p. 817–824 (1982).
See also *1.6*, p. *43*; 17.1, p. 258; 21, p. 338.

M. J. McGetrick, C. F. Coleman, A. E. Hughes, P. Rice-Evans, and F. A. Smith
*Positron Trapping in Mechanically Deformed Titanium
(a) Vol. **72**, No. 2, p. K185–K188 (1982).
See also 21.1, p. 352.

K. D. Moore, P. L. Jones, and F. H. Cocks
*A. Positron Annihilation Study of the Precipitation Hardening Effect in a Maraging Steel
(a) Vol. **72**, No. 2, p. K223–K227 (1982).
See also 21.1.1, p. 361.

V. I. Zubov
*On the Theory of Vacancies in Simple Metals
(b) Vol. **113**, No. 1, p. K73–K76 (1982).
See also 21, p. 339.

O. Simmich
*Theoretical Considerations on the Diffusion-Controlled Particle Growth in an Isotropic Medium
(a) Vol. **73**, No. 1, p. 107–115 (1982).
See also *1.2*, p. *21*; 21, p. 339.

B. Gueffroy-Oettel and H. Löffler
*The Dependence of the Reversion of Guinier-Preston Zones in an Al–4.5 at% Zn–2.0 at% Mg Alloy on the Pre-Ageing Conditions
(a) Vol. **73**, No. 1, p. 153–156 (1982).
See also 21, p. 339; 21.1, p. 353; 21.3, p. 368.

R. Tognato
*On Internal Premelting in Metals
(a) Vol. **73**, No. 1, p. 237–241 (1982).
See also 21, p. 339.

S. A. Ivanov, A. L. Kolesnikova, and A. E. Romanov
*Elastic Stresses and Elastic Energy of a Flux Line in a Half-Space
(a) Vol. **73**, No. 1, p. K31–K34 (1982).
See also *14.2*, p. *200*.

R. A. Varin
*On the Validity of Interpretation of Some Grain Boundary Processes Based on the Assumption about the Existence of Unrelaxed Extrinsic Grain Boundary Dislocations
(a) Vol. **73**, No. 1, p. K51–K55 (1982).
See also 21, p. 339; 21.1, p. 354; 21.1.1, p. 362.

G. Haneczok and J. W. Moroń
*The Analysis of Internal Friction Curves for a Temperature Gradient Existing along the Sample
(a) Vol. 73, No. 1, p. K123–K127 (1982).
See also *12.1*, p. *132*; 21, p. 340.

A. J. Fendrik, A. M. Monti, and E. J. Savino
*Self-Interstitial in Hexagonal Close-Packed Metals
(b) Vol. 113, No. 2, p. 709–714 (1982).
See also 21.3, p. 368.

A. M. Roshchupkin and V. Ya. Platkov
*The Static and Dynamic Hysteresis Responsible for the Dislocational Amplitude Dependent Internal Friction (I): Theory
(a) Vol. 73, No. 2, p. 303–313 (1982).
See also 12.1, p. 132; 21, p. 340.

V. G. Chudinov, N. V. Moseev, B. N. Goshchitskii, and V. I. Protasov
*Atomic Configurations of Point Defects in Intermetallic Nb_3Sn Compound
(a) Vol. 73, No. 2, p. 325–331 (1982).
See also 21, p. 340; 21.7, p. 380.

G. Said, H. H. Hassan, G. A. Hassan, and A. A. Youssef
*Effect of Type and Amount of Pre-Torsional Deformation on the Creep Characteristics of Zinc Wire Samples
(a) Vol. 73, No. 2, p. K175–K177 (1982).
See also 21.1, p. 355.

I. N. Kuzmenko, O. A. Palagina, and V. V. Pustovalov
*Study of Plastic Deformation of Aluminium near the Superconducting Transition
(a) Vol. 74, No. 1, p. 51–57 (1982).
See also 14.2, p. 201; 21, p. 341.

B. Gueffroy-Oettel, R. Ramlau, and H. Löffler
*On the Relation between Precipitation Density and Size and Hardness during the Reversion Treatment of an Al–4.5 at%Zn–2.0 at%Mg Alloy Pre-Aged at 100 °C
(a) Vol. 74, No. 1, p. 149–157 (1982).
See also 21, p. 341; 21.1, p. 356; 21.3, p. 368.

H. J. Blythe, F. Walz, and F. Dworschak
*Magnetic Relaxation in Dilute Iron–Nickel Alloys Following Low-Temperature Electron Irradiation
(a) Vol. 74, No. 1, p. 181–184 (1982).
See also 11, p. 128; *18.2*, p. *273*; 21.1, p. 356; 21.1.1, p. 363.

J. Degauque and B. Astie
*Evolution of the Configuration of Magnetic Domains Interacting with Structural Defects in High Purity Iron
(a) Vol. 74, No. 1, p. 201–210 (1982).
See also *18.2*, p. *273*; 21.1.1, p. 363.

M. Čerňanský
*Broadening and Shift of Diffraction Lines Due to Partical Coherence of X-Rays
(b) Vol. 114, No. 2, p. 365–372 (1982).
See also *1.4*, p. *37*; 21.1, p. 357.

10.2 Non-Metals

Review Articles

A. B. Doktorov and E. A. Kotomin
*Theory of Tunneling Recombination of Defects Stimulated by Their Motion (I): General Formalism
(b) Vol. **114**, No. 1, p. 9–34 (1982).
See als *9*, p. *91*.

E. A. Kotomin and A. B. Doktorov
*Theory of Tunneling Recombination of Defects Stimulated by Their Motion (II): Three Recombination Mechanisms
(b) Vol. **114**, No. 2, p. 287–318 (1982).
See also *9*, p. *91*; 22.5.2, p. 454.

Original Papers and Short Notes

E. A. Kotomin and A. L. Shluger
*Semiempirical Calculations of Defect Properties in LiF Crystal (II): Electron and Hole Centres and Their Recombination
(b) Vol. **109**, No. 1, p. 75–81 (1982).
See also *13.4*, p. *162*; 22.5.2, p. 454.

Zh. Egemberdiev, S. Zazubovich, and V. Nagirnyi
*Polarized Luminescence of Lead Ions Associated with Interstitial Radiation Defects in $KCl:PbCl_2$ Crystals
(b) Vol. **109**, No. 1, p. 173–182 (1982).
See also *20.3*, p. *318*; 22.5.2, p. 454.

S. Milshtein and A. Senderichin
*Model of Single Charged Dislocation in Tetravalent Semiconductors
(b) Vol. **109**, No. 1, p. 429–436 (1982).
See also *13.4*, p. *163*; 22.1.1, p. 395; 22.1.2, p. 400.

I. E. Bolotov and V. Yu. Kolosov
Electron Microscope Investigation of Crystals Based on Bend-Contour Arrangement (I): Relationship between Bend-Contour Arrangement and Bend Geometry
(a) Vol. **69**, No. 1, p. 85–96 (1982).
See also *1.5*, p. *38*; 22.1.3, p. 417.

D. Mergel and R. Labusch
*Optical Excitations of Dislocation States in Silicon
(a) Vol. **69**, No. 1, p. 151–158 (1982).
See also *13.4*, p. *163*; 16, p. 249; 22.1.2, p. 400.

K. S. Arntson and M. B. Langley
*A Phenomenological Theory for Hysteresis in the Ionic Conductivity of Solid Electrolytes
(a) Vol. **69**, No. 1, p. 185–192 (1982).
See also 8, p. 85; *14.4*, p. *234*; 22.5, p. 450.

Ch. Schnittler, G. Teichmann, W. Kaufmann, and H. Scheider
*A Thermodynamic Model for the Formation and Growth of B-Defects in Silicon
(a) Vol. **69**, No. 1, p. 227–236 (1982).
See also 22.1.2, p. 400.

S. K. Maksimov and E. N. Nagdaev
Electron Microscopy Images of Composition-Modulated Crystals (II): Dilatational Contrast Images
(a) Vol. **69**, No. 2, p. 505–512 (1982).
See also *1.4*, p. *32*; 22.2.4, p. 430.

T. M. Bulatova and E. M. Nadgornyi
*Effect of the Polycrystalline Structure on the Relaxation Processes and Dislocation Motion in NaCl Crystals
(a) Vol. **69**, No. 2, p. 527–533 (1982).
See also 22.5.2, p. 455.

M. N. Avasthi, M. T. El-Gemal, and M. Saleem
*High Ionic Conductivity of $Ag_{19}I_{15}P_2O_7$ Solid Electrolyte
(a) Vol. **69**, No. 2, p. 535–539 (1982).
See also *14.4*, p. *234*; 22.8, p. 483.

B. S. Acharya and M. L. Mukherjee
*Role of Tin Centres on the Thermoluminescence of KCl: $SnCl_2$ Single Crystals
(a) Vol. **69**, No. 2, p. 649–655 (1982).
See also 20.1, p. 301; *20.3*, p. *320*; 22.5.2, p. 455.

L. Castellani, P. Gondi, C. Patuelli, and R. Berti
*On the EBIC Contrast of Dislocations in Si
(a) Vol. **69**, No. 2, p. 677–685 (1982).
See also *1.4*, p. *32*; 22.1.2, p. 402.

H. Föll
*Lattice Imaging of Silicide Silicon Interfaces and the Interpretation of Interfacial Defects
(a) Vol. **69**, No. 2, p. 779–788 (1982).
See also *1.6*, p. *42*; 22.1.2, p. 402.

D. N. Popov, N. M. Nancheva, and M. R. Minev
*Positron Annihilation in Sr^{2+} and Ba^{2+} Doped KCl
(a) Vol. **69**, No. 2, p. K143–K146 (1982).
See also 22.5.2, p. 456.

A. Goltzene, B. Meyer, and C. Schwab
On the Generation of Paramagnetic Defects by the Abrahams-Buiocchi Etchant on Semi-insulating GaAs
(a) Vol. **69**, No. 2, p. K151–K156 (1982).
See also *19*, p. *284*; 22.2.1, p. 420.

L. Pasemann
*EBIC Contrast of Surface-Parallel Dislocations in a Planar Semiconductor Diode with Large Surface Recombination
(a) Vol. **69**, No. 2, p. K199–K203 (1982).
See also 14.3.3, p. 220; 22, p. 384.

P. A. Cherdantsev, I. P. Chernov, and A. P. Mamontov
*Chain Annihilation of Defects in Semiconductor Crystals
(a) Vol. **69**, No. 2, p. K215–K218 (1982).
See also *11*, p. *119*; 22.1.2, p. 402; 22.2.1, p. 420.

M. L. Dalal, S. Sivaraman, and Y. V. G. S. Murti
*Thermal Dissociation Energy of F-Centers in Alkali Halides
(b) Vol. **110**, No. 1, p. 349–356 (1982).
See also 9, p. 91; 22.5.2, p. 456.

P. KUKK, Ô. PALMRE, and E. MELLIKOV
*The Structure of Recombination Centres in Activated ZnSe Phosphors
(a) Vol. **70**, No. 1, p. 35–42 (1982).
See also *13.4*, p. *166*; 20.3, p. 321; 22.4.2, p. 442.

L. N. SKUJA and A. R. SILIN
*A Model for the Non-Bridging Oxygen Center in Fused Silica. The Dynamic Jahn-Teller Effect
(a) Vol. **70**, No. 1, p. 43–49 (1982).
See also *13.4*, p. *166*; 20.1, p. 302; 20.3, p. 321; 22.6, p. 469.

M. RENNINGER
*Imaging of Lattice Defects by Double Diffractometric Transmission Topography
(a) Vol. **70**, No. 1, p. 183–188 (1982).
See also 22.1.2, p. 403.

F. CALLENS, W. MAENHOUT-VAN DER VORST, and L. W. KETELLAPPER
*The Effect of the Solution pAg and pH on the Space Charge Characteristics of Silver Bromide Emulsion Grains
(a) Vol. **70**, No. 1, p. 189–195 (1982).
See also *14.4*, p. *235*; 22.5.1, p. 453.

L. PASEMANN, H. BLUMTRITT, and R. GLEICHMANN
*Interpretation of the EBIC Contrast of Dislocations in Silicon
(a) Vol. **70**, No. 1, p. 197–209 (1982).
See also *1.4*, p. *33*; 14.3.3, p. 220; 22.1.2, p. 403.

P. J. RATCLIFFE
*X-Ray Topography and Laser Tomography of Crystalline TGSe
(a) Vol. **70**, No. 1, p. 211–218 (1982).
See also 22.9, p. 504.

K. ISHIDA, Y. MATSUMOTO, and K. TAGUCHI
*Lattice Defects in LPE InP–InGaAsP–InGaAs Structure Epitaxial Layers on InP Substrates
(a) Vol. **70**, No. 1, p. 277–286 (1982).
See also *1.5*, p. *39*; 22.2.4, p. 430.

K. N. REDDY, M. L. RAO, and V. H. BABU
*Thermoluminescence and Optical Absorption Studies of Z_1-Centres in NaCl Crystals Doped with Samarium
(a) Vol. **70**, No. 1, p. 335–341 (1982).
See also 20.1, p. 303; *20.3*, p. *321*; 22.5.2, p. 456.

M. STOJIĆ
*The Origin of Iron Interstitial in Quenched Silicon Single Crystals
(a) Vol. **70**, No. 1, p. K39–K42 (1982).
See also *19*, p. *285*; 22.1.2, p. 404.

E. WOLF
*The Nature and Formation Mechanism of Anomalous Defects in Dislocation Free FZ-Silicon Crystals Caused by Hydrogen Doping
(a) Vol. **70**, No. 1, p. K59–K61 (1982).
See also 3, p. 57; 22.1.2, p. 404.

P. Paufler, O. Boehnke, G. Teichert, and K. Löschke
*On the Influence of Plastic Deformation upon the Mass Density of GaP
(a) Vol. **70**, No. 1, p. K81 – K83 (1982).
See also 22.2.2, p. 424.

S. Bhan
*Luminescence and Optical Investigation of NaF: C^{3+} Single Crystals
(b) Vol. **110**, No. 2, p. 399 – 405 (1982).
See also 20.1, p. 303; *20.3*, p. *321*; 22.5.2, p. 456.

Fat Duen Ho
*The F^+ Center in Lithium Tantalate
(b) Vol. **110**, No. 2, p. 437 – 443 (1982).
See also *13.4*, p. *166*; 20.1, p. 303; 22.8.1, p. 494.

I. A. Parfianovich and V. N. Salomatov
*The F_A(II) Center Relaxed Excited State and Models of F_A(II)-Like Defects in Alkali Halide Crystals
(b) Vol. **110**, No. 2, p. 513 – 520 (1982).
See also 20.1, p. 303; 22.5.2, p. 456.

D. Vignaud, J. L. Farvacque, and D. Ferre
*Anisotropy of the Free Carrier Piezoelectric Scattering by Dislocations in III–V Compounds
(b) Vol. **110**, No. 2, p. 601 – 609 (1982).
See also *14.3*, p. *205*; 22.2, p. 418.

M. Mladenova, A. Malinovski, and M. Georgiev
*Conversion of Impurity–Vacancy Complexes under γ-Irradiation and Subsequent F Band Bleaching in $KCl{:}Eu^{2+}$
(b) Vol. **110**, No. 2, p. 643 – 652 (1982).
See also 19, p. 285; 22.5.2, p. 456.

R. R. Mijangos, J. L. Marín, A. Clark, A. Navarro, and C. Ruíz Mejía
*Effect of Surface Polarization on the Absorption Spectrum of an F-Center
(b) Vol. **110**, No. 2, p. 703 – 708 (1982).
See also *13.4*, p. *167*; 22.5.2, p. 457.

V. B. Ptashnik, T. Yu. Dunaeva, and Yu. M. Baikov
*The Electrical Conductivity of Solid Lithium Hydride Doped with Sulphur
(b) Vol. **110**, No. 2, p. K121 – K124 (1982).
See also *14.4*, p. *235*; 22, p. 385.

S. Bhan
*F_{Z_1} Centres in $NaF{:}Ca^{2+}$ Single Crystals
(b) Vol. **110**, No. 2, p. K125 – K127 (1982).
See also *20.1*, p. *304*; 22.5.2, p. 457.

Z. G. Liu and W. Skrotzki
*Slip on {100} Planes in LiF
(a) Vol. **70**, No. 2, p. 433 – 440 (1982).
See also 22.5.2, p. 457.

K. Tempelhoff and N. van Sung
Formation of Self-Disorder Agglomerates in Dislocation-Free Silicon during Crystal Growth
(a) Vol. **70**, No. 2, p. 441 – 449 (1982).
See also *3*, p. *57*; 22.1.2, p. 404.

V. Holý
The Coherence Description of the Dynamical X-Ray Diffraction from Randomly Disordered Crystals (I): General Formalism
(b) Vol. **111**, No. 1, p. 341 – 351 (1982).
See also *1.1*, p. *11*; 22, p. 386.

L. A. Kazakevich and P. F. Lugakov
*The Effect of Dislocations on the Formation of Radiation Defects in Silicon
(a) Vol. **71**, No. 1, p. 99 – 108 (1982).
See also 11, p. 122; 22.1.2, p. 406.

U. Balachandran and N. G. Eror
*Oxygen Nonstoichiometry of Tantalum-Doped Calcium Titanate
(a) Vol. **71**, No. 1, p. 179 – 184 (1982).
See also 22.8.2, p. 500.

N. D. Zakharov, M. Pasemann, and V. N. Rozhanski
*Observations of Point Defects in Silicon by Means of Dark-Field Lattice Plane Imaging
(a) Vol. **71**, No. 1, p. 275 – 281 (1982).
See also 22.1.2, p. 407.

B. R. Sood
Suitability of NaF as a Phaser Material
(a) Vol. **71**, No. 1, p. K117 – K119 (1982).
See also *6*, p. *74*; *20.2*, p. *317*; 22.5.2, p. 457.

R. K. Sinha and M. L. Mukherjee
*Lattice Parameter and F-Band Position in $KCl_{1-x}Br_x$ Crystals
(b) Vol. **111**, No. 2, p. K117 – K121 (1982).
See also *1.1*, p. *12*; 20.1, p. 306; 22.5.4, p. 467.

B. Escaig, J. L. Farvacque, and D. Ferré
*On the Measurement of Thermally Activated Slip Parameters in Tetrahedrally Coordinated Semiconductors
(a) Vol. **71**, No. 2, p. 329 – 334 (1982).
See also 22.2.3, p. 427.

V. B. Dudnikova and V. S. Urusov
*Solvus of $KCl:Sr^{2+}$ Solid Solution by Density Measurements of Single Crystals with the Flotation Method
(a) Vol. **71**, No. 2, p. 409 – 414 (1982).
See also *1.2*, p. *18*; 22.5.2, p. 458.

G. Kowalski and J. Gronkowski
*On the Intermediary Image in X-Ray Section Topography
(a) Vol. **71**, No. 2, p. 611 – 617 (1982).
See also *1.4*, p. *34*; 22.1.2, p. 408.

M. T. Sebastian, D. Pandey, and P. Krishna
X-Ray Diffraction Study of the 2H to 3C Solid State Transformation in Vapour Grown Single Crystals of ZnS
(a) Vol. **71**, No. 2, p. 633 – 640 (1982).
See also *1.2*, p. *18*; 22.4.1, p. 441.

G. A. Andreev, M. Hartmanová, and V. A. Klimov
*Precipitation-Induced Dislocation Loops and Helices in NaCl–Ni Crystals
(a) Vol. **71**, No. 2, p. K137 – K138 (1982).
See also 22.5.2, p. 458.

E. V. Kučys, O. E. Panchuk, J. V. Kameneckas, and D. F. Junevičius
*On the Kinetics of the Process of Decay of Supersaturated Solid Copper or Silver Solution in Cadmium Telluride
(a) Vol. **71**, No. 2, p. K157–K160 (1982).
See also 9, p. 92; 14.3, p. 207; 22.4.3, p. 446.

F. Callens and W. Maenhout-van der Vorst
*The Influence of EDTA on the Space Charge Characteristics of AgBr and β-AgI Emulsion Grains
(a) Vol. **71**, No. 2, p. K161–K163 (1982).
See also *14.4*, p. *236*; 22.5.1, p. 453.

G. K. Padam, B. Singh, S. K. Gupta, and G. D. Sootha
*Effect of Impurities on the Oxygen Radicals and F^+ Centres in CdS
(a) Vol. **71**, No. 2, p. K169–K172 (1982).
See also *19*, p. *287*; 22.4.1, p. 441.

S. R. Singh and S. Singh
*Phonon Scattering in Plastically Deformed Ge
(b) Vol. **112**, No. 1, p. 51–54 (1982).
See also *6*, p. *75*; 22.1.1, p. 397.

Y. M. M. Hornos, H. Panepucci, C. J. Magon, and M. C. Terrile
*Discrete Saturation Effects of the ESR Spectrum of H^0 in CaF_2
(b) Vol. **112**, No. 1, p. 91–96 (1982).
See also *19*, p. *288*; 22.5.3, p. 465.

V. Holý
*The Coherence Description of the Dynamical X-Ray Diffraction from Randomly Disordered Crystals (II): Some Numerical Results
(b) Vol. **112**, No. 1, p. 161–169 (1982).
See also *1.1*, p. *12*; 22, p. 387.

M. Chinnusamy and T. Nagarajan
*Positron Annihilation in Mn^{2+}-Doped KCl Single Crystals
(b) Vol. **112**, No. 1, p. 297–301 (1982).
See also 22.5.2, p. 458.

A. K. Gerasyuk, V. F. Pisarenko, and V. V. Popov
*On the Theory of Dipolar Thermally Stimulated Depolarization in Doped Ionic Crystals
(b) Vol. **112**, No. 1, p. K11–K14 (1982).
See also *14.4*, p. *236*; 22.5.3, p. 465.

B. M. Loginov and A. A. Predvoditelev
*Computer Simulation of Dislocation Motion through a Flexible and Reactionable Dislocation Forest of Different Density in NaCl and Mg Crystals
(a) Vol. **72**, No. 1, p. 69–77 (1982).
See also *10.1*, p. *98*; 21.3, p. 367; 22.5.2, p. 458.

D. A. Golopentia and L. M. Slifkin
*Dielectric Relaxation in Manganese-Doped Silver Chloride
(a) Vol. **72**, No. 1, p. 123–129 (1982).
See also *14.4*, p. *236*; 22.5.1, p. 453.

S. K. Maksimov and E. N. Nadgaev
*Electron Microscopy Images of Composition-Modulated Crystals (III): Moiré Fringe Images
(a) Vol. **72**, No. 1, p. 135–145 (1982).
See also *1.4*, p. *35*; 22.2.4, p. 430.

J. Echigoya and J. W. Edington
*Stacking Faults in Recrystallized Cadmium Sulphide
(a) Vol. **72**, No. 1, p. K1–K4 (1982).
See also 22.4.1, p. 441.

N. Erben, P. Paufler, K. Löschke, S. Däbritz, and K. Kleinstück
*Kossel Line Broadening near a Scratch on GaP
(a) Vol. **72**, No. 1, p. K15–K18 (1982).
See also 22.2.2, p. 425.

D. E. Ioannou
*Comments on the $E_V + 0.45$ eV Quenched-in Level in Silicon
(a) Vol. **72**, No. 1, p. K33–K36 (1982).
See also *13.4*, p. *171*; 22.1.2, p. 409.

K. G. Nikiforov, A. G. Gurevich, S. I. Radautsan, L. M. Emiryan, and V. E. Tezlevan
*Effect of Nonstoichiometry on Ferromagnetic Resonance in $CdCr_2S_4$ Crystals
(a) Vol. **72**, No. 1, p. K37–K39 (1982).
See also *19*, p. *288*; 22.8, p. 487.

P. K. Khabibullaev, L. P. Khiznichenko, N. S. Tulyaganova, and L. G. Zotova
*Transformation of Self-Interstitials in Silicon under Electronic and Thermal Stimulation
(a) Vol. **72**, No. 1, p. K69–K72 (1982).
See also 11, p. 124; 22.1.2, p. 409.

S. J. Pearton
*Hydrogen Passivation of γ-Induced Point Defects in Silicon
(a) Vol. **72**, No. 1, p. K73–K75 (1982).
See also *14.3*, p. *208*; 22.1.2, p. 410.

S. Milshtein
**C–U* Measurements on Heavily Dislocated Si
(a) Vol. **72**, No. 1, p. K99–K103 (1982).
See also *13.4*, p. *171*; 14.3, p. 208; 22.1.2, p. 410.

J. Kikas and M. Rätsep
*Correlation of Transition Frequencies in Impurity Spectra of Inhomogeneous Solids
(b) Vol. **112**, No. 2, p. 409–415 (1982).
See also *20.1*, p. *309*; 22.5.2, p. 459.

H. Opyrchał, K. D. Nierzewski, and B. Macalik
*Effect of γ-Irradiation on Eu^{2+} Ions in KCl Crystals
(b) Vol. **112**, No. 2, p. 429–434 (1982).
See also 11, p. 124; 19, p. 288; 20.1, p. 309; 20.3, p. 324; 22.5.2, p. 459.

A. Dauletbekova, A. Akilbekov, and A. Elango
*Thermo- and Photostimulated Recombinations of $F–H_A$ and $\alpha–I_A$ Centres in KBr with Large Na Concentration
(b) Vol. **112**, No. 2, p. 445–452 (1982).
See also 20.3, p. 324; 22.5.2, p. 459.

V. I. Kuznetsov and P. F. Lugakov
*Study of Rearrangements of Intrinsic Defects at Annealing of Proton-Irradiated Silicon
(b) Vol. **112**, No. 2, p. 457–462 (1982).
See also 11, p. 124; *13.4*, p. *172*; 22.1.2, p. 410.

S. Bhan
*Thermoluminescence of NaF:Ca^{2+} Single Crystals
(b) Vol. **112**, No. 2, p. 507–514 (1982).
See also *20.3*, p. *324*; 22.5.2, p. 459.

B. Pfalzgraf, O. Roder, and J.-M. Spaeth
*Optical Investigation of Ag^+ Centers in Cesium Halides
(b) Vol. **112**, No. 2, p. 565–574 (1982).
See also 13.5.1, p. 184; *20.1*, p. *309*; 22.5.2, p. 459.

M. Siu Li, M. de Souza, and S. E. Kapphan
*Ionic Thermal Currents under Uniaxial Stress. A New Method to Determine Electric and Elastic Dipole Properties
(b) Vol. **112**, No. 2, p. 685–693 (1982).
See also *14.4*, p. *237*; 22.5.2, p. 459.

W. van Puymbroeck, N. Schrijvers, A. Bouwen, and D. Schoemaker
*Electron Spin Resonance Study of H_D-Type $BrCl^-$ Defects in KCl Doped with Divalent Cation Impurities
(b) Vol. **112**, No. 2, p. 725–733 (1982).
See also *19*, p. *289*; 22.5.2, p. 459.

F. Mariani and J. Vignolo
*Effect of γ-Irradiation on I–V Dipoles in KCl Crystals
(b) Vol. **112**, No. 2, p. K85–K87 (1982).
See also *11*, p. *124*; 22.5.2, p. 459.

N. M. Olekhnovich, A. L. Karpei, V. M. Skidan, and L. D. Puzenkova
*Dynamical Oscillations of X-Ray Integrated Intensity in Laue Diffraction and Their Temperature Dependences for Si Dislocation Crystals
(a) Vol. **72**, No. 2, p. 463–467 (1982).
See also 22.1.2, p. 410.

V. V. Aristov, V. G. Kohn, V. I. Polovinkina, and A. A. Snigirev
*Change of Interference Pattern in the X-Ray Section Topography of Single Crystals with Increasing Dimensions of an Incoherent Source
(a) Vol. **72** No. 2, p. 483–491 (1982).
See also *1.1*, p. *12*; 22.1.1, p. 397; 22.1.2, p. 410.

K. Tempelhoff and N. van Sung
*Formation of Vacancy Agglomerates in Quenched Dislocation-Free Silicon
(a) Vol. **72**, No. 2, p. 617–622 (1982).
See also 22.1.2, p. 411.

I. E. Bolotov, V. Yu. Kolosov, and A. V. Kozhyn
Electron Microscope Investigation of Crystals Based on Bend-Contour Arrangement (II): Bending Phenomena of the Crystal Growth in an Amorphous Film
(a) Vol. **72**, No. 2, p. 645–654 (1982).
See also *1.5*, p. *40*; 22.1.3, p. 417.

V. V. Kveder, Yu. A. Osipyan, W. Schröter, and G. Zoth
On the Energy Spectrum of Dislocations in Silicon
(a) Vol. **72**, No. 2, p. 701–713 (1982).
See also *13.4*, p. *173*; 22.1.2, p. 411.

E. M. LAWSON and S. J. PEARTON
*Hydrogen Passivation of Laser-Induced Acceptor Defects in p-Type Silicon
(a) Vol. **72**, No. 2, p. K155–K158 (1982).
See also *14.3*, p. *209*; 22.1.2, p. 411.

M. R. BROZEL, E. J. FOULKES, and B. TUCK
*The Electrical Properties of InP Diffused with Chromium or Iron
(a) Vol. **72**, No. 2, p. K159–K163 (1982).
See also *14.3*, p. *209*; 22.2.2, p. 426.

T. SUZUKI, Y. HIROSE, and H. HIROSE
*A Model of the Photodoping Mechanism in the Ag/Ag_2S_3 System
(a) Vol. **72**, No. 2, p. K165–K168 (1982).
See also 14.3.2, p. 217; *16*, p. *253*; 22, p. 389.

L. BOSI and M. NIMIS
*Luminescence Properties of an M_A-Type Center in KCl:$SrCl_2$ and KCl:$CaCl_2$
(b) Vol. **113**, No. 1, p. 347–352 (1982).
See also *20.3*, p. *325*; 22.5.2, p. 460.

T. GOWOROEK, C. RYBKA, R. WASIEWICZ, and J. WAWRYSZCZUK
*Positronium Trapping at Thermally Produced Defects in Naphthalene
(b) Vol. **113**, No. 1, p. K9–K13 (1982).
See also 22.9, p. 506.

R. K. GARTIA, R. SINGH, and K. Y. SINGH
*Thermoluminescence Studies of Post-Irradiation Deformation in Ca-Doped NaCl
(b) Vol. **113**, No. 1, p. K57–K60 (1982).
See also *20.3*, p. *325*; 22.5.2, p. 460.

R. CASES, R. ALCALÁ, and V. M. ORERA
*F′-Centers in CsF
(b) Vol. **113**, No. 1, p. K61–K65 (1982).
See also 20.1, p. 311; 20.3, p. 325; 22.5.2, p. 460.

H.-G. BRÜHL and J. BĄK
Determination of Nitrogen Concentrations in (111) Oriented VPE-GaP Epitaxial Layers by Measurement of the Precision Lattice Parameter
(a) Vol. **73**, No. 1, p. 139–144 (1982).
See also *1.5*, p. *40*; 22.2.2, p. 426.

DA YU WANG and A. S. NOWICK
A Grain-Boundary-Type Maxwell-Wagner Peak in the Thermally-Stimulated Depolarization of Doped Ceria Ceramics
(a) Vol. **73**, No. 1, p. 165–172 (1982).
See also *14.4*, p. *238*; 22.6, p. 474.

Y. NISHINO and T. IMURA
*Generation Process of Dislocations in Precipitate-Containing Silicon Crystals
(a) Vol. **73**, No. 1, p. 173–182 (1982).
See also 22.1.2, p. 412.

L. L. REGEL, V. R. REGEL, S. E. BORISKIN, G. G. KNAB, A. A. URUSOVSKAYA, L. I. ALEKSEEVA, and V. V. KLECHKOVSKAYA
*Influence of Irradiation with Heavy Ions on the Defect Structure and Mechanical Properties of LiF Crystals
(a) Vol. **73**, No. 1, p. 255–266 (1982).
See also 11, p. 126; *12.2*, p. *136*; 22.5.2, p. 460.

P. Zaumseil and U. Winter
*Triple Crystal Diffractometer Investigations of Imperfections in Silicon Crystals with Laue Case Diffraction
(a) Vol. **73**, No. 2, p. 455–466 (1982).
See also 22.1.2, p. 413.

P. Veeresham, U. V. Subba Rao, and V. Hari Babu
*Thermoluminescence of Z_1-Centres in $KCl_xBr_{1-x}:Ca^{2+}$ Crystals
(a) Vol. **73**, No. 2, p. 565–572 (1982).
See also 20.1, p. 313; *20.3*, p. *326*; 22.5.2, p. 461.

A. J. Cohen and L. N. Makar
*Models for Color Centers in Smoky Quartz
(a) Vol. **73**, No. 2, p. 593–596 (1982).
See also *20.1*, p. *313*; 22.6, p. 475.

S. P. Kuila and V. V. Ratnam
*Thermal Annealing of V-Bands and Their Correlation with Thermoluminescence Spectra in X-Irradiated Li-Doped KBr Crystals
(a) Vol. **73**, No. 2, p. 597–601 (1982).
See also *20.3*, p. *326*; 22.5.2, p. 462.

A. Castaldini, A. Cavallini, P. Gondi, and E. Bonetti
*Generation and Recombination Images of Dislocations in Si by Scanning Microscopy
(a) Vol. **73**, No. 2, p. 617–624 (1982).
See also 13.4, p. 175; 16, p. 253; 22.1.2, p. 414.

C. Battaglia, S. Bonora, M. Cevolani, G. C. Degli Esposti, and S. Petralia
Thermoluminescence Sensitization in LiF (TLD 100) at Different Post-Annealing Temperatures
(a) Vol. **73**, No. 2, p. K169–K173 (1982).
See also *20.3*, p. *327*; 22.5.2, p. 462.

W. Skrotzki and Z. G. Liu
*Analysis of the Cross Slip Process in Alkali Halides
(a) Vol. **73**, No. 2, p. K225–K229 (1982).
See also 22.5.2, p. 462.

J. Stępień-Damm and E. Mugeński
*Effect of F–F' Conversion on the Lattice Parameter of Electrolytically Coloured KCl, KBr, and KI Crystals
(a) Vol. **73**, No. 2, p. K263–K265 (1982).
See also 22.5.2, p. 462.

J. Prakash
*Electronic Excited State and the Phenomena of Photo-Induced Polarization and Depolarization
(b) Vol. **114**, No. 1, p. 61–66 (1982).
See also 14.4, p. 238; 22.5.2, p. 462.

Ch. Lushchik, J. Kolk, A. Lushchik, N. Lushchik, M. Taiirov, and E. Vasilchenko
*Decay of Excitons into Long-Lived F, H and α, I Pairs in KCl
(b) Vol. **114**, No. 1, p. 103–111 (1982).
See also *20.3*, p. *327*; 22.5.2, p. 462.

D. Mergel and R. Labusch
*Variable Reconstruction of Dislocation Cores in Si
(b) Vol. **114**, No. 2, p. 545–551 (1982).
See also *13.4*, p. *177*; 14.3, p. 211; 16, p. 254; 22.1.2, p. 415.

I. Tale, V. Kortov, and V. Popov
*Trap Spectroscopy in Alkali Halides at Synchronous Measurements of Exoelectron Emission and Luminescence by Fractional Glow Technique
(a) Vol. **74**, No. 2, p. 395–401 (1982).
See also 17, p. 257; 20.3, p. 328; 22.5.2, p. 464.

A. I. Öpik and J. A. Varvas
*Defect Structure of Cu-Doped Cadmium Selenide
(a) Vol. **74**, No. 2, p. 467–473 (1982).
See also *14.3*, p. *212*; 22.4.2, p. 444.

J. Wood, M. J. Howes, and D. V. Morgan
The Effect of Dislocations on the Electron Mobility of GaAs
(a) Vol. **74**, No. 2, p. 493–496 (1982).
See also *14.3.3*, p. *226*; 22.2.1, p. 423.

S. P. Kuila and V. V. Ratnam
*Thermoluminescence Studies of F_A-Centres in X-Irradiated Li-Doped KBr Single Crystals
(a) Vol. **74**, No. 2, p. 615–619 (1982).
See also 20.1, p. 316; *20.3*, p. *328*; 22.5.2, p. 464.

F. S. Nasredinov and A. V. Ermolaev
*Study of Compensation of Excess Charge of Tin and Iron Impurity Atoms in Cobalt Monoxide by Mössbauer Spectroscopy
(a) Vol. **74**, No. 2, p. 631–636 (1982).
See also 5, p. 69; 22.6, p. 476.

C. W. Allen and K. C. Liao
*Dislocation Models for Shear Transformations
(a) Vol. **74**, No. 2, p. 673–681 (1982).
See also *1.2*, p. *24*; 22.3, p. 435.

T. Suzuki and H. Koizumi
*Anomalous Behaviour of Plasticity of CsI
(a) Vol. **74**, No. 2, p. K101–K103 (1982).
See also 22.5.2, p. 464.

K. Narasimha Reddy and V. Hari Babu
*The Study of Z_2-Centres in NaCl Crystals Doped with Samarium
(a) Vol. **74**, No. 2, p. K127–K130 (1982).
See also *20.1*, p. *316*; 22.5.2, p. 464.

J. Leroueille
*Relationship between Carbon and Oxygen Nucleation in CZ Silicon
(a) Vol. **74**, No. 2, p. K159–K163 (1982).
See also 22.1.2, p. 416.

11. Irradiation Effects
(X-Rax Diffraction Investigations See 1 and 10)

Original Papers and Short Notes

V. V. Lasukov, Yu. L. Pivovarov, and O. G. Kostareva
*Spontaneous Radiation of Relativistic Electrons under Planar Channeling and Band Structure of Transverse Energy
(b) Vol. **109**, No. 2, p. 761–772 (1982).
See also *13.1*, p. *149*; 20, p. 294.

K. D. Glinchuk, K. Lukat, and V. I. Vovnenko
*Interaction of Copper Atoms with Radiation Defects in Gallium Arsenide
(a) Vol. **69**, No. 2, p. 521–525 (1982).
See also *20.3*, p. *320*; 22.2.1, p. 420.

H. J. Blythe, F. Walz, I. M. Richardson, and F. Dworschak
*Low-Temperature Magnetic Relaxation in a Dilute Iron–Silicon Alloy after Electron Irradiation at 44 K
(a) Vol. **69**, No. 2, p. 563–567 (1982).
See also *18.2*, p. *268*; 21.1.1, p. 359.

P. K. Madden and S. B. Fisher
Void Swelling Studies of Stainless Steel Containing Silicon and Titanium
(a) Vol. **69**, No. 2, p. 569–588 (1982).
See also *1.4*, p. *32*; 21.1.1, p. 359.

E. Glaser, G. Götz, N. Sobolev, and W. Wesch
*Investigations of Radiation Damage Production in Ion Implanted Silicon
(a) Vol. **69**, No. 2, p. 603–614 (1982).
See also 14.3, p. 204; 19, p. 284; 20.1, p. 301; 22.1.2, p. 402.

S. Loualiche, C. Lucas, P. Baruch, J. P. Gailliard, and J. C. Pfister
Theoretical Model for Radiation Enhanced Diffusion and Redistribution of Impurities. Comparison with Experiments
(a) Vol. **69**, No. 2, p. 663–676 (1982).
See also *9*, p. *91*; 22.1.2, p. 402.

L. Brossard, H. Bernas, L. Thomé, A. Traverse, and P. Nedellec
*Influence of Disorder on Phonon Resistivity of Ion-Implanted Nickel Hybride
(a) Vol. **69**, No. 2, p. 727–732 (1982).
See also *14.1*, p. *193*; 21.1, p. 344.

A. Subrahmanyam
Dielectric Properties of NaCl and KCl Single Crystals X-Irradiated under Different DC Fields
(a) Vol. **69**, No. 2, p. 773–778 (1982).
See also *14.4*, p. *234*; 22.5.2, p. 456.

H. Fischer, H.-D. Geiler, G. Götz, A. V. Dvurechenskii, B. P. Kashnikov, and L. S. Smirnov
*Antiannealing Effects in Ion-Implanted and Electron Pulse Irradiated Quartz Crystals
(a) Vol. **69**, No. 2, p. K139–K142 (1982).
See also 1.2, p. 15; 22.6, p. 469.

A. V. Dvurechenskii, B. P. Kashnikov, L. S. Smirnov, R. Grötzschel, and R. Klabes
*Synthesis of Solid Solutions at Electron Pulse Processing of Ge–Si Structures
(a) Vol. **69**, No. 2, p. K183–K187 (1982).
See also 1.5, p. 38; *3*, p. *57*; 22.1.1, p. 395; 22.1.2, p. 402.

P. A. Cherdantsev, I. P. Chernov, and A. P. Mamontov
*Chain Annihilation of Defects in Semiconductor Crystals
(a) Vol. **69**, No. 2, p. K215–K218 (1982).
See also *10.2*, p. *105*; 22.1.2, p. 402; 22.2.1, p. 420.

V. I. Kuznetsov and P. F. Lugakov
*Study of Radiation Defect Clusters, Their Structure and Properties in Proton-Irradiated Silicon
(a) Vol. **70**, No. 2, p. 387–395 (1982).
See also 14.3, p. 205; 22.1.2, p. 404.

W. R. Fahrner, K. Heidemann, and P. Schöttle
*Results of Ion Implantation into Silicon in the 100 MeV Range (I): Oxygen and Boron Implantation
(a) Vol. **70**, No. 2, p. 463–472 (1982).
See also 14.3.3, p. 221; 22.1.2, p. 405.

H. Kerkow, B. Lukasch, and R. Pippig
*Disorder Production in Ion Implanted Silicon
(a) Vol. **70**, No. 2, p. 489–495 (1982).
See also 22.1.2, p. 405.

M. Servidori, A. Zani, and G. Garulli
Residual Lattice Disorder in Self-Implanted Silicon after Pulsed Laser Irradiation
(a) Vol. **70**, No. 2, p. 691–701 (1982).
See also *10.2*, p. *108*; 22.1.2, p. 405.

V. S. Lysenko, R. N. Litovskii, A. N. Nazarov, and V. P. Kulichkov
Temperature Dependence of Capacitance–Voltage Characteristics in Implanted MOS Structures
(a) Vol. **70**, No. 2, p. 719–728 (1982).
See also *14.3.3*, p. *221*; 22.1.2, p. 405.

A. A. Groza, L. G. Nikolaeva, M. I. Starchik, G. G. Shmatko, and Yu. I. Zaslavskii
Defects in Silicon Crystals after the High Temperature Treatment
(a) Vol. **70**, No. 2, p. 763–768 (1982).
See also *10.2*, p. *108*; 20.1, p. 304; 22.1.2, p. 406.

T. I. Voronina, Yu. A. Goldberg, and T. V. Lvova
*Electrical Properties of GaP Crystals Bombarded with High Energy Protons
(a) Vol. **70**, No. 2, p. K121–K124 (1982).
See also *14.3*, p. *206*; 22.2.2, p. 424.

S. P. Kathuria and C. M. Sunta
TL Response of LiF TLD-100 at High Irradiation Temperatures
(a) Vol. **70**, No. 2, p. K125–K128 (1982).
See also *20.3*, p. *322*; 22.5.2, p. 457.

H.-D. Geiler, F. Thrum, and G. Götz
*Pd-Silicide Reactions Induced by a Millisecond Laser Pulse
(a) Vol. **70**, No. 2, p. K159–K162 (1982).
See also *1.5*, p. *39*; 22.1.2, p. 406.

A. I. Koifman, R. A. Muminov, and A. Ya. Yafasov
*Investigation of the Effect of Large Doses on the Si(Li) Detector Parameters
(a) Vol. **71**, No. 1, p. 59–64 (1982).
See also *14.3.3*, p. *221*; 22.1.2, p. 406.

S. Bahn
Thermoluminescence of NaF Harshaw Single Crystals Irradiated to Very Low and High Doses
(a) Vol. **71**, No. 1, p. 73–78 (1982).
See also *20.3*, p. *323*; 22.5.2, p. 457.

V. S. LYSENKO, T. N. SYTENKO, V. I. ZIMENKO, and O. V. SNITKO
Investigation of Traps in the Transition Region of Si–SiO_2 Structures at Cryogenic Temperatures
(a) Vol. 71, No. 2, p. 619–626 (1982).
See also *13.4*, p. *170*; *14.3.3*, p. *222*; 22.1.2, p. 408.

A. V. DVURECHENSKII, I. A. RYAZANTSEV, and V. A. DRAVIN
*Doping of Amorphous Silicon by Manganese Ion Implantation
(a) Vol. 71, No. 2, p. K133–K136 (1982).
See also 2, p. 49; 13.4, p. 170; *14.3.1*, p. *215*; 22.1.2, p. 408.

K. D. GLINCHUK and K. LUKAT
Non-Trivial Annealing-Induced Increase of the Near-Intrinsic Emission Intensity in γ-Irradiated GaAs
(a) Vol. 71, No. 2, p. K187–K190 (1982).
See also *13.4*, p. *171*; *20.3*, p. *323*; 22.2.1, p. 421.

D. L. SASTRY and K. V. S. RAMA RAO
ESR and Optical Absorption Studies of ClO_3 in $Ba(ClO_4)_2 \cdot 3\,H_2O$ Single Crystals
(b) Vol. 112, No. 1, p. 133–142 (1982).
See also *19*, p. *288*; *20.1*, p. *307*; 22.8.1, p. 495.

V. D. AKHMETOV and V. V. BOLOTOV
*Kinetics of Accumulation of Radiation Defects and Annihilation of Vacancies and Interstitials in Carbon- and Boron-Containing Silicon
(a) Vol. 72, No. 1, p. 61–68 (1982).
See also 22.1.2, p. 409.

W. R. FAHRNER, D. BRÄUNIG, and E. BORSCHERT
*MOS and Thyristor Damage Parameters after 2.5 and 15 MeV High Temperature Electron Irradiation
(a) Vol. 72, No. 1, p. 79–88 (1982).
See also *14.3.3*, p. *223*; 22.1.2, p. 409.

R. V. NANDEDKAR, K. VARATHARAJAN, S. PANCHAPAKESAN, and A. K. TYAGI
*Microhardness, Microstructure, and Blistering of Helium Irradiated Nickel, Alloy Nickel 201, and Nimonic 90
(a) Vol. 72, No. 1, p. 89–94 (1982).
See also *12.1*, p. *131*; 21.1, p. 350.

A. V. DVURECHENSKII, R. GRÖTZSCHEL, N. M. IGONINA, B. P. KASHNIKOV, and N. I. KOMOLOVA
*Impurity Profiles at Multi-Pulse Electron-Beam Annealing of Ion-Implanted Silicon
(a) Vol. 72, No. 1, p. 301–304 (1982).
See also 22.1.2, p. 409.

L. PELIKÁN, V. RYBKA, P. KREJČÍ, V. HNATOWICZ, and J. KVÍTEK
*Study of Boron Implantation in Ag–Si Layer Structures
(a) Vol. 72, No. 1, p. 369–373 (1982).
See also 22.1.2, p. 409.

L. S. WIELUŃSKI, B. M. PAINE, B. X. LIU, C.-D. LIEN, and M.-A. NICOLET
*Ion Mixing of Ni and Pt Layers on Si
(a) Vol. 72, No. 1, p. 399–404 (1982).
See also 9, p. 92; 22.1.2, p. 409.

P. K. Khabibullaev, L. P. Khiznichenko, N. S. Tulyaganova, and L. G. Zotova
Transformation of Self-Interstitials in Silicon under Electronic and Thermal Stimulation
(a) Vol. **72**, No. 1, p. K69–K72 (1982).
See also *10.2*, p. *111*; 22.1.2, p. 409.

Th. Wirth, V. Atzrodt, and H. Lange
*Composition Changes of NiSi and PtSi Due to Ar^+ Ion Bombardment Determined from AES Measurements
(a) Vol. **72**, No. 1, p. K89–K93 (1982).
See also *1.6*, p. *43*; 21, p. 337.

E. Yu. Brailovskii, I. G. Megela, N. H. Pambuchchyan, and V. V. Teslenko
*EPR Study of Electron-Irradiated InP:Fe
(a) Vol. **72**, No. 1, p. K109–K111 (1982).
See also *19*, p. *288*; 22.2.2, p. 425.

H. Opyrchał, K. D. Nierzewski, and B. Macalik
*Effect of γ-Irradiation on Eu^{2+} Ions in KCl Crystals
(b) Vol. **112**, No. 2, p. 429–434 (1982).
See also *10.2*, p. *111*; 19, p. 288; 20.1, p. 309; 20.3, p. 324; 22.5.2, p. 459.

V. I. Kuznetsov and P. F. Lugakov
*Study of Rearrangements on Intrinsic Defects at Annealing of Proton-Irradiated Silicon
(b) Vol. **112**, No. 2, p. 457–462 (1982).
See also *10.2*, p. *111*; *13.4*, p. *172*; 22.1.2, p. 410.

A. D. Pogrebnyak, V. A. Kuzminikh, and K. P. Arefiev
*Positron Annihilation and Profiles of Displaced Atoms in Electron Irradiated Ge
(b) Vol. **112**, No. 2, p. K79–K83 (1982).
See also 22.1.1, p. 397.

F. Mariani and J. Vignolo
*Effects of γ-Irradiation on I–V Dipoles in KCl Crystals
(b) Vol. **112**, No. 2, p. K85–K87 (1982).
See also *10.2*, p. *112*; 22.5.2, p. 459.

S. P. Mhaskar and P. K. Naik
*Effect of Surface Relaxation on Thermal Desorption of Helium from Polycrystalline Molybdenum
(a) Vol. **72**, No. 2, p. 515–520 (1982).
See also *1.6*, p. *43*; 21, p. 338.

V. N. Brudnyi, S. A. Vorobiev, and A. A. Tsoi
*Positron Annihilation and Electrical Properties of Electron-Irradiated ($\approx$ 2 MeV) InAs Crystals
(a) Vol. **72**, No. 2, p. 529–534 (1982).
See also *13.4*, p. *173*; 14.3, p. 209; 22.2.1, p. 422.

N. V. Vershinina and N. N. Gerasimenko
*Some Peculiarities of Radiation-Induced Changes in the Properties of Si–SiO_2 Structures Irradiated at Elevated Temperatures
(a) Vol. **72**, No. 2, p. 569–574 (1982).
See also 13.4, p. 173; *14.3.2*, p. *217*; 22.1.2, p. 411.

B. RAUSCHENBACH and K. HOHMUTH
*Bildung amorpher Metall–Metalloid-Verbindungen durch Ionenimplantation
(Formation of Amorphous Metal–Metalloid Compounds by Ion Implantation)
(a) Vol. **72**, No. 2, p. 667–678 (1982).
See also *2*, p. *51*; 21, p. 338; 21.1, p. 352; 21.1.1, p. 361.

L. K. VODOPYANOV and S. P. KOZYREV
Infrared Reflection Spectroscopy of Ion-Implanted n-$Hg_{0.8}Cd_{0.2}Te$
(a) Vol. **72**, No. 2, p. 737–744 (1982).
See also 14.3, p. 209; *20.1*, p. *309*; 22.4.4, p. 448.

J. M. LÓPEZ and J. A. ALONSO
*Semiempirical Study of Metastable Alloys Obtained by Ion Implantation in Metals and Semiconductors
(a) Vol. **72**, No. 2, p. 777–781 (1982).
See also *1.3*, p. *30*; 21, p. 338; 22, p. 389.

V. A. MOKRITSKII, E. M. KURITSIN, B. A. DANILCHENKO, and V. I. SHAKHOVTSOV
*Epitaxial Germanium Layers Grown by Liquid Phase in γ-Radiation Field
(a) Vol. **72**, No. 2, p. K121–K124 (1982).
See also 1.5, p. 40; *3*, p. *58*; 22.1.1, p. 397.

L. K. VODOPYANOV and S. P. KOZYREV
*Ion Implantation of B^+ in n-$Hg_{0.8}Cd_{0.2}Te$
(a) Vol. **72**, No. 2, p. K133–K136 (1982).
See also 22.4.4, p. 448.

V. E. BORISENKO and V. A. LABUNOV
*Annealing of Antimony Implanted Silicon with Halogen Lamp Irradiation
(a) Vol. **72**, No. 2, p. K173–K176 (1982).
See also *14.3*, p. *209*; 22.1.2, p. 411.

E. V. NIDAEV and A. I. POPOV
*Influence of the Surface on the Stability of Radiation Defects Introduced by Electron Irradiation
(a) Vol. **72**, No. 2, p. K193–K195 (1982).
See also 13.4, p. 173; 14.3.3, p. 224; 22.1.2, p. 411.

V. D. AKHMETOV, V. V. BOLOTOV, L. S. SMIRNOV, O. I. ZHUMAEVA, P. GAWORZEWSKI, and K. SCHMALZ
*Accumulation of Radiation Defects in Oxygen-Containing Silicon Thermally Treated at 600 °C
(a) Vol. **72**, No. 2, p. K211–K216 (1982).
See also 13.4, p. 173; 20.1, p. 310; 22.1.2, p. 411.

B. HENDERSON and A. H. GARRISON
*Ion Implantation in Alkaline Earth Oxides
(b) Vol. ***113***, No. 1, p. 267–276 (1982).
See also *19*, p. 289; 20.1, p. 310; 20.3, p. 325; 22.6, p. 473.

S. KISHIDA, K. MATSUURA, H. FUKUMA, F. TAKEDA, and I. TSURUMI
*Optical Absorption Bands in Neutron Irradiated ZnSe and $ZnS_{0.5}Se_{0.5}$ Crystals
(b) Vol. **113**, No. 1, p. K31–K33 (1982).
See also *20.1*, p. *310*; 22.4.1, p. 441; 22.4.2, p. 443; 22.4.4, p. 448.

A. P. Tyutnev, V. S. Saenko, E. D. Pozhidaev, and A. F. Akkerman
*Transient Radiation-Induced Conductivity in Polymers
(a) Vol. **73**, No. 1, p. 81–89 (1982).
See also *14.3*, p. *210*; 22.9, p. 507.

Bai-Xin Liu, M.-A. Nicolet, and S. S. Lau
*Ti–Au Metallic Glasses Formed by Ion Mixing
(a) Vol. **73**, No. 1, p. 183–188 (1982).
See also *1.5*, p. *41*; 2, p. 52; 21.1, p. 353; 21.6, p. 377.

J. Narayan and O. W. Holland
*Solid-Phase-Epitaxial Growth in Ion-Implanted Silicon
(a) Vol. **73**, No. 1, p. 225–236 (1982).
See also 2, p. 52; *3*, p. *59*; 22.1.2, p. 412.

L. L. Regel, V. R. Regel, S. E. Boriskin, G. G. Knab, A. A. Urusovskaya, L. I. Alekseeva, and V. V. Klechkovskaya
Influence of Irradiation with Heavy Ions on the Defect Structure and Mechanical Properties of LiF Crystals
(a) Vol. **73**, No. 1, p. 255–266 (1982).
See also *10.2*, p. *113*; *12.2*, p. *136*; 22.5.2, p. 460.

S. Nojima
*An Anomalous Aspect of GaAs Surface Dissociation by Q-Switch Pulsed Ruby Laser Irradiation
(a) Vol. **73**, No. 1, p. K7–K9 (1982).
See also 1.6, p. 44; 22.2.1, p. 422.

H.-D. Geiler, K. Hehl, and D. Stock
*A Model of Energy Deposition into Semiconductors during Laser Annealing
(a) Vol. **73**, No. 1, p. K57–K62 (1982).
See also 22, p. 389.

A. D. Avesov, E. M. Gasanov, A. I. Novozhilov, and V. I. Samoilovich
*Some Structure Defects in Neutron Irradiated Fluorphlogopite
(a) Vol. **73**, No. 1, p. K67–K68 (1982).
See also *20.3*, p. *326*; 22.8, p. 489.

B. A. S. de Barros, Jr., C. A. dos Santos, and I. J. R. Baumvol
*Search for Intermetallic Phases in Radiation-Enhanced Diffusion of Tin into Steels
(a) Vol. **73**, No. 1, p. K137–K140 (1982).
See also 5, p. 68; *9*, p. *92*; 21.1.1, p. 362.

K. Ellmer
*Analytical Solution for Total Nuclear Energy Loss Distributions at Low Energies
(b) Vol. **113**, No. 2, p. K97–K102 (1982).

Yu. V. Bulgakov, K. Lenkeit, and G. I. Gorbunova
*Resonance Dechanneling of Hydrogen and Helium Ions with Hundreds of keV Energies in the ⟨110⟩–(110) Axial-to-Planar Transitions in Silicon
(b) Vol. **113**, No. 2, p. K103–K105 (1982).
See also 22.1.2, p. 413.

H. Hedler, W. Andrä, and G. Götz
Activation of High As and Sb Concentrations in Silicon by Laser Irradiation
(a) Vol. **73**, No. 2, p. 333–338 (1982).
See also *1.2*, p. *22*; 14.3, p. 210; 22.1.2, p. 413.

A. P. Tyutnev, G. S. Mingaleev, V. S. Saenko, E. D. Pozhidaev, and A. F. Akkerman
*Bulk Charging of Dielectric Films by Low Energy Electrons
(a) Vol. **73**, No. 2, p. 361–366 (1982).
See also *14.4*, p. *238*; 22.9, p. 507.

C. Donolato and P. Venturi
Influence of the Generation Distribution on the Calculated EBIC Contrast of Line Defects
(a) Vol. **73**, No. 2, p. 377–387 (1982).
See also *10.2*, p. *114*; 22.1.2, p. 413.

B. V. R. Chowdari, A. Rajaratnam, and Y. Ravi Sekhar
*Radiation Damage Studies of Dichromate Doped Alums
(a) Vol. **73**, No. 2, p. K231–K235 (1982).
See also 19, p. 290; 20.1, p. 313; 22.8.1, p. 497.

A. D. Avesov and E. M. Gasanov
Some Radiation-Optical Properties of Fluorphlogopite with Admixture of Vanadium
(a) Vol. **73**, No. 2, p. K259–K261 (1982).
See also *20.1*, p. *313*; 22.8, p. 490.

K. D. Glinchuk, A. V. Prokhorovich, and N. S. Zayats
Comparison of Irradiation- and Annealing-Induced Changes of the Intrinsic Emission Intensity in Electron-Bombarded Undeformed and Plastically Deformed GaAs
(a) Vol. **73**, No. 2, p. K267–K270 (1982).
See also *20.3*, p. *327*; 22.2.1, p. 423.

S. K. Maksimov, V. L. Egorov, V. V. Kryuk, D. I. Piskunov, E. A. Pitirmova, and V. F. Veselov
*Effect of Stoichiometric Disturbances in $A^{III}B^{V}$ Compounds during Ion Implantation
(a) Vol. **73**, No. 2, p. K283–K285 (1982).
See also 22.2.1, p. 423; 22.2.3, p. 429.

H. J. Blythe
*Magnetic Disaccommodation in Ordered Ni_3Fe Following Either Cold-Work or Low-Temperature Electron Irradiation
(a) Vol. **73**, No. 2, p. K311–K315 (1982).
See also *18.2*, p. *273*; 21.1, p. 356; 21.1.1, p. 363.

N. Nishimaki and S. Shimanuki
New Luminescence from M and N Centers Perturbed by Divalent and Monovalent Impurities in γ-Irradiated NaCl
(b) Vol. **114**, No. 1, p. K59–K62 (1982).
See also *10.2*, p. *116*; *20.3*, p. *328*; 22.5.2, p. 463.

L. A. Kazakevich and P. F. Lugakov
*The Effect of Dislocations on the Radiation Defect Annealing Processes in Silicon
(a) Vol. **74**, No. 1, p. 113–122 (1982).
See also *10.2*, p. *116*; 13.4, p. 176; 14.3, p. 211; 22.1.2, p. 414.

H. Endert and W. Melle
*Laser-Induced Damage in KDP Crystals. The Influence of Growth Ghosts and Growth Bands
(a) Vol. **74**, No. 1, p. 141–148 (1982).
See also *1.4*, p. *37*; 20.2, p. 317; 22.8.1, p. 498.

H. J. Blythe, F. Walz, and F. Dworschak
Magnetic Relaxation in Dilute Iron–Nickel Alloys Following Low-Temperature Electron Irradiation
(a) Vol. **74**, No. 1, p. 181–184 (1982).
See also *10.1*, p. *101*; *18.2*, p. *273*; 21.1, p. 356; 21.1.1, p. 363.

V. Hnatowicz, J. Kvítek, P. Krejči, L. Pelikán, and V. Rybka
Determination of Arsenic, Antimony, and Bismuth in Silicon Using 200 keV α-Particle Backscattering
(a) Vol. **74**, No. 1, p. 323–328 (1982).
See also *1.6*, p. *44*; 22.1.2, p. 414.

T. S. Shi, S. N. Sahu, G. S. Oehrlein, A. Hiraki, and J. W. Corbett
Models for the Hydrogen-Related Defect–Impurity Complexes and Si–H Infrared Bands in Crystalline Silicon
(a) Vol. **74**, No. 1, p. 329–341 (1982).
See also *4*, p. *65*; *10.2*, p. *116*; 20.1, p. 315; 22.1.2, p. 414.

R. Klabes, J. Matthäi, M. Voelskow, and S. Mutze
*Pulsed Incoherent Light Annealing of Arsenic and Phosphorus Implanted Polycrystalline Silicon
(a) Vol. **74**, No. 1, p. K5–K7 (1982).
See also 9, p. 93; *14.3.1*, p *216*; 22.1.2,p. 415.

N. Sieber, R. Klabes, M. Voelskow, F. Fenske, and K. H. Stegemann
*Annealing Characteristics of SiO_2–Si Structures after Incoherent Light Pulse Processing
(a) Vol. **74**, No. 1, p. K9–K12 (1982).
See also 13.3, p. 161; *14.3.2*, p. *218*; 22.1.2, p. 415.

A. P. Pathak
*Superposition Criteria in Charge Densities and Potentials
(b) Vol. **114**, No. 2, p. K177–K181 (1982).
See also 13, p. 147.

W. R. Fahrner, H. G. Bach, D. Bräunig, and M. Knoll
*Implantation Profiles of Buried Layers (MOS pn) from HF or LF $C(U)$ Curves
(a) Vol. **74**, No. 2, p. 421–429 (1982).
See also *14.3.3*, p. *225*; 22.1.2, p. 415.

P. F. Lugakov, T. A. Lukashevich, and V. V. Shusha
Nature of the Defect Determining the Fermi Level Stabilization in Irradiated Silicon
(a) Vol. **74**, No. 2, p. 445–452 (1982).
See also *13.4*, p. *177*; *14.3*, p. *212*; 22.1.2, p. 416.

W. Scharff, J.-W. Erben, A. Wolf, K. Breuer, Chr. Weissmantel, R. Klabes, H. Woittennek, K.-H. Heinig, M. Voelskov, J. Matthäi, and A. Schmidt
Flash-Lamp-Induced Crystal Growth of Siicon on Amorphous Substrates Using Artificia Surface-Relief Structures
(a) Vol. **74**, No. 2, p. 545–552 (1982).
See also *2*, p. *55*; *3*, p. *60*; 22.1.2, p. 416.

12. Mechanical Properties (Plastic Deformations See 10)

Original Papers and Short Notes

E. M. Anastassakis
*Generalized Internal Strains
(b) Vol. **110**, No. 1, p. 169–174 (1982).
See also *6*, p. *72*.

W. Andrä and H. Danan
*Interface Energy as an Origin of Intrinsic Stress in Thin Films
(a) Vol. **70**, No. 2, p. K145–K149 (1982).
See also *1.5*, p. *39*.

E. Mann
On the Diffusion Theory of Kink-Pair Formation on Dislocations
(b) Vol. **111**, No. 2, p. 541–553 (1982).
See also *9*, p. *91*; *10*, p. *94*.

J. Kubát
*A Cooperative Mechanism Limiting the Spectral Width of Stress Relaxaion Processes in Solids
(b) Vol. **111**, No. 2, p. 599–606 (1982).
See also *10*, p. *95*.

J. Lothe, V. L. Indenbom, and V. A. Chamrov
*Elastic Field and Self-Force of Dislocations Emerging at the Free Surfaces of an Anisotropic Halfspace
(b) Vol. **111**, No. 2, p. 671–677 (1982).
See also *10*, p. *95*.

I. V. Stasyuk and R. A. Grigorchuk
The Theory of the Strain Effects in the Crystals Described by the s(d)–f Model
(b) Vol. **112**, No. 1, p. 327–338 (1982).
See also *1*, p. *10*; *13*, p. *144*.

E. K. Naimi
*The Amplitude Dependence of Internal Friction and the Young's Modulus Defect in Solids Due to Anharmonism of Dislocation Motion at Low Vibration Amplitudes
(a) Vol. **72**, No. 2, p. 825–832 (1982).
See also *10*, p. *95*; 21, p. 338; 22.5, p. 452.

H. M. Ledbetter and R. D. Kriz
Elastic-Wave Surfaces in Solids
(b) Vol. **114**, No. 2, p. 475–480 (1982).
See also *1.6*, p. *44*.

I. T. Suprun
Forced Pipe Diffusion of Mobile Pinning Points
(a) Vol. **74**, No. 2, p. 411–419 (1982).
See also *9*, p. *93*; 10, p. 95.

12.1 Metals

Review Article

T. A. PARKHOMENKO and V. V. PUSTOVALOV
The Low-Temperature Yield Stress Anomaly in Metals and Alloys
(a) Vol. **74**, No. 1, p. 11–42 (1982).
See also *10.1*, p. *96*; 21, p. 330.

Original Papers and Short Notes

YU. YA. BOGUSLAVSKII
Equation for the Melting Curve of Solids under High Pressure
(b) Vol. **109**, No. 1, p. 145–151 (1982).
See also *1.2*, p. *14*; *8*, p. *84*; 21.2, p. 365; 23, p. 509.

H.-J. KAUFMANN and P. P. PAL-VAL
*Peculiarities of High-Amplitude Dislocation Internal Friction in Molybdenum Single Crystals of High Purity in the Temperature Range 5.9 to 300 K
(a) Vol. **69**, No. 1, p. 127–132 (1982).
See also *10.1*, p. *96*; 21, p. 331.

H. MÜHLBACH
*Dislocation Relaxation Peaks in High-Purity Molybdenum Single Crystals
(a) Vol. **69**, No. 2, p. 615–624 (1982).
See also *10.1*, p. *96*; 21, p. 331.

L. M. LITYAGINA, Z. V. MALYUSHITSKAYA, T. A. PASHKINA, and S. S. KABALKINA
*Isothermal Compression of Al to 10 GPa at 673 K
(a) Vol. **69**, No. 2, p. K147–K150 (1982).
See also *1.1*, p. *11*; 21, p. 332.

G. S. MURTHY and D. H. SASTRY
Impression Creep of Zinc and the Rate-Controlling Dislocation Mechanism of Plastic Flow at High Temperatures
(a) Vol. **70**, No. 1, p. 63–71 (1982).
See also *10.1*, p. *97*; 21.1, p. 345.

J. X. ZHANG, T. S. KÊ, G. Y. LI, G. FANTOZZI, P. F. GOBIN, and M. WELLER
*Low-Temperature Cold Work Internal Friction Peaks in Al–0.5 wt% (0.21 at%) Cu
(a) Vol. **70**, No. 1, p. 159–165 (1982).
See also *10.1*, p. *97*; 21, p. 333; 21.1, p. 346.

H. MIZUBAYASHI and S. OKUDA
*Low Temperature Internal Friction in Deformed Ta
(a) Vol. **70**, No. 1, p. 257–268 (1982).
See also *10.1*, p. *97*; 21, p. 333.

S. A. MAHMOUD, M. A. SEMARY, and Z. FARID
*The Effect of Plastic Deformation on the Coherency of the θ′-Phase
(a) Vol. **70**, No. 1, p. K77–K79 (1982).
See also *10.1*, p. *97*; 21, p. 333; 21.1, p. 346.

J. Heinrichs and P. Magain
Density-Functional Theory of the Pressure Distribution at Metal Surfaces
(b) Vol. **110**, No. 2, p. 495–498 (1982).
See also *13.3*, p. *158*; 21, p. 334.

R. M. Agrawal and R. P. S. Rathore
Interactions in Non-Simple F.C.C. Metals
(b) Vol. **110**, No. 2, p. 677–686 (1982).
See also *4*, p. *62*; *6*, p. *73*; 21, p. 334; 21.1, p. 347.

R. Kumar, Vishwamittar, K. Chandra, and S. K. Date
Mössbauer and Magnetic Studies of $Fe_{0.968-x}Mn_xSi_{0.032}$ Ternary Systems
(a) Vol. **70**, No. 2, p. 545–553 (1982).
See also *1.3*, p. *27*; 5, p. 67; 18, p. 260; 21.1.1, p. 360.

D. Lepski and P. Burck
Precipitate Structure Analysis by Diffraction Contrast Image Matching for a Metastable Semicoherent Carbide Phase in Molybdenum (II): Image Matching
(a) Vol. **70**, No. 2, p. 571–581 (1982).
See also *1.2*, p. *16*; *1.4*, p. *33*; 10.1, p. 98; 21.1, p. 347.

J. D. Livingston
*Magnetomechanical Properties of Amorphous Metals
(a) Vol. **70**, No. 2, p. 591–596 (1982).
See also 2, p. 47; *18.2*, p. *269*; 21, p. 334.

T. Soma, Y. Kimura, and H.-Matsuo Kagaya
Elastic-Stiffness Constants under Pressure of Cs–K, Cs–Rb, and Rb–K Solid Solutions
(b) Vol. **111**, No. 2, p. 679–683 (1982).
See also *1.3*, p. *28*; 21.2, p. 365.

C. Minier, J. Lauzier, C. Esnouf, and G. Fantozzi
*A Study of the Peaking Effects in a Hexagonal Structure Metal: Magnesium
(a) Vol. **71**, No. 2, p. 381–386 (1982).
See also *10.1*, p. *98*; 11, p. 122; 21.3, p. 367.

W. C. Johnson and J. K. Lee
*An Integral Equation Aproach to the Elastic Interaction of Two Precipitates
(a) Vol. **71**, No. 2, p. 589–602 (1982).
See also *10.1*, p. *98*; 21, p. 335.

T. Soma, Y. Kimura, and H.-M. Kagaya
Compression Effect on the Lattice Dynamics of Alkali Metals
(b) Vol. **112**, No. 1, p. 151–154 (1982).
See also *6*, p. *76*; 21.2, p. 366.

R. V. Nandedkar, K. Varatharajan, S. Panchapakesan, and A. K. Tyagi
*Microhardness, Microstructure, and Blistering of Helium Irradiated Nickel, Alloy Nickel 201, and Nimonic 90
(a) Vol. **72**, No. 1, p. 89–94 (1982).
See also *11*, p. *123*; 21.1, p. 350.

J. P. Jernot, M. Coster, and J. L. Chermant
*Model to Describe the Elastic Modulus of Sintered Materials
(a) Vol. **72**, No. 1, p. 325–332 (1982).
See also *14.1*, p. *196*; 21.4, p. 371.

A. del Moral, J. F. D. Montenegro, J. S. Abell, and M. R. Ibarra
On the Magnetization and Magnetostriction of ErZn
(a) Vol. **72**, No. 1, p. 353–360 (1982).
See also *18.1*, p. *266*; 18.2, p. 271; 21.1, p. 351; 21.4, p. 371.

O. H. Prasad and M. Suryanarayana
*Fourth-Order Elastic Constant, C_{1111}, of Cubic Crystals
(b) Vol. **112**, No. 2, p. 627–631 (1982).
See also *12.2*, p. *136*; 21, p. 337; 21.1, p. 351; 21.6, p. 377; 22.1, p. 393; 22.5.2, p. 459.

T. Soma, Y. Kimura, and H. M. Kagaya
Equation of State of the Alloy System Composed of Alkali Metal Elements
(b) Vol. **112**, No. 2, p. K107–K109 (1982).
See also *1.3*, p. *29*; 21.2, p. 366.

M. Pstrokoński and W. Chomka
*Measurements of the Internal Friction and the Shear Modulus in Cold-Worked Ordered and Disordered Ag_3Mg Alloys
(a) Vol. **72**, No. 2, p. 475–482 (1982).
See also *10.1*, p. *99*; 21.3, p. 368; 21.6, p. 377.

B. Kumar and K. S. Sharma
A Pseudopotential Calculation of Phonons in Copper
(b) Vol. **113**, No. 1, p. 73–78 (1982).
See also *6*, p. *78*; 13.1, p. 153; 21.1, p. 352.

C. R. Brooks, P. J. Meschter, and T. G. Kollie
The Magnetic Heat Capacity of the Configurationally Disordered Ni–25at% Fe Alloy
(a) Vol. **73**, No. 1, p. 189–198 (1982).
See also *8*, p. *88*; 14.1, p. 197; 18.2, p. 272; 21.1, p. 353; 21.1.1, p. 362.

G. Haneczok and J. W. Moroń
*The Analysis of Internal Friction Curves for a Temperature Gradient Existing along the Sample
(a) Vol. **73**, No. 1, p. K123–K127 (1982).
See also *10.1*, p. *101*; 21, p. 340.

T. Soma, Y. Iwayama, and H. Matsuo Kagaya
*Equation of State of Cu-Based Binary Alloys
(b) Vol. **113**, No. 2, p. K131–K133 (1982).
See also *1.2*, p. *21*; 21.1, p. 355.

A. M. Roshchupkin and V. Ya. Platkov
The Static and Dynamic Hysteresis Responsible for the Dislocational Amplitude Dependent Internal Friction (I): Theory
(a) Vol. **73**, No. 2, p. 303–313 (1982).
See also *10.1*, p. *101*; 21, p. 340.

J. P. Jernot and J. L. Chermant
*Model to Predict the Tensile Strength of Sintered Materials
(a) Vol. **73**, No. 2, p. 579–586 (1982).
See also 7, p. 83; 14.1, p. 197; 21.1.1, p. 363.

T. C. R. Penna and R. E. Medrano
*On Stress Relaxation
(a) Vol. **74**, No. 1, p. 265–272 (1982).
See also 21.1, p. 356.

L. Takácz, A. Vértes, and H. Leidheiser, Jr.
*Effect of Stresses on the Mössbauer Linie-Intensities in Nickel Electrodeposits of [100] Texture
(a) Vol. **74**, No. 1, p. K45–K48 (1982).
See also *5*, p. *69*; 18.2, p. 274; 21.1, p. 357.

V. A. Melik-Shakhnazarov, I. A. Naskidashvili, V. I. Serdobintsev, and V. I. Savin
*Anelastic Effects in the β-Phase of Niobium Hybride
(a) Vol. **74**, No. 2, p. 593–602 (1982).
See also 1.4, p. 37; *7*, p. *83*; 21, p. 342.

A. Baldan and D. R. F. West
*The Strength of Certain Ni–Al–Ta and Ni–Al–Hf Alloys Containing γ′-and β-Phases
(a) Vol. **74**. No. 2, p. 705–713 (1982).
See also *1.3*, p. *31*; 21, p. 342; 21.1, p. 358.

12.2 Non-Metals

Original Papers and Short Notes

J. SHANKER and J. P. SINGH
*Third-Order Elastic Constants and Pressure Derivatives of Second-Order Elastic Constants of Fluorite-Structure Crystals
(b) Vol. **110**, No. 2, p. 697 – 702 (1982).
See also *4*, p. *62*; 22.5.3, p. 465.

T. SOMA, Y. KATABIRA, and H. MATSUO
*Compression Effect on the Elastic Constants of Si–Ge Solid Solution
(b) Vol. **110**, No. 2, p. K167 – K169 (1982).
See also 22.1.1, p. 396; 22.1.2, p. 404.

J. SHANKER and J. P. SINGH
*Third- and Fourth-Order Elastic Constants for Silver Halides, Alkaline Earth Oxides, and Chalcogenides of Pb and Sn
(a) Vol. **70**, No. 2, p. 677 – 681 (1982).
See also 22, p. 385; 22.5.1, p. 453 ; 22.6, p. 470.

K. NARASIMHA REDDY and V. HARI BABU
*Study of Hardness Anisotropy in Doped NaCl Crystals
(a) Vol. **70**, No. 2, p. K97 – K100 (1982).
See also 22.5.2, p. 457.

T. SOMA, H. IWANAMI, and H.-MATSUO KAGAYA
Pressure–Volume Relation under High Pressure for Si–Ge Solid Solution
(b) Vol. **111**, No. 1, p. K23 – K26 (1982).
See also *1.2*, p. *17*; 4, p. 62; 22.1.1, p. 396; 22.1.2, p. 406.

L. G. KHVOSTANTSEV, A. I. ORLOV, N. KH. ABRIKOSOV, T. E. SVECHNIKOVA, and S. N. CHIZHEVSKAYA
Thermoelectric Properties and Phase Transitions in Bi_2Te_3 under Hydrostatic Pressure up to 9 GPa and Temperature up to 300 °C
(a) Vol. **71**, No. 1, p. 49 – 53 (1982).
See also *1.2*, p. *17*; 14.3, p. 206; *15*, p. *246*; 22.7, p. 479.

K. RAMANI and A. M. GHODGAONKAR
*Lattice Parameters of Alkali Halides by the Simons-Parr-Finlan Technique
(a) Vol. **71**, No. 2, p. 345 – 350 (1982).
See also 8, p. 87; 22.5.2, p. 458.

J. F. J. JORDAN, A. AXMANN, H. EGGER, and J. KALUS
Temperature Dependence of the Compressibility of d_8-Naphthalene
(a) Vol. **71**, No. 2, p. 457 – 462 (1982).
See also *1*, p. *10*; 6, p. 75; 22.9, p. 505.

J. K. KRÜGER, M. PIETRALLA, and H.-G. UNRUH
The Elastic Properties of Hexatriacontane Single Crystals at Their Various Phase Transitions
(a) Vol. **71**, No. 2, p. 493 – 504 (1982).
See also *1.2*, p. *18*; *6*, p. *75*; 22.9, p. 505.

I. Gorczyca
Effect of Hydrostatic Pressure on the Band Structure of InSb. A Pseudopotential Calculation
(b) Vol. **112**, No. 1, p. 97–104 (1982).
See also *13.1*, p. *152*; 22.2.3, p. 428.

J. Shanker and K. Singh
Analysis of the Grüneisen Parameter and Its Volume Dependence in Alkali Halide Crystals
(b) Vol. **112**, No. 2, p. 615–620 (1982).
See also 4, p. 63; *6*, p. *77*; 22.5.2, p. 459.

O. H. Prasad and M. Suryanarayana
*Fourth-Order Elastic Constant, C_{1111}, of Cubic Crystals
(b) Vol. **112**, No. 2, p. 627–631 (1982).
See also *12.1*, p. *132*; 21, p. 337; 21.1, p. 351; 21.6, p. 377; 22.1, p. 393; 22.5.2, p. 459.

R. K. Singh and D. K. Neb
Role of Zero-Point Motion and Three-Body Effects in the Study of Crystal Properties of Rare Gas Solids
(b) Vol. **112**, No. 2, p. 735–746 (1982).
See also *4*, p. *63*; 6, p. 77; 8, p. 88; 23, p. 509.

T. Soma, Y. Katabira, and H. M. Kagaya
Mode Grüneisen Parameter for Long-Wave Phonons of Si–Ge Solid Solution
(b) Vol. **112**, No. 2, p. K99–K102 (1982).
See also *6*, p. *77*; *8*, p. *88*; 22.3, p. 434.

F. Vávra
*Formation of Wavy Slip Bands in AgCl Crystals
(a) Vol. **72**, No. 2, p. K125–K128 (1982).
See also 22.5.2, p. 453.

F. Prokert
Neutron Scattering Studies on Strontium Barium Niobate
(b) Vol. **113**, No. 1, p. 239–252 (1982).
See also *6*, p. *78*; 14.4.1, p. 242; 22.8.1, p. 496.

R. S. Narayanan, J. Shankar, H. C. Gupta, and B. B. Tripathi
Theory of Interionic Forces in Alkali Halide Crystals
(b) Vol. **113**, No. 1, p. 339–346 (1982).
See also *1.1*, p. *13*; *4*, p. *64*; 22.5.2, p. 460.

L. L. Regel, V. R. Regel, S. E. Boriskin, G. G. Knab, A. A. Urusovskaya, L. I. Alekseeva, and V. V. Klechkovskaya
*Influence of Irradiation with Heavy Ions on the Defect Structure and Mechanical Properties of LiF Crystals
(a) Vol. **73**, No. 1, p. 255–266 (1982).
See also *10.2*, p. *113*; 11, p. 126; 22.5.2, p. 460.

G. Sorge, H. Beige, A. Almeida, and L. A. Shuvalov
Elastic Nonlinearity of $KH_3(SO_3)_2$ and $KD_3(SeO_3)_2$ Crystals
(a) Vol. **73**, No 1, p. K63–K66 (1982).
See also *1.2*, p. *21*; *14.4.1*, p. *243*; 22.8, p. 489.

J. Grammatikakis
Measurements of the Ionic Conductivity of NaI and Calculation of the Bulk Modulus and Specific Heat of Vacancies
(b) Vol. **113**, No. 2, p. 721–726 (1982).
See also 8, p. 89; *10.2*, p. *114*; *14.4*, p. *238*; 22.5.2, p. 461.

K. R. Allakhverdiev, Sh. G. Gasymov, T. G. Mamedov, E. Yu. Salaev, and I. K. Efendieva
*Shift of the Fundamental Absorption Edge of TlSe under Hydrostatic Pressure
(b) Vol. **113**, No. 2, p. K127–K129 (1982).
See also 13.1, p. 155; *20.1*, p. *312*; 22.3, p. 434.

U. Straube, G. Sorge, and N. R. Ivanov
*Temperature-Dependent Electrostriction Coefficient and Pseudoproper Phase Transition in TSCC
(a) Vol. **73**, No. 2, p. 467–474 (1982).
See also 1.2, p. 22; *14.4.1*, p. *243*; 22.8, p. 490.

C. Patel, W. F. Sherman, and G. R. Wilkinson
Lattice Dynamics of InP under Hydrostatic Pressures
(b) Vol. **114**, No. 1, p. 169–174 (1982).
See also *6*, p. *80*; 20.1, p. 314; 22.2.2, p. 426.

C. Basu, A. Kanrar, and U. S. Ghosh
*Elastic Behaviour of Mixed Alkali Halide Crystals of Rocksalt Structure
(b) Vol. **114**, No. 1, p. 221–225 (1982).
See also 22.5.4, p. 467.

R. K. Singh and R. D. Singh
Study of Lattice Statics and Allied Properties of Zincblende-Type Compound Semiconductors
(b) Vol. **114**, No. 1, p. 235–242 (1982).
See also *4*, p. *64*; 6, p. 80; 22.2, p. 419; 22.4, p. 438.

V. P. Gupta and C. P. Agrawal
Rydberg Potential in Diatomic Ionic Crystals
(b) Vol. **114**, No. 1, p. 265–270 (1982).
See also *4*, p. *65*; *6*, p. *80*; 8, p. 89; 22.5.2, p. 462; 22.6, p. 475.

V. K. Jain and J. Shanker
Analysis of Interionic Potentials in Alkali Chalcogenide Crystals
(b) Vol. **114**, No. 1, p. 271–275 (1982).
See also *4*, p. *65*; 22, p. 391.

J. Shanker and R. K. Varshney
*Temperature Dependence of Elastic Constants of Alkali Halide Crystals
(b) Vol. **114**, No. 1, p. K71–K74 (1982).
See also 6, p. 80; 22.5.2, p. 463.

G. Sorge, A. Almeida, and L. A. Shuvalov
*Influence of Deuteration and Defects on Elastic Properties of $KH_3(SeO_3)_2$
(a) Vol. **74**, No. 1, p. 91–96 (1982).
See also 14.4.1, p. 244; 22.8.1, p. 497.

13. Electron States

Review Article

H. Böttger and V. V. Bryksin
Hopping Conductivity in Ordered and Disordered Systems (III)
(b) Vol. **113**, No. 1, p. 9–49 (1982).
See also 2, p. 45; 6, p. 70; 13.5.3, p. 188; *14.3*, p. *202*; 22, p. 382.

Original Papers and Short Notes

P. Palanichamy and S. P. Singh
*Change in Electron Mass and Collision Frequency Due to Acoustic Wave Propagation in Piezoelectric Semiconductors
(b) Vol. **109**, No. 1, p. 119–123 (1982).
See also *7*, p. *82*; 22.4.1, p. 439.

A. K. Das
*Some Properties of Stochastic Diffusion in Constant Potential Fields
(b) Vol. **109**, No. 1, p. 153–159 (1982).

M. Błaszak
Spin–Orbit Interaction in Superexchange with Orbital Degeneracy (I): Coupling Coefficients and Operator Algebra for the Chain 0⊂SU2
(b) Vol. **109**, No. 1, p. 195–210 (1982).
See also *18*, p. *259*.

K. Elk
On the Susceptibility of the Periodic Anderson Model
(b) Vol. **109**, No. 1, p. 217–222 (1982).
See also *18.1*, p. *265*; 21, p. 330.

W. Tews and R. Gründler
Electron-Energy-Loss Spectroscopy of Different Al_2O_3 Modifications (I): Energy Loss Function, Dielectric Function, Oscillator Strength Sum Rule and the Quantity $\varepsilon_2 E$
(b) Vol. **109**, No. 1, p. 255–264 (1982).
See also *14.4*, p. *234*; *20.1*, p. *298*; 22.6, p. 468.

C. Pisani
*An Expansion Technique for Evaluating Densities of States and Related Integrals
(b) Vol. **109**, No. 1, p. 353–360 (1982).

V. A. Kozlov and I. M. Nefedov
*Modified Self-Scattering Procedure for Monte-Carlo Simulation of Hot Electrons
(b) Vol. **109**, No. 1, p. 393–401 (1982).
See also *14.3.4*, p. *277*; 22, p. 382.

H. van Cong, S. Brunet, and S. Charar
*Fermi-Dirac Integrals and Fermi Energy for Degenerate Narrow-Gap Semiconductors
(b) Vol. **109**, No. 1, p. K1–K5 (1982).
See also *14.3*, p. *202*; 22, p. 383.

H. KOSTIAL and L. F. KURTENOK
*Temperature Dependence of the Multivalued Sasaki Effect
(b) Vol. **109**, No. 2, p. K109 – K113 (1982).
See also *14.3.4*, p. *227*; 22.1.2, p. 402.

T. SOMA, H. IWANAMI, and H. MATSUO
*Equation of State of Si–Ge Solid Solution
(b) Vol. **110**, No. 1, p. 75 – 79 (1982).
See also *1.3*, p. *26*; 12.2, p. 134; 22.3, p. 433.

W. WELLER, V. N. PRIGODIN, and YU. A. FIRSOV
*Localization in a Random System of N Coupled Chains (I): Microscopic Model, General Scheme
(b) Vol. **110**, No. 1, p. 143 – 155 (1982).
See also *14*, p. *190*.

V. V. NEMOSHKALENKO, A. G. LESNIK, YU. V. KUDRYAVTSEV, I. V. LEZHNENKO, L. I. NIKOLAEV, and L. S. VOSKREKASENKO
*Electronic Structure of Fe–45 Al Films
(b) Vol. **110**, No. 1, p. 239 – 243 (1982).
See also *1.5*, p. *38*; 20, p. 294; 21.1.1, p. 360.

C. E. BIELSCHOWSKY and B. MAFFEO
*Cluster Calculations of Alkaline-Earth Fluorides by the MSXα Method
(b) Vol. **110**, No. 1, p. 263 – 271 (1982).
See also 22.5.3, p. 465.

L. SKÁLA
*The CNDO/BW Study of Lithium Clusters (III): Different Crystallographic Structures
(b) Vol. **110**, No. 1, p. 299 – 306 (1982).
See also *4*, p. *61*; 21.2, p. 365.

G. B. BACHELET, C. TANNER, and M. SCHLUTER
*Static Atomic Polarizabilities from Norm-Conserving Pseudopotentials
(b) Vol. **110**, No. 1, p. 313 – 322 (1982).
See also *4*, p. *62*; 22.1, p. 393; 22.1.2, p. 403.

S. K. LAI and S. WANG
*Calculation of the Electronic Density of States in Disordered Metallic Systems
(b) Vol. **110**, No. 1, p. 331 – 340 (1982).
See also *2*, p. *46*; 21, p. 332.

M. I. AUSLENDER, N. G. BEBENIN, A. B. ZOLOTOVITSKII, and V. P. KALASHNIKOV
*On the Density of States of a Broad-Band Ferromagnetic Semiconductor at $T \gtrsim T_c$
(b) Vol. **110**, No. 1, p. 369 – 377 (1982).
See also 18.2, p. 269; 22, p. 384.

S. GHOSH and V. K. AGARWAL
Excitation of Acoustohelicon Waves Due to Modulation of a Laser Beam in Longitudinally Magnetised Semiconductors
(b) Vol. **110**, No. 1, p. K37 – K40 (1982).
See also *7*, p. *82*; 18, p. 260; *20*, p. *294*; 22, p. 384.

F. BROSENS and J. T. DEVRESSE
*Effective Interactions in Jellium Including Exchange
(b) Vol. **111**, No. 1, p. 365–374 (1982).
See also 21, p. 334.

T. M. GASYMOV, A. A. KATANOV, and F. K. KOCHARLI
*Propagation of Strong Electromagnetic Waves in Kane Semiconductors under Conditions of Electron and Phonon Heating
(b) Vol. **111**, No. 1, p. 383–395 (1982).
See also *6*, p. *73*; 14.3.4, p. 228; 22, p. 386.

U. PIETSCH
*The Influence of Anharmonic Core Vibrations in the X-Ray Bond Charge Analysis of $A^N B^{8-N}$ Compounds
(b) Vol. **111**, No. 1, p. K7–K12 (1982).
See also *4*, p. *62*; 22.2, p. 418; 22.4, p. 437; 22.5, p. 450.

N.-C. CHAO
*A Phenomenological Model for a System of Spins with Coupling to the Charge Field
(b) Vol. **111**, No. 1, p. K27–K31 (1982).
See also *18*, p. *260*.

I. M. DYKMAN and L. A. PROKHNITSKII
*Influence of Electron Heating on Reflection and Attenuation Coefficients of Electromagnetic Waves in Semiconductors with Superlattice
(b) Vol. **111**, No. 2, p. 449–454 (1982).
See also *14.3*, p. *206*; 20, p. 295; 22, p. 387.

T. KANEKO and Y. H. OHTSUKI
*Calculation of the Equilibrium Mean Charge of Helium Ions
(b) Vol. **111**, No. 2, p. 491–500 (1982).
See also *11*, p. *122*.

R. H. LANGLEY
*Bond Hybrids and the Structures of the Lanthanide Metals
(b) Vol. **111**, No. 2, p. 501–505 (1982).
See also *4*, p. *62*; 21.4, p. 370.

G. S. DUBEY
*Calculation of the Static Structure Factor of the Classical Two-Dimensional Plasma
(b) Vol. **111**, No. 2, p. K99–K101 (1982).
See also *1*, p. *10*.

S. ROBASZKIEWICZ and M. BARTKOWIAK
*Ground State Phase Diagram of the Extended Hubbard Model with Ising-Like Exchange Interactions
(b) Vol. **111**, No. 2, p. K107–K111 (1982).
See also *1.2*, p. *18*; 14, p. 191; 18, p. 261.

I. BOSE and S. CHATTERJEE
*A Migdal-Kadanoff Type of Recursion Scheme for Two-Dimensional Square Lattice
(b) Vol. **111**, No. 2, p. K137–K140 (1982).
See also *18*, p. *261*.

S. Ghosh and V. K. Agrawal
*Effect of Magnetostatic Field on the Stimulated Raman Scattering of Transversely Polarized Electromagnetic Waves and Raman Instability in Semiconductor Plasmas
(b) Vol. **112**, No. 1, p. 119 – 131 (1982).
See also **6**, p. **76**; *20*, p. *295*; **22.2.3**, p. **428**.

A. Nath and U. S. Ghosh
*Ligand Field Theory of the Magnetic Behaviours of Neodymium Ethylsulphate Nonahydrate
(b) Vol. **112**, No. 1, p. 187 – 192 (1982).
See also *18*, p. *261*; 22.8.1, p. **495**.

C. Pisani and P. Ugliengo
*Spectral Point Schemes for Evaluating the Expansion Coefficients of State Densities into Orthogonal Functions
(b) Vol. **112**, No. 1, p. 251 – 259 (1982).

I. V. Stasyuk and R. A. Grigorchuk
*The Theory of the Strain Effects in the Crystals Described by the s(d)–f Model
(b) Vol. **112**, No. 1, p. 327 – 338 (1982).
See also *1*, p. *10*; 12, p. **129**.

B. Lorenz
On the Stability of the Wigner Crystallized Ground State
(b) Vol. **112**, No. 1, p. K47 – K49 (1982).

E. Heiner
*The Dielectric Response of a Charge Density Wave (CDW) Distorted Electron–Phonon System in the Precritical Region (I): The Self-Consistent Method
(b) Vol. **112**, No. 2, p. 489 – 500 (1982).
See also 1.2, p. **19**; 14.4, p. **237**; 22, p. **388**.

P. Blaudeck
*Embedded Cluster Approach to the Density of Tail States for Disordered s-Scatterers
(b) Vol. **112**, No. 2, p. 525 – 531 (1982).
See also *2*, p. *50*.

F. A. Khan and D. P. Bhattacharya
*Hot-Carrier Galvanomagnetic Effects in High Magnetic Fields
(b) Vol. **112**, No. 2, p. 651 – 657 (1982).
See also *14.3.4*, p. *230*; 22, p. **388**.

K. Elk
Temperature Dependence of the Electrical Conductivity of the Periodic Anderson Model
(b) Vol. **112**, No. 2, p. 659 – 665 (1982).
See also *14.1*, p. *196*; 21, p. **337**.

K. C. Sharma and D. R. Sharma
*On the Gunn Instability near Threshold
(a) Vol. **72**, No. 2, p. K137 – K140 (1982).
See also *14.3.4*, p. *230*; 22, p. **389**.

U. Behn
*Tight Binding Quasiparticle Motion in a Poisson Correlated Discrete Stochastic Potential. The Density of States
(b) Vol. **113**, No. 1, p. 219–226 (1982).

E. V. Zarochentsev and S. V. Teplov
Theory of an Adiabatic Potential of Simple Metals (I): Exchange and Correlation Effects of Electrons
(b) Vol. **113**, No. 1, p. 259–266 (1982).
See also *4*, p. *64*; 21, p. 338.

V. M. Volkov, E. F. Kustov, T. K. Maketov, and G. Steczko
*Invariant Expansion of Energy of the Induced Magnetic Anisotropy
(b) Vol. **113**, No. 1, p. 301–305 (1982).
See also *18*, p. *262*; 22.8, p. **488**.

S. Katsura and T. Shirakura
*Statistical Mechanics of the Random Anisotropic Planar Heisenberg Model. Discussion on the Spin Glass
(b) Vol. **113**, No. 1, p. 327–338 (1982).
See also *18*, p. *262*.

Mariana Weissmann and Norah V. Cohan
*Charge Transfers in Structurally Disordered Alloys
(b) Vol. **113**, No. 1, p. 395–399 (1982).
See also *2*, p. *52*; 21.2, p. 366; **21.6**, p. **377**.

I. S. Zheludev
*Conditions of the Second-Order Phase Transition by Landau and by Lifschits. Magnetogyration
(b) Vol. **113**, No. 1, p. K77–K81 (1982).
See also *1.2*, p. *20*; 18, p. **262**.

F. Brouers and J. Franz
*Monte-Carlo Calculation of Density of States and Localization in Alloys Including Short-Range Order
(b) Vol. **113**, No. 2, p. 431–437 (1982).
See also *2*, p. *52*; 21.2, p. 366; **21.6**, p. **377**.

V.-L. Bonch-Bruevich and To Thi Bich Nguyet
*On the Theory of the Conductivity of Disordered Semiconductors
(b) Vol. **113**, No. 2, p. 467–469 (1982).
See also 2, p. 52; *14.3*, p. *210*; 22, p. 390.

V. L. Bonch-Bruevich
*On the Problem of a Power-Law Localization of Charge Carriers in a Two-Dimensional Disordered Semiconductor
(b) Vol. **113**, No. 2, p. 493–496 (1982).
See also 2, p. 52; *14.3*, p. *210*; 22, p. 390.

B. Lorenz and D. Ihle
*Band Effects on Charge Ordering in Fe_3O_4
(b) Vol. **113**, No. 2, p. 601–607 (1982).
See also *1.3*, p. *30*; 22.6, p. 474.

M. Asche, Z. S. Gribnikov, V. M. Ivastchenko, H. Kostial, and V. V. Mitin
*Conductivity and Transverse Fields in n-Si for Currents in the {110} and {100} Planes
(b) Vol. **114**, No. 2, p. 429–438 (1982).
See also *14.3.4*, p. *232*; 22.1.2, p. 415.

D. J. González, J. A. Alonso, and M. P. Iñiguez
*Cellular Density Functional Theory of the Heat of Formation of Disordered Simple Alloys. Influence of the Cell Partitioning
(b) Vol. **114**, No. 2, p. 495–501 (1982).
See also *2*, p. *55*; 21.3, p. 368.

G. Heber, D. Żymierska, and H. Beyer
*Dynamics of Two-Spin Cluster in Spin Glass (III): Magnetization Process at Finite, but Low Temperature
(b) Vol. **114**, No. 2, p. 571–580 (1982).
See also *18*, p. *264*; 22.8, p. 491.

V. M. Fomin and E. P. Pokatilov
Conductivity of Semiconductors Influenced by Strong Electromagnetic Radiation
(b) Vol. **114**, No. 2, p. 615–625 (1982).
See also 6, p. 81; *14.3.4*, p. *232*; 22, p. 391.

E. Ya. Glushko
Nonhomogeneous Kinetics of Quasiparticles in Steady-State Packet Representation
(b) Vol. **114**, No. 2, p. 685–694 (1982).
See also 6, p. 81; *13.5.1*, p. *186*.

V. L. Bonch-Bruevich and A. G. Mironov
*On the Problem of Electron Localization in a Two-Dimensional System
(b) Vol. **114**, No. 2, p. 715–719 (1982).
See also 14.3.4, p. 232.

A. Freudenhammer and C. Heber
*Nonlinear Dynamics of a Coupled Pair of Spins
(b) Vol. **114**, No. 2, p. K125–K130 (1982).
See also *18*, p. *264*.

M. I. Auslender, N. G. Bebenin, and V. P. Kalashnikov
*The Asymptotics of the Electron Density of States in the s–f Exchange Model for Ferromagnetic Semiconductors near T_C
(b) Vol. **114**, No. 2, p. K147–K152 (1982).
See also 14.3, p. 211; *18.2*, p. *274*.

A. P. Pathak
Superposition Criteria in Charge Densities and Potentials
(b) Vol. **114**, No. 2, p. K177–K181 (1982).
See also *11*, p. *128*.

13.1 Band Structure

Review Articles

V. V. Nemoshkalenko, V. N. Antonov, Vl. N. Antonov, W. John, H. Wonn, and P. Ziesche
*Electronic Structure and Soft X-Ray Emission Spectra of 5d Transition Metals
(b) Vol. **111**, No. 1, p. 11–52 (1982).
Erratum
(b) Vol. **114**, No. 2, p. 741 (1982).
See also *20*, p. *293*; 21, p. 330; 21.1, p. 343; 21.6, p. 374.

Z. M. Jarzębski
Preparation and Physical Properties of Transparent Conducting Oxide Films
(a) Vol. **71**, No. 1, p. 13–41 (1982).
See also 1.5, p. 38; 3, p. 57; *14.3.1*, p. *213*; *20.1*, p. *298*; 22.6, p. 468; 22.6.1, p. 477.

F. Bechstedt
*Electronic Relaxation Effects in the Core Level Spectra of Solids
(b) Vol. **112**, No. 1, p. 9–49 (1982).
See also 4, p. 61; 17, p. 255; 20, p. 293; 21, p. 330; 22, p. 382; 23, p. 509.

Original Papers and Short Notes

A. M. Sandauer and P. Byszewski
*Evidence of the Conduction–Valence Band Overlap Induced by a Magnetic Field in Zero-Gap $Hg_{1-x}Mn_xTe$
(b) Vol. **109**, No. 1, p. 167–171 (1982).
See also *14.3*, p. *202*; 18, p. 259; 22.8, p. 481.

H. Sørum, O. M. Weng, and J. Bremer
The CrK Series X-Ray Emission Spectra
(b) Vol. **109**, No. 1, p. 335–342 (1982).
See also *20*, p. *293*; 21.1, p. 343.

F. Henneberger, N. S. Averkiev, and R. J. Rasulov
On a New Photogalvanic Effect Due to Free-Carrier Absorption
(b) Vol. **109**, No. 1, p. 343–351 (1982).
See also 6, p. 70; *16*, p. *249*; 22.1.3, p. 417.

I. V. Bodnar, A. P. Bologa, and B. V. Korzun
*Composition Dependence of the Band Gap of $CuGa_xIn_{1-x}Se_2$ Solid Solutions
(b) Vol. **109**, No. 1, p. K31–K33 (1982).
See also 20.1, p. 299; 22.8, p. 481.

M. Załużny
*On the Cyclotron Resonance in Size-Quantized Films
(b) Vol. **109**, No. 1, p. K69–K73 (1982).
See also *19*, p. *283*.

V. A. Trefakov, L. N. Amitin†, A. I. Kuznetsov, and A. L. Kholkin
*Vaccum Ultraviolet Reflection Spectra of Pb-Containing ABO_3 Perovskites. Application to Electrooptic Properties
(a) Vol. **69**, No. 1, p. 425–429 (1982).
See also *20.1*, p. *299*; 22.8.1, p. 492.

I. I. Gegusin, V. N. Datsyuk, and R. V. Vedrinskii
*Electron Structure and X-Ray Absorption Spectra of Potassium Chloride (I): Band Structure Approach
(b) Vol. **109**, No. 2, p. 563–572 (1982).
See also *20*, p. *293*; 22.5.2, p. 455.

V. Christoph
On the Phonon Contribution to the Electrical Resistivity of Intermediate Valence Systems
(b) Vol. **109**, No. 2, p. 727–732 (1982).
See also *14*, p. *190*.

V. V. Lasukov, Yu. L. Pivovarov, and O. G. Kostareva
*Spontaneous Radiation of Relativistic Electrons under Planar Channeling and Band Structure of Transverse Energy
(b) Vol. **109**, No. 2, p. 761–772 (1982).
See also *11*, p. *119*; 20, p. 294.

V. D. Krevchik and A. Ya. Yafasov
Two-Photon Interband Absorption in Quantized Semiconductor Films with Participation of Deep Impurity Centers
(b) Vol. **109**, No. 2, p. K97–K101 (1982).
See also *13.4*, p. *164*; *20.1*, p. *300*; 22, p. 384.

T. G. Kerimova, Sh. S. Mamedov, and E. Yu. Salaev
*Special Features of the Electronic Spectrum of $CdGa_2S_4$ in the Region of the Fundamental Absorption Edge
(b) Vol. **109**, No. 2, p. K119–K122 (1982).
See also *20.1*, p. *300*; 22.8, p. 482.

B. S. Sundersheshu and T. Kendelewicz
Temperature Dependence of the Absorption Edge of $Cd_{1-x}Mn_xTe$ Mixed Crystals
(a) Vol. **69**, No. 2, p. 467–474 (1982).
See also *20.1*, p. *300*; 22.4.4, p. 447.

J. Calas, C. Fau, J. R. Vellas, and M. Royer
Shubnikov-de-Haas Effect in EDRI-Type Epitaxial HgCdTe Thin Films
(a) Vol. **69**, No. 2, p. 751–756 (1982).
See also *14.3.1*, p. *214*; 22.4.4, p. 447.

A. H. Abou El Ela and S. Mahmoud
Optical Absorption of Thin Zinc Films
(a) Vol. **69**, No. 2, p. K193–K197 (1982).
See also *20.1*, p. *301*; 21.1, p. 345.

S. Baroni, G. Grosso, and G. Pastori Parravicini
*Energy Bands of Krypton and Xenon with the GOPW Method
(b) Vol. **110**, No. 1, p. 227–232 (1982).
See also 23, p. 509.

W. H. LAMFRIED and R. STREHLOW
*Cluster-Bethe-Lattice Approach to the Electronic Density of States of III–V Semiconductors: Application to GaAs
(b) Vol. **110**, No. 1, p. K79–K83 (1982).
See also 22.2.1, p. 420.

R. J. IWANOWSKI
*The Γ-Point Band Structure Parameters of $Cd_xHg_{1-x}Se$ ($x \leqq 0.1$) Mixed Crystals from Thermomagnetic Measurements
(b) Vol. **110**, No. 2, p. 461–470 (1982).
See also 15, p. 246; 22.4.4, p. 447.

V. MILANOVIĆ and D. TJAPKIN
*Energy Band Calculation and Zero Energy Gap Conditions for Semiconductor Superlattices
(b) Vol. **110**, No. 2, p. 687–695 (1982).
See also 14.3.3, p. 221; 22.2.4, p. 430.

E. S. ALEKSEEV, R. G. ARKHIPOV, and S. V. POPOVA
*Band Structure of Hexagonal Tungsten Carbide
(b) Vol. **110**, No. 2, p. K151–K154 (1982).
See also 22, p. 385.

C. RINCOŃ and J. GONZÁLEZ
Influence of Impurities on the Optical Properties of $CuInSe_2$ near the Fundamental Absorption Edge
(b) Vol. **110**, No. 2, p. K171–K174 (1982).
See also 13.4, p. 167; *20.1*, p. *304*; 22.8, p. 483.

H. FINKENRATH, G. FRANZ, and N. UHLE
Optical Absorption in $PbSe_{1-x}Te_x$ Due to Transitions between Band States and Localized Energy Levels
(b) Vol. **111**, No. 1, p. 125–130 (1982).
See also *20.1*, p. *304*; 22, p. 385.

J. CISOWSKI
*Level Ordering in II_3–V_2 Semiconducting Compounds
(b) Vol. **111**, No. 1, p. 289–293 (1982).
See also 22.3, p. 433.

M. SEN and S. CHATTERJEE
*Electronic Energy Bands of Nearly Free Electron Metallic Glasses. An Application to $Mg_{0.7}Zn_{0.3}$ Amorphous Alloy
(b) Vol. **111**, No. 1, p. 353–360 (1982).
See also 2, p. 48; 21.1, p. 348; 21.3, p. 367.

R. J. BARTLETT and C. G. OLSON
Absorption Spectrum of Ni from 40 to 220 eV
(b) Vol. **111**, No. 1, p. K33–K36 (1982).
See also *20.1*, p. *305*; 21.1, p. 348.

A. N. Chakravarti, K. P. Ghatak, K. K. Ghosh, and G. B. Rao
Effect of Band Non-Parabolicity on the Generalized Einstein Relation for a Semiconductor Superlattice
(b) Vol. **111**, No. 1, p. K61 – K65 (1982).
See also *14.3*, p. *206*; 22, p. 386.

R. V. Vedrinskii, I. I. Gegusin, V. N. Datsyuk, A. A. Novakovich, and V. L. Kraizman
*Electron Structure and X-Ray Absorption Spectra of Potassium Chloride (II): Effect of a Core Hole in the Cluster Approach
(b) Vol. **111**, No. 2, p. 433 – 442 (1982).
See also *20*, p. *295*; 22.5.2, p. 457.

L. Komorowski
*Numerical Approach to the Band Structure in TCNQ Stack
(b) Vol. **111**, No. 2, p. 443 – 447 (1982).
See also 22.9, p. 505.

F. López-Aguilar, J. Costa-Quintana, and J. S. Muñoz
*Coloumb (or Strong) Correlation Effects in the Band Structure of Eu_2O_3
(b) Vol. **111**, No. 2, p. 659 – 669 (1982).
See also 22.6, p. 472.

G. A. Bordovskii, M. L. Gordeev, A. N. Ermoshkin, V. A. Izvozchikov, and R. A. Evarestov
*Energy Band Structure of Rhombic Lead Monoxide
(b) Vol. **111**, No. 2, p. K123 – K127 (1982).
See also 22.6, p. 472.

H. Neumann, B. Perlt, B. Schumann, and G. Kühn
*Optical Properties of $CuIn_2Se_{3.5}$ near the Fundamental Absorption Edge
(b) Vol. **111**, No. 2, p. K133 – K136 (1982).
See also *20.1*, p. *306*; 22.8, p. 485.

K. Yasuda, M. Takeda, H. Masuda, and A. Yoshida
*Changes of the Band Gap of SiC Crystals by High Energy Electron Irradiation and Annealing
(a) Vol. **71**, No. 2, p. 549 – 553 (1982).
See also *11*, p. *122*; 20.1, p. 307; 22.3, p. 434.

A. J. McEvoy
ESCA Spectrum and Band Structure of Ruthenium Chloride
(a) Vol. **71**, No. 2, p. 569 – 574 (1982).
See also *17*, p. *256*; 22.5, p. 451.

T. G. Kerimova, F. R. Adyalova, A. Sh. Khidirov, and E. Yu. Salaev
Optical Properties of α- and β-$CdIn_2Se_4$ in the Region of the Fundamental Absorption Edge
(a) Vol. **71**, No. 2, p. K211 – K213 (1982).
See also *20.1*, p. *307*; 22.8, p. 486.

P. B. Allen
*Relaxation of Interband Disturbances. A Phenomenological Theory
(b) Vol. **112**, No. 1, p. 61 – 67 (1982).
See also 14, p. 191.

K. Bange, A. Liebegall, and B. Lorentz
Interband Magnetoreflectivity at $Pb_{1-x}Sn_xSe$ Single Crystals
(b) Vol. **112**, No. 2, p. 543 – 547 (1982).
See also *18*, p. *262*; *20.1*, p. *309*; 22.8, p. 487.

H. Presting and R. von Baltz
Bulk Photovoltaic Effect in a Ferroelectric Crystal. A Model Calculation
(b) Vol. **112**, No. 2, p. 559 – 564 (1982).
See also *14.4.1*, p. *242*; *16*, p. *252*; 22.8.1, p. 496.

K. Germanova, K. Marinova, and A. Trayanov
*On the Electronic Subbands in n-Inversion Layers on (111) Ge
(b) Vol. **112**, No. 2, p. 575 – 580 (1982).
See also *14.3.2*, p. *217*; 22.1.1, p. 397.

T. P. Humphreys and G. P. Srivastava
*Nonlocal Pseudopotential Calculations for Two Isoelectronic Series: Ge–GaAs–ZnSe and α-Sn–InSb–CdTe
(b) Vol. **112**, No. 2, p. 581 – 597 (1982).
See also 22, p. 388.

A. N. Gusatinskii, G. I. Alperovich, and A. V. Soldatov
*X-Ray Spectroscopic Investigation and Calculation of the Energy Band Structure of Rare Earth Monophosphide and Monosulphide
(b) Vol. **112**, No. 2, p. 599 – 607 (1982).
See also 20, p. 296; 22, p. 388.

A. Persson and R. Jones
*First Principles Calculation of the Energy Band Structure of Diamond
(b) Vol. **112**, No. 2, p. 641 – 650 (1982).
See also 22.1, p. 394.

P. Swarup, R. K. Jain, S. N. Verma, S. Charan, and D. M. Tandle
*LPE Growth of GaInAlAs on GaAs
(a) Vol. **72**, No. 2, p. K189 – K192 (1982).
See also *2*, p. *51*; 20.1, p. 310; 20.3, p. 325; 22.2.4, p. 431.

B. Kumar and K. S. Sharma
A Pseudopotential Calculation of Phonons in Copper
(b) Vol. **113**, No. 1, p. 73 – 78 (1982).
See also *6*, p. *78*; 12.1, p. 132; 21.1, p. 352.

C. Ance, J. P. Ferraton, J. M. Berger, and F. de Chelle
*Calculation Model for Valence Electrons and Hydrogen Concentration in a-Si:H
(b) Vol. **113**, No. 1, p. 105 – 111 (1982).
Erratum
(b) Vol. **114**, No. 2, p. 742 (1982).
See also *2*, p. *51*; 14.4, p. 237; 21.1.2, p. 412.

A. P. Dmitriev, M. P. Mikhailova, and I. N. Yassievich
*High Energy Distribution Function in an Electric Field and Electron Impact Ionization in $A^{III}B^{V}$ Semiconductors
(b) Vol. **113**, No. 1, p. 125 – 135 (1982).
See also 22.2, p. 418.

H. Sørum
Intensity Distributions of the K Emission Spectra of Iron
(b) Vol. **113**, No. 1, p. 197–202 (1982).
See also *20*, p. *296*; 21.1.1, p. 362.

U. Pietsch
A Proof of Aspherical Valence Charge Parts in the Rock-Salt Structure Compounds by Means of an X-Ray Bond Charge Model
(b) Vol. **113**, No. 1, p. 203–207 (1982).
See also *4*, p. *64*; 22.5.2, p. 460; 22.6, p. 473.

C. Engler and W. Lorenz
*Calculation of Local Densities of States, Local Charges, and Adsorption Energy on Large Semiconductor Substrate with Single Adsorbate
(b) Vol. **113**, No. 1, p. 227–237 (1982).
See also 4, p. 64; 13.3, p. 160; 22.2.2, p. 426.

T. K. Boletskaya, B. E. Egorushkin, and V. P. Fadin
*Electron Structure and Optical Conductivity of NiTi
(b) Vol. **113**, No. 1, p. 307–311 (1982).
See also 14.1, p. 196; 20, p. 296; 21.1, p. 353.

M. Załużny
*On the Inter-Subband Optical Absorption in Size-Quantized Semiconductor Films
(b) Vol. **113**, No. 1, p. K1–K4 (1982).
See also *20.1*, p. *310*; 22.2.3, p. 428.

S. F. Nee, T. W. Nee, S. F. Fan, and D. W. Lynch
Wavelength Modulated Reflectance Spectrum of ZrS_3
(b) Vol. **113**, No. 1, p. K5–K8 (1982).
See also *20.1*, p. *310*; 22, p. 389.

G. Dräger, F. Werfel, and O. Brümmer
*Polarized OKα Valence Band Spectra of TiO_2 (Rutile)
(b) Vol. **113**, No. 1, p. K15–K17 (1982).
See also *20*, p. *296*; 22.6, p. 473.

K. R. Allakhverdiev, T. G. Mamedov, E. Yu. Salaev, and I. K. Efendieva
*The Fundamental Absorption Edge of $TlInSe_2$
(b) Vol. **113**, No. 1, p. K43–K47 (1982).
See also *20.1*, p. *311*; 22.8, p. 488.

Ch. Kleint, S. Schulze, and R. D. Tomlinson
*Low Energy Electron Loss Peaks of $CuInTe_2$
(b) Vol. **113**, No. 1, p. K53–K56 (1982).
See also 13.5.2, p. 187; 22.8, p. 489.

B. A. Orlowski
*The Initial and Final States for UPS Optical Transitions in $Cd_{0.2}Hg_{0.8}Te$ Crystal
(b) Vol. **113**, No. 2, p. 487–492 (1982).
See also 20.1, p. 311; 22.4.4, p. 448.

J. A. VERGÉS, D. GLÖTZEL, M. CARDONA, and O. K. ANDERSEN
*Absolute Hydrostatic Deformation Potentials of Tetrahedral Semiconductors
(b) Vol. **113**, No. 2, p. 519–534 (1982).
See also 22.1.1, p. 398; 22.1.2, p. 413; 22.2.1, p. 422.

B. JAGANNATHA RAO and A. R. CHETAL
*Study of CuK Absorption Edge in Cu(I) Compounds
(b) Vol. **113**, No. 2, p. 727–730 (1982).
See also *20*, p. *297*; 22.5, p. 452.

H. VAN CONG, S. CHARAR, S. BURNET, M. AVEROUS, and J. L. BIRMAN
*Heavy Doping Effects on the Energy Band Structure of n-Type Crystals
(b) Vol. **113**, No. 2, p. 757–778 (1982).
See also 13.4, p. 175; 22.1.2, p. 413.

A. R. BEATTIE
*The Influence of the L-Conduction Band on the Internal Quantum Efficiency of InSb
(b) Vol. **113**, No. 2, p. K83–K85 (1982).
See also *20.3*, p. *326*; 22.2.3, p. 428.

K. R. ALLAKHVERDIEV, SH. G. GASYMOV, T. G. MAMEDOV, E. YU. SALAEV, and I. K. EFENDIEVA
Shift of the Fundamental Absorption Edge of TlSe under Hydrostatic Pressure
(b) Vol. **113**, No. 2, p. K127–K129 (1982).
See also *12.2*, p. *137*; *20.1*, p. *312*; 22.3, p. 434.

A. L. DAWAR, P. KUMAR, S. K. PARADKAR, T. D. SADANA, and P. C. MATHUR
Electrical Transport Properties of Epitaxial Films of $Pb_{1-x}Sn_xTe$
(b) Vol. **73**, No. 2, p. 503–507 (1982).
See also *14.3.1*, p. *215*; 22, p. 390.

T. KALA
*Study of the Influence of Manganese and Lanthanum Oxides on the Band Structure of $Pb(Zr, Ti)O_3$
(a) Vol. **73**, No. 2, p. 573–578 (1982).
See also *20.1*, p. *313*; 22.8.1, p. 497.

J. HORÁK, P. LOŠŤÁK, L. ŠIŠKA, and M. STORDEUR
The Nature of Silver Impurity Atoms in Antimony Telluride
(b) Vol. **114**, No. 1, p. 39–45 (1982).
See also *14.3*, p. *210*; 15, p. 248; *20.1*, p. *314*; 22.7, p. 479.

B. I. CRAIG
*The Electronic Structure of a Hydrogen Impurity in Aluminium (II): The Bandstructure Model
(b) Vol. **114**, No. 2, p. 337–348 (1982).
See also *13.4*, p. *177*; 21, p. 341.

AL. L. EFROS, L. M. KANSKAYA, S. I. KOKHANOVSKII, and R. P. SEISYAN
*Diamagnetic Excitons and the Indium Antimonide Energy Band Parameters
(b) Vol. **114**, No. 2, p. 373–382 (1982).
See also 13.5.1, p. 185; 20.1, p. 315; 22.2.3, p. 429.

P. Rennert, U. H. Gläser, W. Hergert, and A. N. Timoshevskii
*Electronic Structure of Intermetallic Compounds by Interpolation Schemes
(b) Vol. **114**, No. 2, p. 449–458 (1982).
See also 21.1, p. 357.

M. Singh, J. Cisowski, P. R. Wallace, J. C. Portal, and J. M. Broto
*The Band Structure of Cd_3P_2 in the Presence of a Magnetic Field
(b) Vol. **114**, No. 2, p. 481–486 (1982).
See also 22.3, p. 434.

Ch. Müller and P. Ziesche
*Local Partial Al DOS in FeAl and AlAl
(b) Vol. **114**, No. 2, p. 523–528 (1982).
See also 21, p. 342; 21.1.1, p. 364.

F. López-Aguilar and J. Costa-Quintana
*Coloumb Correlation Effects in the Spin Polarized Band Structure of Eu_2O_3
(b) Vol. **114**, No. 2, p. 599–608 (1982).
See also 22.6, p. 476.

V. Kumar Srivastava
*Plasmon Approach to the Band Gap of Alkali Halides
(b) Vol. **114**, No. 2, p. 667–671 (1982).
See also 13.5.2, p. 187; 14.4, p. 239; 22.5.2, p. 463.

A. Fujimori and N. Tsuda
*XPS and UPS Spectra of Non-Stoichiometric CeH_x
(b) Vol. **114**, No. 2, p. K139–K142 (1982).
See also *17*, p. *257*; 21.4, p. 372.

W. H. Lamfried and R. Strehlow
*On the Electronic Structure of SiO_x–Random Bonding Model
(b) Vol. **114**, No. 2, p. K165–K170 (1982).
See also 22.6, p. 476.

13.2 Fermi Surfaces

Original Papers and Short Notes

I. Mertig, E. Mrosan, R. Zeller, and P. H. Dederichs
*Anisotropic Cyclotron Masses and Electron–Phonon Interaction in Silver
(b) Vol. **109**, No. 1, p. K17–K21 (1982).
See also *19*, p. *283*; 21.6, p. 374.

B. Schmidt, V. N. Antonov, and E. Mrosan
*Cyclotron Masses and Many-Body Effects in Transition Metals
(b) Vol. **109**, No. 2, p. 627–633 (1982).
See also 21.1, p. 344.

E. Donato, B. Ginatempo, E. S. Giuliano, R. Ruggeri, and A. Stancanelli
*Fermi Surfaces in Nb_cMo_{1-c} Random Alloys
(b) Vol. **110**, No. 1, p. 39–46 (1982).
See also 21, p. 332.

W. Braune, B. Fellmuth, N. Kubicki, and R. Herrmann
*Composition Dependence of the Electron Fermi Surfaces Tilt in $Bi_{1-x}Sb_x$ Alloys ($0 \leqq x \lessapprox 0.22$)
(b) Vol. **110**, No. 2, p. 549–556 (1982).
See also 21.7, p. 379.

V. V. Nemoshkalenko, A. V. Zhalko-Titarenko, V. N. Antonov, Vl. N. Antonov, and E. Mrosan
*Anisotropy of Cyclotron Masses in Gold
(b) Vol. **111**, No. 2, p. K81–K83 (1982).
See also *19*, p. *287*; 21.6, p. 376.

W. Gerber, W. Hoppe, O. Brümmer, and G. Dlubek
*Estimation of Fermi Surface Parameters from the One-Dimensional Fourier Transform of the Positron Annihilation Angular Correlation Curves
(b) Vol. **112**, No. 1, p. 231–235 (1982).
See also 21, p. 336; 21.2, p. 366; 21.3, p. 367.

L. Weiss and H. Eschrig
Phonon Anomalies in Copper
(b) Vol. **114**, No. 2, p. 419–427 (1982).
See also *6*, p. *80*; 21.1, p. 357.

H. Bross
*Conduction Spin Density and Knight Shift of Copper
(b) Vol. **114**, No. 2, p. 529–535 (1982).
See also 19, p. 291; 21.1, p. 357.

13.3 Surface and Interface States

Original Papers and Short Notes

M. SEILENTHAL and K.-S. REBANE
*Photovoltage Spectroscopy of Fast Surface States in Silicon MOS Structures
(a) Vol. **69**, No. 1, p. K121–K123 (1982).
See also 16, p. 250; 22.1.2, p. 401.

V. CHÁB, J. ZEMEK, S. KOC, and P. MIKUŠÍK
*Photoemission Study of Oxygen Adsorbed on Sputtered a-Si(H) and a-Si
(b) Vol. **109**, No. 2, p. 495–501 (1982).
See also 2, p. 46; *17*, p. *255*; 22.1.2, p. 401.

T. D. DZHAFOROV, A. SH. MEKHTIEV, and M. S. SADIGOV
The Effect of γ-Irradiation on Photoelectric Properties of Sb–GaSe Structures
(a) Vol. **69**, No. 2, p. 643–648 (1982).
See also *14.3.3*, p. *220*; *16*, p. *250*; 22.3, p. 433.

G. I. ROSENMAN, E. I. BOIKOVA, and YU. L. CHEPELOV
*Emission Study of Surface States in Ferroelectrics
(a) Vol. **69**, No. 2, p. K173–K177 (1982).
See also 14.4.1, p. 240; *17*, p. *255*; 22.8.1, p. 493.

H. SOLBRIG
*The Excitation Map of Surface Barrier Resonances (I): The Excitation Map for Ni(001) in the (10) Azimuth
(b) Vol. **110**, No. 1, p. 191–201 (1982).
See also 21.1, p. 345.

K. B. KADYRAKUNOV, E. V. NIDAEV, and L. S. SMIRNOV
*Behaviour of a Built-in Charge in a Dielectric at Pulsed Annealing of $Si–SiO_2$ Structures
(a) Vol. **70**, No. 1, p. K15–K17 (1982).
See also *14.4*, p. *235*; 22.1.2, p. 403; 22.6, p. 469.

J. HEINRICHS and P. MAGAIN
*Density-Functional Theory of the Pressure Distribution at Metal Surfaces
(b) Vol. **110**, No. 2, p. 495–498 (1982).
See also 12.1, p. 131; 21, p. 334.

H. SOLBRIG
*The Excitation Map of Surface Barrier Resonances (II): Some Qualitative Arguments Concerning the 2D NFE Behaviour of Surface Barrier Resonances
(b) Vol. **110**, No. 2, p. 543–548 (1982).
See also 21.1, p. 346.

H. UEBA and S. ICHIMURA
*Enhanced Infrared Absorption of Adsorbed Molecules
(b) Vol. **110**, No. 2, p. K161–K165 (1982).
See also *20.1*, p. *304*.

J. Marczewski, R. Nitecki, and M. Liberadzka
*The Influence of the Negative Bias–Temperature Test on the Properties of the $Si–SiO_2$ Interface
(a) Vol. **70**, No. 2, p. 555–561 (1982).
See also *14.3.2*, p. *217*; 20.1, p. 304; 22.1.2, p. 405.

Y. Hasegawa and Y. Abe
*Electrical and Optical Characteristics of a Schottky Barrier on a Cleaved Surface of the Layered Semiconductor InSe
(a) Vol. **70**, No. 2, p. 615–621 (1982).
See also 13.4, p. 167; *14.3.3*, p. *221*; 16, p. 250; 22.3, p. 433.

N. M. Ravindra, K. Sunil Kumar, and V. K. Srivastava
*Temperature and Voltage Dependence of the Barrier Height in SnO_2/Si Solar Cells
(a) Vol. **70**, No. 2, p. 623–630 (1982).
See also 14.3.3, p. 221; 16, p. 251; 22.1.2, p. 405.

F. Schauer, M. Jedlička, and J. Polcer
Amorphous–Crystalline Heterojunction with Built-in Interface Charge (I): Steady-State Transport
(a) Vol. **70**, No. 2, p. 755–762 (1982).
See also 2, p. 48; *14.3.3*, p. *221*; 22, p. 385; 22.4.2, p. 443.

H. Solbrig
*The Excitation Map of Surface Barrier Resonances (III): Resonances at Tilted Incidence on the Ni(001) Surface
(b) Vol. **111**, No. 1, p. 203–211 (1982).
See also 21.1, p. 348.

I. Kamohara and Y. H. Ohtsuki
*Light Scattering by the Surface Wake
(b) Vol. **111**, No. 2, p. 619–624 (1982).
See also *20*, p. *295*.

A. I. Bobrysheva, S. A. Moskalenko, and V. T. Zyukov
*The Angular Dependences of Two-Photon Transitions to Surface States
(b) Vol. **111**, No. 2, p. K75–K80 (1982).
See also *13.5.1*, p. *184*; 22.2.1, p. 421.

M. Tomášek and Š. Pick
*About the Influence of Certain Perturbation Effects on Surface States
(b) Vol. **111**, No. 2, p. K89–K91 (1982).
See also 21, p. 335; 21.1, p. 349.

F. Schauer, M. Jedlička, and J. Macků
*Amorphous–Crystalline Heterojunction with Built-in Interface Charge (II): Non-Steady-State Transport
(a) Vol. **71**, No. 2, p. 335–343 (1982).
See also 2, p. 49; *14.3.3*, p. *222*; 22, p. 387; 22.4.2, p. 443.

K. Sen
The Effect of Interface States on the Performance of MIS Solar Cells
(a) Vol. **71**, No. 2, p. 641–648 (1982).
See also *14.3.3*, p. *223*; *16*, p. *252*; 22.1.2, p. 408.

T. Jung and W. Titel
*Application of Auger Depth Profiling on the Al–SiO_2 Interface
(a) Vol. **74**, No. 1, p. 85–90 (1982).
See also *1.6*, p. *44*; 22.6, p. 475.

N. Sieber, R. Klabes, M. Voelskow, F. Fenske, and K. H. Szegemann
Annealing Characteristics of SiO_2–Si Structures after Incoherent Light Pulse Processing
(a) Vol. **74**, No. 1, p. K9–K12 (1982).
See also *11*, p. *128*; *14.3.2*, p. *218*; 22.1.2, p. 415.

J. Szuber
*Photoemission Yield Investigations of the Electronic Properties of UHV Annealed NiO Single Crystal Surfaces
(a) Vol. **74**, No. 1, p. K83–K87 (1982).
See also 17, p. 257; 22.6, p. 476.

K. Masuda
*Electronic States of Ordered Overlayers on a Tight-Binding Metal Surface. Two-Level Adsorbate
(b) Vol. **114**, No. 2, p. 393–398 (1982).
See also 21, p. 342.

A. Persson
*Electron State Associated with the Unreconstructed (111) Surface in Diamond
(b) Vol. **114**, No. 2, p. 581–583 (1982).
See also 22.1, p. 394.

G. B. Demidovich, S. N. Karyagin, and V. F. Kiselev
*Spin Centres on Clean Germanium and Silicon Surfaces and Their Interaction with Adsorbed Molecules
(b) Vol. **114**, No. 2, p. 705–713 (1982).
See also 19, p. 291; 22.1.1, p. 398; 22.1.2, p. 415.

13.4 Impurity and Defects States

Review Articles

M. PRASAD
Quantum-Statistical Theories of the Cyclotron Resonance Lineshape in Parabolic Semiconductors
(b) Vol. **109**, No. 1, p. 11–35 (1982).
See also *19*, p. *282*; 22, p. 382.

P. P. SEREGIN, A. R. REGEL, A. A. ANDREEV, and F. S. NASREDINOV
*A Study of Local Structure of Chalcogenide Vitreous Semiconductors and of States of Impurity Atoms in Them by Means of Mössbauer and Electron Spectroscopies
(a) Vol. **74**, No. 2, p. 373–394 (1982).
See also 1.2, p. 14; *2*, p. *45*; 5, p. 66; 22, p. 382.

Original Papers and Short Notes

V. NEKVASIL
*Crystal-Field Analysis of Electronic Raman Scattering Data in YbAG and YbGG
(b) Vol. **109**, No. 1, p. 67–74 (1982).
See also *20.1*, p. *298*; 22.8.2, p. 499.

E. A. KOTOMIN and A. L. SHLUGER
*Semiempirical Calculations of Defect Properties in LiF Crystal (II): Electron and Hole Centres and Their Recombination
(b) Vol. **109**, No. 1, p. 75–81 (1982).
See also *10.2*, p. *103*; 22.5.2, p. 454.

E. G. SIEVERTS, S. H. MULLER, C. A. J. AMMERLAAN, R. L. KLEINHENZ, and J. W. CORBETT
*EPR Observation of an Au–Fe Complex in Silicon (II): Electronic Structure
(b) Vol. **109**, No. 1, p. 83–94 (1982).
See also *19*, p. *282*; 22.1.2, p. 400.

V. I. LITVINOV, K. A. VALUEV, and K. D. TOVTSYUK
*Redistribution of Impurities in the Field of a Strong Electromagnetic Waves
(b) Vol. **109**, No. 1, p. 189–194 (1982).
See also 14.3, p. 202; 20, p. **293**; 22, p. 382.

D. M. DARASELIA and A. BRÄUER
*Investigation of Superhyperfine Interactions of Mn^{2+} in α-$LiIO_3$ and Evaluation of Lithium Quadrupolar Interactions
(b) Vol. **109**, No. 1, p. 223–231 (1982).
See also *19*, p. *282*; 22.8.1, p. 492.

U. HAHN, R. HAENSEL, and N. SCHWENTNER
*Kr and Xe Guest Atoms in Ar Matrix-: Emission Spectra, Excitation Spectra, and Lifetimes
(b) Vol. **109**, No. 1, p. 233–244 (1982).
See also 20.1, p. 298; *20.3*, p. *318*; 23, p. 509.

B. I. Craig and P. V. Smith
*Model Hamiltonian Wannier Functions
(b) Vol. **109**, No. 1, p. 303–313 (1982).

B. V. R. Chowdari, Y. Ravisekhar and C. Muralikrishna
*Radiation Damage of Ni^{2+}-Doped $NH_4MgPO_4 \cdot 6\,H_2O$
(b) Vol. **109**, No. 1, p. 315–324 (1982).
See also 19, p. **282**; 20.1, p. 298; 22.8.1, p. 492.

Y. V. G. S. Murti and N. Sucheta
Thermoluminescence of Coloured Alkali Halides
(b) Vol. **109**, No. 1, p. 325–334 (1982).
See also *20.3*, p. *318*; 22.5.2, p. 454.

E. Kartheuser, A. Babcenco, and R. Evrard
*Electron–Hole Correlation Effects on Donor–Acceptor Pairs
(b) Vol. **109**, No. 1, p. 375–386 (1982).
See also 22.2.2, p. 424; 22.4.2, p. 442.

S. Milshstein and A. Senderichin
*Model of Single Charged Dislocation in Tetravalent Semiconductors
(b) Vol. **109**, No. 1, p. 429–436 (1982).
See also *10.2*, p. *103*; 22.1.1, p. 395; 22.1.2, p. 400.

A. V. Vaisleib, V. P. Oleinikov, and Yu. B. Rozenfeld
*Local and Pseudolocal Jahn-Teller States of Crystals with T Level Impurity
(b) Vol. **109**, No. 1, p. K35–K38 (1982).
See also 22, p. 383.

D. Mergel and R. Labusch
*Optical Excitations of Dislocation States in Silicon
(a) Vol. **69**, No. 1, p. 151–158 (1982).
See also *10.2*, p. *103*; 16, p. 249; 22.1.2, p. 400.

P. Krispin
*Thermochemical Model Applied to a Deep-Level Defect in GaP
(a) Vol. **69**, No. 1, p. 193–200 (1982).
See also 14.3.3, p. **219**; 22.2.2, p. **424**.

A. H. Abou El Ela, N. Abdelmohsen, and H. H. A. Labib
Dielectric Relaxation Study of $CuTlSe_2$ in the Solid and Liquid Phases
(a) Vol. **69**, No. 1, p. 377–384 (1982).
See also 2, p. 45; *14.4*, p. *234*; 22.8, p. 481.

S. G. Aldullaeva and V. A. Aliev
Special Features of Negative Photoconductivity in $TlGaSe_2$
(a) Vol. **69**, No. 1, p. K33–K36 (1982).
See also *16*, p. *249*; 22.8, p. 481.

J. González, J. Alberto Torres, and G. Sanchez Pérez
Photoconductivity Spectrum of p-Type $CuInS_2$ Single Crystals
(a) Vol. **69**, No. 1, p. K37–K41 (1982).
See also *16*, p. *250*; 22.8, p. 481.

K. D. Glinchuk, K. Lukat, and V. I. Vovnenko
*Effect of Electron Irradiation on Recombination Characteristics of Deep Radiative Centres in GaAs
(a) Vol. **69**, No. 1, p. K43–K47 (1982).
See also *11*, p. *118*; 20.3, p. 319; 22.2.1, p. 420.

Nguyen Ba An, Nguyen Van Hieu, Nguyen Toan Thang, and Nguyen Ai Viet
*On the Role of the Coulomb Interaction of Charge Carriers in the Electronic Raman Scattering on Donor Levels in Direct Band Gap Semiconductors
(b) Vol. **109**, No. 2, p. 463–472 (1982).
See also 13.5.1, p. 181; *20.1*, p. *300*; 22, p. 383.

M. Höhne
Electron Paramagnetic Resonance of Gold in Silicon (II): Cluster Centres
(b) Vol. **109**, No. 2, p. 525–534 (1982).
See also *19*, p. *283*; 22.1.2, p. 401.

E. Gutsche
*Non-Condon Approximations and the Static Approach in the Theory of Non-Radiative Multiphonon Transitions
(b) Vol. **109**, No. 2, p. 583–597 (1982).
See also *6*, p. *71*; 20.3, p. 319; 22, p. 383.

K. Peuker, R. Enderlein, A. Schenk, and E. Gutsche
*Theory of Non-Radiative Multiphonon Capture Processes. Solution of Old Controversies
(b) Vol. **109**, No. 2, p. 599–606 (1982).
See also *6*, p. *71*; 20.3, p. 319; 22, p. 383.

E. Molva, J. P. Chamonal, and J. L. Pautrat
*Shallow Acceptors in Cadmium Telluride
(b) Vol. **109**, No. 2, p. 635–644 (1982).
See also 20.3, p. 319; 22.4.3, p. 445.

A. N. Chakravarti, K. P. Ghatak, S. Ghosh, and A. K. Chowdhury
Effect of Heavy Doping on the Electronic Heat Capacity in Semiconductors
(b) Vol. **109**, No. 2, p. 705–710 (1982).
See also *8*, p. *85*; 22.2.1, p. 420.

V. D. Krevchik and A. Ya. Yafasov
*Two-Photon Interband Absorption in Quantized Semiconductor Films with Participation of Deep Impurity Centers
(b) Vol. **109**, No. 2, p. K97–K101 (1982).
See also 13.1, p. 149; *20.1*, p. *300*; 22, p. 384.

S. J. Pearton
*Energy Levels of Some Rare-Earth Related Impurities in Germanium
(b) Vol. **109**, No. 2, p. K135–K138 (1982).
See also 22.2.1, p. 395.

A. Sodeika, Z. Silevičius, Z. Januškevičius, and A. Sakalas
*The Influence of Intrinsic Defects on the Electrical Properties of Single Crystals of $CdSiP_2$ and $ZnGeP_2$
(a) Vol. **69**, No. 2, p. 491–495 (1982).
See also *14.3*, p. *203*; 22.8, p. 482.

A. N. Georgobiani, S. I. Radautsan, and I. M. Tiginyanu
*Electroabsorption and Non-Equilibrium Carrier Recombination in $CdGa_2S_4$ Single Crytals
(a) Vol. **69**, No. 2, p. 513–520 (1982).
See also **16**, p. 250; 20.1, p. 301; *20.3*, p. *320*; 22.8, p. 483.

J. Pichon, J. Salardenne, and A. S. Barrière
Study of Electronic Localized States Responsible of the Polarization of FeF_3 Thin Films
(a) Vol. **69**, No. 2, p. 699–706 (1982).
See also *14.3.1*, p. *213*; *14.3.3*, p. *220*; 21.1.1, p. 359.

S. A. Kulagin and I. S. Osadko
*Relaxation and Optical Processes in Impurity Centres with Two-Well Adiabatic Potential
(b) Vol. **110**, No. 1, p. 57–68 (1982).
See also 20.1, p. 301; 20.3, p. 320.

N. G. Romanov, Yu. P. Veshchunov, V. A. Vetrov, and P. G. Baranov
*EPR Study of the Relaxed Excited State of F-Centres in Alkali Halides
(b) Vol. **110**, No. 1, p. 89–96 (1982).
See also *19*, p. *284*; 22.5.2, p. 456.

S. Myhra
*Acceptor Point Defects in n-Type α-Sn
(b) Vol. **110**, No. 1, p. 97–103 (1982).
See also *14.3*, p. *204*; 15, p. 245; 22.1, p. 393.

K. Weber, M. Grünewald, W. Fuhs, and P. Thomas
Field Effect in a-Si:H Films. Influence of Annealing and Light Exposure
(b) Vol. **110**, No. 1, p. 133–142 (1982).
See also *2*, p. *46*; *14.3.1*, p. *214*; 22.1.2, p. 403.

A. Haufe, K. Henneberger, J. Röseler, and H.-J. Wünsche
*Local Density Approximation Applied to Bound Excitons in Polar Semiconductors
(b) Vol. **110**, No. 1, p. 157–167 (1982).
See also *13.5.1*, p. *182*; 22.4, p. 437; 22.4.1, p. 439; 22.4.2. p. 442; 22.4.3, p. 445.

H. J. Fuchs, W. Karthe, and R. Böttcher
EPR and ENDOR Investigations of an Oxygen Centre in CaF_2
(b) Vol. **110**, No. 1, p. 203–209 (1982).
See also *19*, p. *284*; 22.5.3, p. 465.

L. Palermo and X. A. da Silva
*Induced Magnetism in a System Consisting of van Vleck Ions Coupled to an Electron Gas in the Narrow Band Limit
(b) Vol. **110**, No. 1, p. 307–311 (1982).
See also *18.2*, p. *268*; 21.4, p. 370.

T. D. Clark, K. K. Bajaj, W. M. Theis, and D. E. Phelps
*Theory of Donor–Bound Exciton Complexes in Semiconductors
(b) Vol. **110**, No. 1, p. 341–347 (1982).
See also *13.5.1*, p. *182*; 22, p. 384.

H. Haberlandt
*The E_1' Centre in SiO_2. MO Calculation of the ^{17}O, ^{29}Si, and ^{1}H Hyperfine Splittings
(b) Vol. **110**, No. 2, p. 531–535 (1982).
See also *19*, p. *285*; 22.6, p. 470.

W. Schmid and P. J. Dean
*Bound Exciton Time Decay and the Auger Effect in Zinc Telluride
(b) Vol. **110**, No. 2, p. 591–600 (1982).
See also *13.5.1*, p. *183*; 20.3, p. 321; 22.4.3, p. 445.

S. Benzi, F. Fermi, and R. Reverberi
Double Structure of A_X and A_T Emissions in Low-Doped KI:Tl^+ Crystals
(b) Vol. **110**, No. 2, p. 611–617 (1982).
See also *20.3*, p. *322*; 22.5.2, p. 456.

W. G. Proctor, P. Lawaetz, Y. Marfaing, and R. Triboulet
*On the S-Shaped Characteristic of Semiconductors at Low Temperature
(b) Vol. **110**, No. 2, p. 637–641 (1982).
See also *14.3.4*, p. *228*; 22.4.3, p. 445.

R. R. Mijangos, J. L. Marín, A. Clark, A. Navarro, and C. Ruíz Mejía
*Effect of Surface Polarization on the Absorption Spectrum of an F-Center
(b) Vol. **110**, No. 2, p. 703–708 (1982).
See also *10.2*, p. *107*; 22.5.2, p. 457.

A. Sh. Chavchanidze, I. S. Shpotin, and V. S. Stepanyuk
*The Perturbation of Electron Density around Mn in α-Fe-Based Dilute Solid Solution
(b) Vol. **110**, No. 2, p. K155–K159 (1982).
See also 21.1.1, p. 360.

C. Rincoń and J. González
Influence of Impurities on the Optical Properties of $CuInSe_2$ near the Fundamental Absorption Edge
(b) Vol. **110**, No. 2, p. K171–K174 (1982).
See also 13.1, p. 150; *20.1*, p. *304*; 22.8, p. 483.

E. Ziegler, A. Heinrich, H. Oppermann, and G. Stöver
Growth and Electrical Properties of Non-Stoichiometric ZnO Single Crystals Doped with Co
(a) Vol. **70**, No. 2, p. 563–570 (1982).
See also 3, p. 57; *14.3*, p. *205*; 22.4, p. 437.

Y. Hasegawa and Y. Abe
Electrical and Optical Characteristics of a Schottky Barrier on a Cleaved Surface of the Layered Semiconductor InSe
(a) Vol. **70**, No. 2, p. 615–621 (1982).
See also *13.3*, p. *159*; *14.3.3*, p. *221*; 16, p. 250; 22.3, p. 433.

T. G. Abdel-Malik, A. M. Abdeen, and A. A. Aly
*Thermally Stimulated Current in β-Metal-Free Phthalocyanine Single Crystals
(a) Vol. **70**, No. 2, p. 703–712 (1982).
See also *14.3*, p. *205*; 22.9, p. 504.

J. Rosenzweig and U. Birkholz
Photo-Hall Effect and Imperfection Analysis in CdTe in the Case of Bipolar Photoconductivity
(a) Vol. **70**, No. 2, p. K171–K174 (1982).
See also *16*, p. *251*; 22.4.3, p. 446.

J. Gazsó
*A Refinement upon the Method of Evaluating the Density of the Localized States from the Field Effect
(a) Vol. **70**, No. 2, p. K181–K183 (1982).
See also 2, p. 48; 14.3.1, p. 214; *14.3.4*, p. *228*; 22, p. 385.

D. Backs and H. J. Freyer
*Photosensitive EPR of Ge^{3+} and Sn^{3+} in Twinned Cubic Zinc Sulfide
(b) Vol. **111**, No. 1, p. 117–124 (1982).
See also *19*, p. *286*; 22.4.1, p. 440.

P. Schlottmann
*f-Level Green's Function for an Intermediate-Valence Impurity
(b) Vol. **111**, No. 1, p. 161–170 (1982).

H. Zeile, U. Harten, and K. Lassmann
*Ultrasonic Investigation of the Acceptor Ground State of Si(B) (I): The Longitudinal and Transverse Relaxation Rates of the Strain-Split Two-Level System
(b) Vol. **111**, No. 1, p. 213–220 (1982).
See also 7, p. 83; 22.1.2, p. 406.

S. K. Misra, P. Mikolajczak, and M. Jalochowski
*Host-Lattice Effects in the EPR Spectra of Gd^{3+} Doped Hydrated Single Crystals of Rare-Earth Chlorides, Nitrates, and Sulphates
(b) Vol. **111**, No. 1, p. 227–234 (1982).
See also 18.1, p. 266; *19*, p. *286*; 22.8, p. 484.

D. Ball
*The Paramagnetic Resonance of Nd^{3+} and Yb^{3+} in Zircon Structure Silicates
(b) Vol. **111**, No. 1, p. 311–320 (1982).
See also *19*, p. *286*; 22.8.1, p. 494.

J. R. Morante, J. E. Carceller, P. Cartujo, and J. J. Barbolla
*Analysis of Thermal Capture of the Acceptor Level of Gold in Silicon
(b) Vol. **111**, No. 1, p. 375–382 (1982).
See also 22.1.2, p. 406.

K. D. Glinchuk, N. M. Litovchenko, and Z. A. Salnik
*Structure and Formation Kinetics of Oxygen-Induced Recombination Centres in Heat-Treated Silicon
(a) Vol. **71**, No. 1, p. 83–87 (1982).
See also 22.1.2, p. 406.

K. D. Glinchuk, K. Lukat, and A. V. Prokhorovich
*Photoluminescence of Copper-Compensated p-Type GaAs after Irradiation by 2.2 MeV Electrons
(a) Vol. **71**, No. 1, p. 233–238 (1982).
See also *20.3*, p. *323*; 22.2.1, p. 421.

R. M. Bilbé, J. E. Nicholls, B. Lunn, and D. H. J. Totterdell
*α-Particle Polarization Effects in Cadmium Telluride Radiation Detectors
(a) Vol. **71**, No. 1, p. 239–243 (1982).
See also *14.3.3*, p. *222*; 22.4.3, p. 446.

G. B. Abdullaev, O. B. Tagiev, G. M. Niftiev, and T. Kh. Azizov
*Injection and Thermoactivation Currents in $EuGa_2S_4$ Single Crystals
(a) Vol. **71**, No. 1, p. K45–K48 (1982).
See also *14.3.4*, p. *229*; 22.8, p. 484.

N. I. Akulovich, V. V. Petrov, and V. I. Utenko
On the Electrical Activity of Transmutated Phosphorus in Neutron Transmutation Doped Silicon
(a) Vol. **71**, No. 1, p. K63–K67 (1982).
See also *11*, p. *122*; *14.3*, p. *206*; 22.1.2, p. 407.

S. M. Wasim, G. Sanchez Porras, and R. D. Tomlinson
Some Electrical Characteristics of Zn and Cd Doped $CuInTe_2$
(a) Vol. **71**, No. 1, p. K73–K78 (1982).
See also *14.3*, p. *206*; 22.8, p. 484.

I. Thurzo and T. Lalinský
*On the Relationship between Charge- and Current-Based DLTS
(a) Vol. **71**, No. 1, p. K83–K87 (1982).
See also *14.3.3*, p. *222*; 22.2.1, p. 421.

W. Korczak and M. Subotowicz
Analysis of Temperature Variation of EPR Linewidths for Gd^{3+} Ion in CeF_3 Single Crystals
(b) Vol. **111**, No. 2, p. 455–460 (1982).
See also *19*, p. *287*; 22.5, p. 450.

A. Andreeff, Th. Frauenheim, E. A. Goremychkin, H. Griessmann, B. Lippold, W. Matz, O. D. Chistyakov, and E. M. Savitskii
*Crystal Field on the Laves Phase Compound $PrNi_2$
(b) Vol. **111**, No. 2, p. 507–512 (1982).
See also *1*, p. *10*; 8, p. 87; 15, p. 246; 18.1, p. 266; 21.1, p. 349; 21.4, p. 370.

M. Ikeya
*ESR Studies of Manganese Centres in LiH X-Rayed at Low Temperature
(b) Vol. **111**, No. 2, p. 525–530 (1982).
See also *19*, p. *287*; 22.5.2, p. 458.

H. Zeile and K. Lassmann
*Ultrasonic Investigation of the Acceptor Ground State of Si(B) (II): The Distribution of Splittings
(b) Vol. **111**, No. 2, p. 555–564 (1982).
See also 7, p. 83; 22.1.2, p. 407.

J. Horák, L. Koudelka, J. Klikorka, and L. Šiška
*Transport Properties and Plasma Resonance of Bismuth Selenide Crystals Doped with Cadmium
(b) Vol. **111**, No. 2, p. 575–580 (1982).
See also *14.3*, p. *207*; 15, p. 247; 22.7, p. 479.

K. D. Glinchuk and K. Lukat
*Non-Trivial Annealing-Induced Increase of the Near-Intrinsic Emission Intensity in γ-Irradiated GaAs
(a) Vol. **71**, No. 2, p. K187 – K190 (1982).
See also **11**, p. **123**; *20.3*, p. *323*; **22.2.1**, p. **421**.

H. Lemke
*Anomalie von Ionenpaaren flacher Akzeptoren mit Mangan in Silizium
(Anomaly of Ion Pairs of Shallow Acceptors with Manganese in Silicon)
(a) Vol. **71**, No. 2, p. K215 – K217 (1982).
See also 22.1.2, p. 408.

B. A. Lombos, M. Averous, C. Fau, and J. Calas
*Conduction Processes in High Acceptor Concentration HgTe
(b) Vol. **112**, No. 1, p. 237 – 242 (1982).
See also *14.3*, p. *208*; **22.4.3**, p. **446**.

N. A. Kulagin and D. T. Sviridov
*Change of the Effective Occupation Numbers of 3d-Shells of Mn^{2+} Ions in Manganese Halide Crystals
(b) Vol. **112**, No. 1, p. K61 – K66 (1982).
See also *4*, p. *63*; 22.5, p. 451.

T. G. Abdel-Malik, A. M. Abdeen, H. M. El-Labany, and A. A. Aly
*Bulk Trapping States in β-Zinc Phthalocyanine Single Crystals
(a) Vol. **72**, No. 1, p. 99 – 104 (1982).
See also *14.3.4*, p. *230*; 22.9, p. 506.

H. Lemke
*Eigenschaften einiger Störstellenkomplexe von Zink in Silizium
(Properties of Some Impurity Complexes of Zinc in Silicon)
(a) Vol. **72**, No. 1, p. 177 – 187 (1982).
See also **14.3**, p. **208**; 22.1.2, p. 409.

D. E. Ioannou
*Comments on the $E_V + 0.45$ eV Quenched-in Level in Silicon
(a) Vol. **72**, No. 1, p. K33 – K36 (1982).
See also *10.2*, p. *111*; 22.1.2, p. 409.

J. P. Fillard, J. Bonnafe, and M. Castagne
Thermal Spectroscopy of Impurity Levels by Optical Absorption in Bulk GaAs
(a) Vol. **72**, No 1, p. K65 – K68 (1982).
See also *20.1*, p. *308*; 22.2.1, p. 422.

S. Milshtein
**C–U* Measurements on Heavily Dislocated Si
(a) Vol. **72**, No. 1, p. K99 – K103 (1982).
See also *10.2*, p. *111*; **14.3**, p. **208**; 22.1.2, p. 410.

I. Sh. Akhmadullin and R. V. Saburova
*Phonon Scattering by Paraelectrical Centers in Crystals (I): General
(b) Vol. **112**, No. 2, p. 367 – 378 (1982).
See also *6*, p. *76*.

T. O. Gegechkori, V. G. Yakeli, and Z. S. Kachlishvili
Low-Temperature Conductivity and Mobility in Semiconductors
(b) Vol. **112**, No. 2, p. 379–390 (1982).
See also *14.3*, p. *208*; 22, p. 388.

M. Korkmaz, Ö. Korkmaz, and S. Kospançali
*EPR of Copper-Doped $(NH_4)_2C_2O_4 \cdot H_2O$ Single Crystals
(b) Vol. **112**, No. 2, p. 423–427 (1982).
See also *19*, p. *288*; 22.8.1, p. 496.

G. G. Fedoruk and I. Z. Rutkovskii
Free Induction of E_1'-Centres in Crystalline Quartz
(b) Vol. **112**, No. 2, p 453–456 (1982).
See also *19*, p. *288*; 22.6, p. 472.

V. I. Kuznetsov and P. F. Lugakov
*Study of Rearrangements of Intrinsic Defects at Annealing of Proton-Irradiated Silicon
(b) Vol. **112**, No. 2, p. 457–462 (1982).
See also *10.2*, p. *111*; 11, p. 124; 22.1.2, p. 410.

L. Stauffer and G. Munschy
*Energy of the Exciton–Ionized Donor Complex in Anisotropic Semiconductors
(b) Vol. **112**, No. 2, p. 501–505 (1982).
See also *13.5.1*, p. *184*; 22.4.2, p. 443.

R. Podloucky, J. Deutz, R. Zeller, and P. H. Dederichs
*Electronic Structure of 3d and 4d Impurities in Mo and Nb
(b) Vol. **112**, No. 2, p. 515–524 (1982).
See also 21, p. 337.

K. Masuda
*Calculation of Binding Energies of Small Vacancy Clusters in B.C.C. Transition Metals
(b) Vol. **112**, No. 2, p. 609–614 (1982).
See also *10.1*, p. *99*; 21, p. 337; 21.1.1, p. 361.

J. Kreissl and W. Gehlhoff
*EPR Investigation of Manganese–Boron Pairs in Silicon
(b) Vol. **112**, No. 2, p. 695–704 (1982).
See also *19*, p. *289*; 22.1.2, p. 410.

A. J. Wojtowicz and H. J. Łożykowski
*Unknown Excited States of the Self-Activated Center in ZnSe
(b) Vol. **112**, No. 2, p. K103–K106 (1982).
See also *20.3*, p. *324*; 22.4.2, p. 443.

Yu. E. Perlin, A. A. Kaminskii, M. G. Blazha, and V. N. Enakii
*Multiphonon Nonradiative Transitions of TR^{3+} Ions in $Y_3Al_5O_{12}$ and $Lu_3Al_5O_{12}$ Garnets
(b) Vol. **112**, No. 2, p. K125–K130 (1982).
See also 6, p. 77; *20*, p. *296*; 22.8.2, p. 501.

L. E. Kholodenkov and A. G. Makhanek
Two-Photon Luminescence Excitation of Eu^{3+} in Fluorite
(b) Vol. **112**, No. 2, p. K149–K151 (1982).
See also *20.3*, p. *324*; 22.5.3, p. 466.

V. N. Brudnyi, S. A. Vorobiev, and A. A. Tsoi
*Positron Annihilation and Electrical Properties of Electron-Irradiated ($\approx$ 2 MeV) InAs Crystals
(a) Vol. **72**, No. 2, p. 529–534 (1982).
See also *11*, p. *124*; 14.3, p. 209; 22.2.1, p. 422.

N. V. Vershinina and N. N. Gerasimenko
Some Peculiarities of Radiation-Induced Changes in the Properties of Si–SiO_2 Structures Irradiated at Elevated Temperatures
(a) Vol. **72**, No. 2, p. 569–574 (1982).
See also *11*, p. *124*; *14.3.2*, p. *217*; 22.1.2, p. 411.

H. Krause
*DC Characteristics of MOS Structures and Hopping Currents in Thermally Grown SiO_2 Films
(a) Vol. **72**, No. 2, p. 607–616 (1982).
See also *14.3.3*, p. *223*; 22.6, p. 472.

M. Blaschke, H. J. Grundmann, and K. Wandel
Current Dependent Electroluminescence Decay and Degradation Behaviour in Green and Yellow LED's
(a) Vol. **72**, No. 2, p. 687–694 (1982).
See also 14.3.3, p. 224; *20.3*, p. *325*; 22.2.4, p. 431.

V. V. Kveder, Yu. A. Osipyan, W. Schröter, and G. Zoth
*On the Energy Spectrum of Dislocations in Silicon
(a) Vol. **72**, No. 2, p. 701–713 (1982).
See also 10.2, p. 112; 22.1.2, p. 411.

K. D. Glinchuk, A V. Prokhorovich, and N. S. Zayats
*Photoluminescence of Copper-Doped n-Type GaAs after Irradiation by 2.2 MeV Electrons
(a) Vol. **72**, No. 2, p. 715–719 (1982).
See also *20.3*, p. *325*; 22.2.1, p. 422.

E. V. Nidaev and A. I. Popov
Influence of the Surface on the Stability of Radiation Defects Introduced by Electron Irradiation
(a) Vol. **72**, No. 2, p. K193–K195 (1982).
See also *11*, p. *125*; 14.3.3, p. 224; 22.1.2, p. 411.

V. D. Akhmetov, V. V. Bolotov, L. S. Smirnov, O. I. Zhumaeva, P. Gaworzewski, and K. Schmalz
Accumulation of Radiation Defects in Oxygen-Containing Silicon Thermally Treated at 600 °C
(a) Vol. **72**, No. 2, p. K211–K216 (1982).
See also *11*, p. *125*; 20.1, p. 310; 22.1.2, p. 411.

M. A. S. Sweet
*The Analysis of Complex Thermoluminescence from Partial Glow Curves
(a) Vol. **72**, No. 2, p. K217–K222 (1982).
See also 20.3, p. 325; 22.4.1, p. 441.

I. SH. AKHMADULLIN and R. V. SABUROVA
*Phonon Scattering by Paraelectrical Centers in Crystals (II): Lowest Order Approximation
(b) Vol. **113**, No. 1, p. 137–145 (1982).
See also *6*, p. *78*.

S. E. KAPPHAN
*UV Absorption and Electro-Optical Alignment Behavior of Off-Center Ag^+ in RbI
(b) Vol. **113**, No. 1, p. 147–156 (1982).
See also 14.4, p. 237; *20.1*, p. *310*; 22.5.2, p. 460.

A. PERSSON
*Electron States Associated with Partial Dislocations in Diamond
(b) Vol. **113**, No. 1, p. 253–257 (1982).
See also 22.1, p. 394.

T. J. BASTOW
*Temperature and Pressure Dependence of the Electric Field Gradient at the Br Sites in $PbBr_2$ and $SnBr_2$
(b) Vol. **113**, No. 1, p. 361–365 (1982).
See also 19, p. 289; 22.5, p. 452.

M. SEKITA and T. KOBAYASHI
Oscillator Strengths of Transitions between Stark Levels of $Nd_xGd_{1-x}P_5O_{14}$
(a) Vol. **73**, No. 1, p. 61–64 (1982).
See also *20.1*, p. *311*; 22.8.1, p. 496.

G. BLASSE
*Energy Transfer Phenomena in the System $(Y, Ce, Gd, Tb)F_3$
(a) Vol. **73**, No. 1, p. 205–208 (1982).
See also *20.3*, p. *326*; 22.8, p. 489.

S. J. PEARTON and A. J. TAVENDALE
The Motion of Deep Donor Centres in Reverse Biased n-GaAs Surface Barrier Diodes
(a) Vol. **73**, No. 1, p. K75–K78 (1982).
See also *14.3.3*, p. *224*; *14.3.4*, p. *231*; 22.1.1, p. 398; 22.2.1, p. 422.

J. KATRIEL and R. BAR-ADON
*Spectrum of the F_2^+-Center
(b) Vol. **113**, No. 2, p. 481–486 (1982).
See also 10.2, p. 114; 22.5.2, p. 461.

E. I. GRANCHAROVA, J. P. LASCARAY, J. DIOURI, and J. ALLÈGRE
Comments on the Behaviour of the Absorption Edge of $Cd_{1-x}Mn_xTe$ for $x > 0.4$
(b) Vol. **113**, No. 2, p. 503–508 (1982).
See also *20.1*, p. *311*; 22.4.4, p. 448.

D. GOURIER, J. P. PARANT, and D. VIVIEN
*ESR Investigation of a $[Cr(V)O_4Pb_8]$ Center in $PbMoO_4$. Origin of the Yellow Coloration of Lead Molybdate
(b) Vol. **113**, No. 2, p. 509–518 (1982).
See also *19*, p. *290*; 22.8.1, p. 497.

K. SCHMITT
Absorption and Stress-Induced Linear Dichroism of Pb^{++} Centres in KCl and RbCl
(b) Vol. **113**, No. 2, p. 559–568 (1982).
See also *20.1*, p. *311*; 22.5.2, p. 461.

J. MERTSCHING
*Kinks in a Quasi-One-Dimensional System with a Nearly 1/3-Filled Band
(b) Vol. **113**, No. 2, p. 609–621 (1982).
See also 21, p. 340.

B. I. CRAIG and P. V. SMITH
*The Electronic Structure of a Hydrogen Impurity in Aluminium (I): The Jellium Model
(b) Vol. **113**, No. 2, p. 747–756 (1982).
See also 21, p. 340.

H. VAN CONG, S. CHARAR, S. BRUNET, M. AVEROUS, and J. L. BIRMAN
Heavy Doping Effects on the Energy Band Structure of n-Type Crystals
(b) Vol. **113**, No. 2, p. 757–778 (1982).
See also *13.1*, p. *155*; 22.1.2, p. 413.

M. VELTER-ŞTEFĂNESCU and S. V. NISTOR
*Spin-Lattice Relaxation of Yb^{3+} Ions in SrF_2 Single Crystals
(b) Vol. **113**, No. 2, p. K115–K117 (1982).
See also *19*, p. *290*; 22.5.3, p. 466.

N. S. BUINOV, V. V. MIKHNEVICH, and S. R. SYRTSOV
Phase Transition in Narrow-Band Compensated Ferroelectric Semiconductors
(b) Vol. **113**, No. 2, p. K123–K126 (1982).
See also *14.4.1*, p. *243*; 22, p. 390.

D. HENNIG, O. GOEDE, and W. HEIMBRODT
*Koster-Slater Model for Isolated and Paired Isoelectronic Te Impurities in II–VI Semiconductors
(b) Vol. **113**, No. 2, p. K163–K167 (1982).
See also *13.5.1*, p. *185*; 22.4.4, p. 449.

A. JAKUBOWSKI and K. INIEWSKI
*Simple Method for Determination of the Interface Trap Density at the Midgap in MOS Structures
(a) Vol. **73**, No. 2, p. 545–549 (1982).
See also *14.3.2*, p. *218*.

A. CASTALDINI, A. CAVALLINI, P. GONDI, and E. BONETTI
Generation and Recombination Images of Dislocations in Si by Scanning Microscopy
(a) Vol. **73**, No. 2, p. 617–624 (1982).
See also *10.2*, p. *115*; 16, p. 253; 22.1.2, p. 414.

V. A. ZUIKOV and V. V. SAMARTSEV
Reverse Photon Echo as a Method of Investigation of Resonant Medium Parameters
(a) Vol. **73**, No. 2, p. 625–632 (1982).
See also *20.1*, p. *313*; *20.3*, p. *327*; 22.8.1, p. 497.

J. Jimenez, M. A. Gonzalez, L. F. Sanz, and J. Bonnafe
On the Photoconductivity Response of GaAs Doped with Chromium and Oxygen
(a) Vol. **73**, No. 2, p. K189 – K192 (1982).
See also *16*, p. *253*; 22.2.1, p. 423.

G. Rudlof and H. Glaefeke
*Comparison of Isochronal and Time-Linear Thermal Annealing Curves and Conclusions about the Correlation Glow-Peak Nature of Traps and the Mechanism of TSEE for α-Al_2O_3
(a) Vol. **73**, No. 2, p. K253 – K258 (1982).
See also *17*, p. *257*; 22.6, p. 475.

F. J. Lohmeier and F. Fischer
Optical Properties of AsO_2^- Impurities in Alkali Halide Crystals
(b) Vol. **114**, No. 1, p. 95 – 102 (1982).
See also *20.1*, p. *314*; *20.3*, p. *327*; 22.5.2, p. 462.

P. A. Alekseev, I. P. Sadikov, Yu. L. Shitikov, I. A. Markova, O. D. Chistyakov, E. M. Savitskii, and J. Kjems
Crystalline Electric Field Effects in Pr(La, Y)Al_3
(b) Vol. **114**, No. 1, p. 161 – 167 (1982).
See also *1*, p. *10*; 21.4, p. 372.

E. Z. Imamov and V. D. Krevchik
On the Diamagnetism of Deep Impurities in Quantized Semiconductor Films
(b) Vol. **114**, No. 1, p. 201 – 207 (1982).
See also *18*, p. *263*; 22, p. 391.

Fat Duen Ho
*Point-Ion-Lattice Calculations of F^+ Centers in Lithium Niobate and Lithium Tantalate
(b) Vol. **114**, No. 1, p. 209 – 212 (1982).
See also 22.8.1, p. 497.

J. Baumann and K. Sieber
Simple Approach to Temperature and Pressure Dependence of F-Centre Absorption Energies in Alkali Halides
(b) Vol. **114**, No. 1, p. K1 – K4 (1982).
See also *10.2*, p. *116*; *20.1*, p. *314*; 22.5.2, p. 462.

W. Ulrici
*Photoinduced Transient Optical Absorption of Cr^{4+} in Semi-Insulating GaAs:Cr
(b) Vol. **114**, No. 1, p. K87 – K90 (1982).
See also *20.1*, p. *314*; 22.2.1, p. 423.

H. L. Oczkowski
*Thermoluminescence Model with Intra-Pair Metastable State
(a) Vol. **74**, No. 1, p. 65 – 74 (1982).
See also *20.3*, p. *328*; 22.4.2, p. 444.

L. A. Kazakevich and P. F. Lugakov
The Effect of Dislocations on the Radiation Defect Annealing Processes in Silicon
(a) Vol. **74**, No. 1, p. 113 – 122 (1982).
See also *10.2*, p. *116*; *11*, p. *127*; 14.3, p. 211; 22.1.2, p. 414.

S. P. Kuila and V. V. Ratnam
*Monochromatic Thermoluminescence Glow Curve Method for Evaluating Trapping Parameters
(a) Vol. **74**, No. 1, p. 255–263 (1982).
See also 10.2, p. 116; *20.3*, p. *328*; 22.5.2, p. 463.

B. I. Craig
*The Electronic Structure of a Hydrogen Impurity in Aluminium (II): The Bandstructure Model
(b) Vol. **114**, No. 2, p. 337–348 (1982).
See also *13.1*, p. *155*; 21, p. 341.

L. Skála, L. Künne, H.-G. Fritsche, and H. Müller
Cluster Model for Interstitial Hydrogen Bonding in Nickel
(b) Vol. **114**, No. 2, p. 439–448 (1982).
See also *4*, p. *65*; 10.1, p. 102; 21.1, p. 357.

D. Mergel and R. Labusch
*Variable Reconstruction of Dislocation Cores in Si
(b) Vol. **114**, No. 2, p. 545–551 (1982).
See also *10.2*, p. *117*; 14.3, p. 211; 16, p. 254; 22.1.2, p. 415.

V. Drchal and J. Kudrnovský
*Theory of Auger Spectra from Disordered Alloys. The Effect of Electron Correlations in Filled Narrow Bands
(b) Vol. **114**, No. 2, p. 627–635 (1982).
See also 17, p. 257; 21, p. 342.

A. M. Kuznetsov and J. Ulstrup
*On the Theory of Long-Range Electron Hopping in Polar Media
(b) Vol. **114**, No. 2, p. 673–683 (1982).
See also 6, p. 81; 22, p. 392.

P. I. Baranskii, V. V. Baidakov, V. N. Ermakov, N. N. Grigorev, V. V. Kolomoets, and T. A. Kudykina
*The Ionisation of Impurity States of Sb in Ge, Connected with L- and X-Valleys of the Conduction Band
(b) Vol. **114**, No. 2, p. K101–K103 (1982).
See also *14.3.4*, p. *232*; 22.1.1, p. 398.

V. V. Sidorik and A. N. Zhevno
*On the Identification of Paramagnetic Defects: The Nature of the Si-B3 Center
(b) Vol. **114**, No. 2, p. K119–K124 (1982).
See also 22.1.2, p. 415.

E. Litwin-Staszewska, L. Kończewicz, R. Piotrzkowski, and W. Szymańska
*On the Deep Impurity Level in n-InSb
(b) Vol. **114**, No. 2, p. K153–K156 (1982).
See also *14.3*, p. *211*; 22.2.3, p. 429.

P. F. Lugakov, T. A. Lukashevich, and V. V. Shusha
*Nature of the Defect Determining the Fermi Level Stabilization in Irradiated Silicon
(a) Vol. **74**, No. 2, p. 445–452 (1982).
See also 11, p. 128; *14.3*, p. *212*; 22.1.2, p. 416.

13.5 Elementary Excitations (Phonons See 6)

Original Papers and Short Notes

U. Lindner
*Solitary Excitations in 1-D Antiferromagnetic Systems with Finite Band Width
(b) Vol. **109**, No. 1, p. K41–K46 (1982).
See also *18.4*, p. *279*.

I. Balslev and A. Stahl
*Properties of Exciton Polaritons Studied in a One-Dimensional Model
(b) Vol. **111**, No. 2, p. 531–539 (1982).
See also *13.5.1*, p. *184*; 20.1, p. 306.

F. Kh. Abdullaev, A. A. Abdumalikov, and A. S. Rakhmatov
*The Dynamics of Solitons in One-Dimensional Molecular Crystals
(b) Vol. **112**, No. 1, p. K5–K10 (1982).
See also *6*, p. *76*; 22, p. 387.

H.-E. Ponath and S. V. Strizhevskii
*Scattering of Surface Polaritons by the Order Parameter Fluctuations in the Vicinity of a Phase Transition Point
(b) Vol. **112**, No. 2, p. 533–542 (1982).
See also 1.2, p. 19; *13.3*, p. *160*.

V. V. Hizhnyakov and A. V. Sherman
*Self-Trapping of Polaritons
(b) Vol. **113**, No. 2, p. 471–479 (1982).
See also *13.5.1*, p. *185*; 20, p. 296; 22.5.2, p. 461.

A. Stahl and I. Balslev
*Polariton Theory in Electron–Hole Configuration Space. Bulk Solution for a Wannier-Type Model
(b) Vol. **113**, No. 2, p. 583–587 (1982).
See also 22, p. 390.

F. Kh. Abdullaev and A. A. Abdumalikov
*Dynamics of Solitons in the Disordered Anharmonic Chain
(b) Vol. **113**, No. 2, p. 685–689 (1982).

G. Gómez-Santos and F. Yndurain
*New Theoretical Method for Studying Elementary Excitations in Crystalline Disordered Alloys. Applications to Photons in Ni_xPt_{1-x}
(b) Vol. **113**, No. 2, p. 697–707 (1982).
See also 6, p. 79; 21.1, p. 354; 21.6, p. 378.

L. Nedbal
*On the Difference between GME and PME Solutions in the Long Time Region
(b) Vol. **113**, No. 2, p. K139–K142 (1982).

P. I. Khadzhi, S. A. Moskalenko, A. H. Rotaru, and G. D. Shibarshina
*Reflection and Refraction of Wave Packets Forming Polariton Solitons in Crystals
(b) Vol. **114**, No. 1, p. K25–K29 (1982).
See also *20.1*, p. *314*.

13.5.1 Excitons

Original Papers and Short Notes

A. A. GOGOLIN
The Spectrum of an Intermediate Polaron and Its Bound States with Phonons at Strong Coupling
(b) Vol. **109**, No. 1, p. 95–108 (1982).
See also *13.5.3*, p. *188*; 20.1, p. 298.

L. NEDBAL
*Higher-Order Test of the Kenkre-Knox Relation for Excitonic Optical and Transport Properties of Molecular Crystals
(b) Vol. **109**, No. 1, p. 109–118 (1982).
See also 14.3, p. 202; 20.1, p. 298; 22, p. 382.

V. ČÁPEK and R. W. MUNN
*From Band to Hopping Diffusion of Excitons and Charge Carriers in Molecular Crystals (II): Results for Local and Weak Nonlocal Electron–Phonon Coupling
(b) Vol. **109**, No. 1, p. 245–253 (1982).
See also 6, p. 70; 22, p. 382.

U. SCHMID and P. REINEKER
*Coherent Motion of Triplet Excitons in Molecular Crystals with Two Differently Oriented Molecules per Unit Cell. Application to the Charge Transfer Crystal Anthracene/Tetracyanobenzene
(b) Vol. **109**, No. 1, p. 265–273 (1982).
See also 19, p. 282; 22.9, p. 503.

NGUYEN BA AN, NGUYEN VAN HIEU, NGUYEN TOAN THANG, and NGUYEN AI VIET
On the Role of the Coulomb Interaction of Charge Carriers in the Electronic Raman Scattering on Donor Levels in Direct Band Gap Semiconductors
(b) Vol. **109**, No. 2, p. 463–472 (1982).
See also *13.4*, p. *164*; *20.1*, p. *300*; 22, p. 383.

M. YOSIDA, N. OHNO, K. NAKAMURA, and Y. NAKAI
*Exciton Spectra of Orthorhombic Indium Bromide
(b) Vol. **109**, No. 2, p. 503–509 (1982).
See also 20.1, p. 300; 22.5, p. 450.

R. MACH, P. FLÖGEL, L. G. SUSLINA, A. G. ARESHKIN, J. MAEGE, and G. VOIGT
The Influence of Compositional Disorder on Electrical and Optical Properties of ZnS_xSe_{1-x} Single Crystals
(b) Vol. **109**, No. 2, p. 607–615 (1982).
See also *14.3*, p. *203*; 20.1, p. 300; *20.3*, p. *319*; 22.4.4, p. 447.

T. SKETTRUP
*Experimental Check of Different Additional Boundary Conditions in Case of Spatial Dispersion
(b) Vol. **109**, No. 2, p. 663–672 (1982).
See also *20.1*, p. *300*; 22.4, p. 437.

F. HENNEBERGER and V. MAY
*Non-Linear Optical Response and Optical Bistability Due to Excitonic Molecules
(b) Vol. **109**, No. 2, p. K139–K143 (1982).
See also *20.1*, p. *300*; 22.5, p. 450.

B. Bonno, J. L. Laporte et Y. Rousset
*Effets des pièges sur la fusion triplet-triplet dans les cristaux de naphtalène
(Effects of Traps on the Triplet–Triplet Annihilation in Naphthalene Crystals)
(b) Vol. **110**, No. 2, p. 581 – 589 (1982).
See also *20.3*, p. *321*; 22.9, p. 504.

W. Schmid and P. J. Dean
*Bound Exciton Time Decay and the Auger Effect in Zinc Telluride
(b) Vol. **110**, No. 2, p. 591 – 600 (1982).
See also *13.4*, p. *167*; 20.3, p. 321; 22.4.3, p. 445.

G. Kurtze, V. G. Lyssenko, and C. Klingshirn
*The Behaviour of the Biexciton in CdS in Magnetic Fields
(b) Vol. **110**, No. 2, p. K103 – K105 (1982).
See also 22.4.1, p. 440.

R. Stegmann, B. Kloth, and G. Oelgart
The Temperature Dependence of Luminscence Intensity on $GaAs_{1-x}P_x$:N
(a) Vol. **70**, No. 2, p. 423 – 431 (1982).
See also *20.3*, p. *322*; 22.2.4, p. 430.

M. Ganser, M. Seelmann-Eggebert, and R. P. Huebener
*Exciton and Electron–Hole Liquid Luminescence of Germanium in an Electric Field
(b) Vol. **111**, No. 1, p. 131 – 140 (1982).
See also 14.3, p. 206; *20.3*, p. *322*; 22.1.1, p. 396.

Yu. V. Kryuchenko and V. I. Sugakov
*Electrodynamics of a Spatially Dispersive Medium with Spherical Cavity
(b) Vol. **111**, No. 1, p. 177 – 185 (1982).
See also *20.1*, p. *305*; 22.4.1, p. 440.

R. Baltramieyunas and E. Kuokshtis
Collective Phenomena in Edge Emission of ZnTe and ZnSe Crystals
(b) Vol. **111**, No. 1, p. 187 – 193 (1982).
See also 20.1, p. 305; *20.3*, p. *322*; 22.4.2, p. 443; 22.4.3, p. 446.

D. Fröhlich and R. Kenklies
*Polarization Dependence of Two-Photon Magnetoabsorption of the 1S Exciton in Cu_2O
(b) Vol. **111**, No. 1, p. 247 – 253 (1982).
See also *20.1*, p. *305*; 22.6, p. 471.

I. Ya. Fugol, O. N. Grigorashchenko, and E. V. Savchenko
*Polariton and Spatial Dispersion Effects in Exciton Spectra of Xenon Cryocrystals
(b) Vol. **111**, No. 1, p. 397 – 406 (1982).
See also *6*, p. *74*; 20.1, p. 305; 20.3, p. 322; 23, p. 509.

P. Ranson, P. Peretti, and Y. Rousset
*Crystal Heating and Non-Radiative Relaxation Processes of Triplet Excitons in 4–4′ Dichlorobenzophenone
(a) Vol. **71**, No. 1, p. K69 – K71 (1982).
See also 20.3, p. 323; 22.9, p. 505.

V. A. Kiselev
Pseudo-Exciton Effect in the Schottky-Barrier Electroreflectance
(b) Vol. **111**, No. 2, p. 461 – 468 (1982).
See also *14.3.4*, p. *229*; *20.1*, p. *305*; 22.2.1, p. 421.

I. Balslev and A. Stahl
*Properties of Exciton Polaritons Studied in a One-Dimensional Model
(b) Vol. **111**, No. 2, p. 531–539 (1982).
See also *13.5*, p. *179*; 20.1, p. 306.

M. S. Brodin, I. V. Blonskii, B. M. Nitsovich, A. S. Krochuk, and A. V. Franiv
*Dynamical Properties of Excitons in Layer Crystals of PbI_2
(b) Vol. **111**, No. 2, p. 625–630 (1982).
See also 20.1, p. 306; 22.5, p. 451.

A. I. Bobrysheva, S. A. Moskalenko, and V. T. Zyukov
*The Angular Dependences of Two-Photon Transitions to Surface States
(b) Vol. **111**, No. 2, p. K75–K80 (1982).
See also *13.3*, p. *159*; 22.2.1, p. 421.

M. Yamawaki and C. Hamaguchi
Effect of Electric Fields on Luminescence in GaAs
(b) Vol. **112**, No. 1, p. 201–210 (1982).
See also 20.1, p. 307; *20.3*, p. *323*; 22.2.1, p. 422.

V. I. Melnik, K. I. Nelipovich, A. N. Faidysh, M. T. Shpak, and L. B. Yankovskaya
*Influence of Secondary Excitons on Intensity and Decay of Phosphorescence in Doped Benzophenone Single Crystals
(b) Vol. **112**, No. 2, p. 435–443 (1982).
See also *20.3*, p. *324*; 22.9, p. 506.

L. Stauffer and G. Munschy
*Energy of the Exciton–Ionized Donor Complex in Anisotropic Semiconductors
(b) Vol. **112**, No. 2, p. 501–505 (1982).
See also *13.4*, p. *172*; 22.4.2, p. 443.

B. Pfalzgraf, O. Roder, and J.-M. Spaeth
Optical Investigation of Ag^+ Centers in Cesium Halides
(b) Vol. **112**, No. 2, p. 565–574 (1982).
See also *10.2*, p. *112*; *20.1*, p. *309*; 22.5.2, p. 459.

E. Müller, W. Gebhardt, and V. Gerhardt
*Exciton Transfer between the Manganese Ions in the Semiconductor Alloy $Cd_{1-x}Mn_xTe$ with $x = 0.51$
(b) Vol. **113**, No. 1, p. 209–218 (1982).
See also *20.3*, p. *325*; 22.4.4, p. 448.

G. S. Davidović-Ristovski, Lj. M. Ristovski, and B. S. Tošić
*The Diffusion in the Exciton System and the Nonconservation of Excitons
(b) Vol. **113**, No. 1, p. 285–290 (1982).

R. Stegmann, B. Kloth, and G. Oelgart
The Influence of the Doping Level on the Cathodoluminescence in $GaAs_{1-x}P_x$:N, Zn Epitaxial Layers
(a) Vol. **73**, No. 1, p. 129–137 (1982).
See also *20.3*, p. *326*; 22.2.4, p. 431.

A. I. Boboysheva, I. I. Jeru, and S. A. Moskalenko
*The Isotropic Shift of the Excitonic Absorption Lines in Cu_2O Crystals
(b) Vol. **113**, No 2, p. 439–445 (1982).
See also 6, p. 79; *20.1*, p. *311*; 22.6, p. 474.

N. P. Gavaleshko, A. N. Lyakhovich, P. P. Vatamanyuk, and A. I. Savchuk
Faraday Effect in III–VI Layer Compounds
(b) Vol. **113**, No. 2, p. 447–451 (1982).
See also *18*, p. *263*; *20.1*, p. *311*; 22.3, p. 434.

V. V. Hizhnyakov and A. V. Sherman
*Self-Trapping of Polaritons
(b) Vol. **113**, No. 2, p. 471–479 (1982).
See also *13.5*, p. *179*; 20, p. 296; 22.5.2, p. 461.

S. Permogorov, A. Reznitzkii, S. Verbin, G. O. Müller, P. Flögel, and M. Nikiforova
*Localized Excitons in $CdS_{1-x}Se_x$ Solid Solutions
(b) Vol. **113**, No. 2, p. 589–600 (1982).
See also 20.1, p. 311; *20.3*, p. *326*; 22.4.4, p. 448.

S. Schmitt-Rink and H. Haug
*Comment on Nonlinear Optical Response and Optical Bistability Due to Excitonic Molecules
(b) Vol. **113**, No. 2, p. K143–K145 (1982).
See also *20.1*, p. *312*.

F. Henneberger and V. May
*Nonlinear Optical Response and Retarded Photon Self-Energy. Comment on "Non-Linear Optical Response and Optical Bistability Due to Excitonic Molecules"
(b) Vol. **113**, No. 2, p. K147–K151 (1982).
See also *20.1*, p. *312*.

D. Hennig, O. Goede, and W. Heimbrodt
*Koster-Slater Model for Isolated and Paired Isoelectronic Te Impurities in II–VI Semiconductors
(b) Vol. **113**, No. 2, p. K163–K167 (1982).
See also *13.4*, p. *175*; 22.4.4, p. 449.

G. B. Talapatra and T. N. Misra
*Delayed Luminescence of Organic Mixed Crystals. Delayed Excimer Fluorescence of Pyrene in p-Terphenyl
(b) Vol. **114**, No. 1, p. 73–83 (1982).
See also *20.3*, p. *327*; 22.9, p. 507.

A. Twardowski and J. Ginter
*The Influence of Exchange Spin Splitting on Exciton in $Cd_{1-x}Mn_xTe$ Mixed Crystals
(b) Vol. **114**, No. 2, p. 331–336 (1982).
See also 18, p. 264; 20.1, p. 315; 22.4.4, p. 449.

Al. L. Efros, L. M. Kanskaya, S. I. Kokhanovskii, and R. P. Seisyan
Diamagnetic Excitons and the Indium Antimonide Energy Band Parameters
(b) Vol. **114**, No. 2, p. 373–382 (1982).
See also *13.1*, p. *155*; 20.1, p. 315; 22.2.3, p. 429.

D. Fröhlich, K. Reimann, and P. Koidl
*Investigation of Mixed-Mode Polariton Dispersion in $AgGaS_2$
(b) Vol. **114**, No. 2, p. 553–559 (1982).
See also *20.1*, p. *315*; 22.8, p. 491.

13.5.2 Plasmons

Original Papers and Short Notes

F. Sukegawa, M. Kobayashi, and I. Yokota
*Collective Motion in Silver Chalcogenides
(b) Vol. **109**, No. 2, p. 717–725 (1982).
See also *6*, p. *71*; 14, p. 190; 22, p. 383.

F. Brosens, J. T. Devreese, L. M. Kahn, and J. Ruvalds
*Exchange Contributions and Plasmon Spectra of One-Dimensional Systems
(b) Vol. **111**, No. 1, p. 95–109 (1982).

L. E. Oliveira
*On the Long-Wavelength Plasmon in the Alkali Metals. A Nearly-Free-Electron Pseudopotential Calculation
(b) Vol. **111**, No. 2, p. 685–690 (1982).
See also 14.4, p. 236; 21.2, p. 365.

H. D. Lutz and G. Kliche
Far-Infrared Reflection Spectra, Optical and Dielectric Constants, Effective Charges, and Lattice Dynamics of the Skutterudites CoP_3, $CoAs_3$, and $CoSb_3$
(b) Vol. **112**, No. 2, p. 549–557 (1982).
See also *6*, p. *77*; *20.1*, p. *309*; 22, p. 388.

O. P. Burmistrova and V. A. Kosobukin
*Radiative Decay of Surface Collective Excitations in Polarizable Cylinders
(b) Vol. **112**, No. 2, p. 675–683 (1982).
See also 6, p. 77; *13.3*, p. *160*; 20, p. 296.

Ch. Kleint, S. Schulze, and R. D. Tomlinson
Low Energy Electron Loss Peaks of $CuInTe_2$
(b) Vol. **113**, No. 1, p. K53–K56 (1982).
See also *13.1*, p. *154*; 22.8, p. 489.

V. Kumar Srivastava
Plasmon Approach to the Band Gap of Alkali Halides
(b) Vol. **114**, No. 2, p. 667–671 (1982).
See also *13.1*, p. *156*; 14.4, p. 239; 22.5.2, p. 463.

A. Selmi, R. Le Toullec, and R. Faymonville
*Influence of the Spin Disorder on the Plasma Edge in n-Type $HgCr_2Se_4$
(b) Vol. **114**, No. 2, p. K97–K99 (1982).
See also *20.1*, p. *315*; 22.8, p. 491.

13.5.3 Polarons

Review Articles

H. Böttger and V. V. Bryskin
Hopping Conductivity in Ordered and Disordered Systems (III)
(b) Vol. **113**, No. 1, p. 9–49 (1982).
See also 2, p. 45; 6, p. 70; 13, p. 139; *14.3*, p. *202*; 22, p. 382.

Original Papers and Short Notes

A. A. Gogolin
*The Spectrum of an Intermediate Polaron and Its Bound States with Phonons at Strong Coupling
(b) Vol. **109**, No. 1, p. 95–108 (1982).
See also 13.5.1, p. 181; 20.1, p. 298.

J. P. Lascaray, J. Diouri, J. P. Nadai, and J. Ferre
Infrared Absorption Spectra of Eu-Rich EuO. Excitations of the Bound Magnetic Polaron
(b) Vol. **109**, No. 1, p. 417–422 (1982).
See also *20.1*, p. *299*; 22.6, p. 468.

V. K. Fedyanin and C. Rodriguez
*Temperature and Coupling Strength Dependence of Polaron Effective Mass and Radius
(b) Vol. **110**, No. 1, p. 105–113 (1982).

G. I. Tsurkan and G. M. Shmelev
*Carrier Mobility in ZnSe in a Wide Temperature Range
(b) Vol. **110**, No. 1, p. K33–K36 (1982).
See also *14.3*, p. *204*; 22.4.2, p. 442.

F. M. Peeters and J. T. Devreese
*Phase Transition of Fröhlich Polarons, in the Feynman Approximation, Subjected to a Magnetic Field
(b) Vol. **110**, No. 2, p. 631–635 (1982).

F. Brosens and J. T. Devreese
*Time-Dependent Momentum Distribution of Polarons at Arbitrary Temperature and Electric Field
(b) Vol. **111**, No. 2, p. 591–597 (1982).
See also 6, p. 74.

P. Pecheur and M. O. Selme
Polaron Lattices
(b) Vol. **111**, No. 2, p. 691–696 (1982).

F. M. Peeters and J. T. Devreese
*On the Existence of a Phase Transition for the Fröhlich Polaron
(b) Vol. **112**, No. 1, p. 219–229 (1982).

B. Oleś, A. M. Oleś, S. Robaszkiewicz, R. Micnas, and K. A. Chao
*Bipolaron in a Finite System
(b) Vol. **113**, No. 2, p. K159–K162 (1982).

13.5.4 Magnons

Original Papers and Short Notes

O. A. C. Nunes
*Magnon Instability in the Presence of Two Radiation Fields
(b) Vol. **110**, No. 1, p. K27–K31 (1982).

L. Merten and A. Lehmeyer
*Reflectivity of Uniaxial Antiferromagnetic and Ferrimagnetic Crystals
(b) Vol. **110**, No. 2, p. 423–430 (1982).
See also 18, p. 260; *20.1*, p. *303*.

S. V. Tarasenko and G. K. Chepurnykh
*Study of Relaxation Processes in Antiferromagnets Using Sound Absorption Measurements
(b) Vol. **110**, No. 2, p. 507–511 (1982).
See also 7, p. 82; *18.4*, p. *279*.

K. M. Häussler, A. Lehmeyer, and L. Merten
*Far-Infrared Reflectivity of Optical Magnons in FeF_2 and CoF_2
(b) Vol. **111**, No. 2, p. 513–518 (1982).
Erratum
(b) Vol. **114**, No. 1, p. 277 (1982).
See also 18, p. 260; *20.1*, p. *306*; 22.5, p. 451.

T. G. Petrova
Electron-Magnon Impurity States in the Energetic Spectrum of the Ferromagnetic Semiconductor
(b) Vol. **111**, No. 2, p. K141–K145 (1982).
See also *13.4*, p. *170*; *18.2*, p. *270*; 22, p. 387.

Yu. A. Izyumov and V. M. Laptev
*Soliton Magnetoelastic Excitation in the Heisenberg Ferromagnetic Chain
(b) Vol. **112**, No. 1, p. 155–159 (1982).
See also 18.2, p. 271.

F. Kh. Abdullaev and A. A. Abdumalikov
*Dynamics of Solitons in the Disordered Anharmonic Chain
(b) Vol. **112**, No. 2, p. K137–K142 (1982).
See also *6*, p. *78*.

14. Electrical Properties. Transport Phenomena

Original Papers and Short Notes

M. Bartkowiak and S. Robaszkiewicz
Charge Ordering in the Extended Hubbard Model. The Caron-Pratt Approximation Results
(b) Vol. **109**, No. 1, p. K11 – K16 (1982).
See also *1.2*, p. *14*; *13*, p. *140*.

E. Kolley, W. Kolley, and H. Fehske
*CPA Study of the Electrical Conductivity for Various Percolation Models
(b) Vol. **109**, No. 2, p. 551 – 561 (1982).
See also *13*, p. *140*.

R. Lenk
Integral Representation of the Incoherent Field in Disordered Materials (I): Current Balance Density-of-States Test, and New Mobility Formula
(b) Vol. **109**, No. 2, p. 617 – 626 (1982).
See also *2*, p. *46*; 13, p. 140.

F. Sukegawa, M. Kobayashi, and I. Yokota
Collective Motion in Silver Chalcogenides
(b) Vol. **109**, No. 2, p. 717 – 725 (1982).
See also *6*, p. *71*; *13.5.1*, p. *187*; 22, p. 383.

V. Christoph
*On the Phonon Contribution to the Electrical Resistivity of Intermediate Valence Systems
(b) Vol. **109**, No. 2, p. 727 – 732 (1982).
See also *13.1*, p. *149*.

H. Nagasawa
*Mixed Valence State and Metal–Non-Metal Transition of KCP
(b) Vol. **109**, No. 2, p. 749 – 759 (1982).
See also *13*, p. *140*; 19, p. 283; 22.8, p. 482.

W. Kraak, R. Herrmann, and H. Haupt
*Investigation of the Charge Carrier Scattering in Bi under High Hydrostatic Pressure
(b) Vol. **109**, No. 2, p. 785 – 792 (1982).
See also 6, p. 71; 21.7, p. 379.

W. Weller, V. N. Prigodin, and Yu. A. Firsov
*Localization in a Random System of *N* Coupled Chains (I): Microscopic Model, General Scheme
(b) Vol. **110**, No. 1, p. 143 – 155 (1982).
See also *13*, p. *141*.

V. N. Rudko
*Low Frequency Conductivity in Disordered Systems Due to Variable Range Hopping. Experiment Versus Theory
(b) Vol. **110**, No. 1, p. K17 – K20 (1982).
See also *2*, p. *46*.

B. Oleś, A. M. Oleś, K. A. Chao, and Yi-Chen Cheng
A Model for n-Methyl-Phenazinium Tetracyanoquinodimethan
(b) Vol. **110**, No. 2, p. K111 – K116 (1982).
See also *2*, p. *47*; *13*, p. *142*; 18, p. 260; 22.9, p. 504.

H. Kuwamoto
*Domain Effects Accompanying the Metal–Insulator Transition in $(V_{0.99}Cr_{0.01})_2O_3$
(b) Vol. **110**, No. 2, p. K175–K178 (1982).
See also *1.4*, p. *33*; 22.6.1, p. 477.

R. Lenk
Integral Representation of the Incoherent Field in Disordered Materials (II): Hall Current and Thermoelectric Phenomena
(b) Vol. **111**, No. 1, p. 255–260 (1982).
See also *2*, p. *48*; 15, p. 246.

S. Robaszkiewicz and M. Bartkowiak
Ground State Phase Diagram of the Extended Hubbard Model with Ising-Like Exchange Intractions
(b) Vol. **111**, No. 2, p. K107–K111 (1982).
See also *1.2*, p. *18*; *13*, p. *143*; 18, p. 261.

P. B. Allen
Relaxation of Interband Disturbances. A Phenomenological Theory
(b) Vol. **112**, No. 1, p. 61–67 (1982).
See also *13.1*, p. *151*.

E. Gerlach and B. Harbecke
*Electron Transport in a Magnetic Field
(b) Vol. **112**, No. 1, p. 143–150 (1982).
See also *18*, p. *261*.

P. Orlandi, A. Magistris, and A. Rigamonti
*Mössbauer and Conductivity Study of Ferroelastic Ferric Molybdate
(a) Vol. **72**, No. 2, p. 543–549 (1982).
See also *5*, p. *67*; 22.8.1, p. 496.

E. Wieser, G. Krabbes, and E. I. Terukov
Detection of the Metastable Simultaneous Occurrence of High- and Low-Temperature States in $Co_yFe_{1-y}S$ by Mössbauer Spectroscopy
(a) Vol. **72**, No. 2, p. 695–699 (1982).
See also *1.2*, p. *20*; *5*, p. *68*; 18, p. 262; 22.8, p. 488.

J.-P. Crine
*Influence of Electric Field and Low Pressure on the Activation Energy for Conduction in Polyethylene
(a) Vol. **72**, No. 2, p. 789–797 (1982).
See also 22.9, p. 506.

R. Lenk
Integral Representation of the Incoherent Field in Disordered Materials (III): Diffusion, Einstein Relation, and Landauer Formula
(b) Vol. **113**, No. 1, p. 179–183 (1982).
See also *2*, p. *51*; 9, p. 92.

R. Lenk
Integral Representation of the Incoherent Field in Disordered Materials (IV): Relation to Multiple Scattering Theory and Inclusion of Pair Correlations
(b) Vol. **114**, No. 1, p. 113–120 (1982).
See also *2*, p. *53*; 13, p. 146.

14.1 Metals. Semi-Metals

Original Papers and Short Notes

M. REIFFERS, K. FLACHBART, Š. JÁNOŠ, A. B. BEZNOSOV, and G. ESKA
*Transport Properties of $PrNi_5$ at Low Temperatures
(b) Vol. **109**, No. 1, p. 369–373 (1982).
See also *8*, p. *84*; 21.1, p. 343; 21.4, p. 369.

E. I. TERUKOV, N. MATTERN, W. REICHELT, H. OPPERMANN, and M. WOLF
*The Behaviour of Resistance and Thermopower and the Coexistence of Two Phases at Metal–Insulator Transitions
(a) Vol. **69**, No. 1, p. 97–101 (1982).
See also 1.2, p. 14; *14.3*, p. *202*; 15, p. 245; 22.6, p. 468.

M. P. SHEARER, S. K. SEN, and C. L. BAUER
Investigation of Interfacial Reactions in Bimetallic Thin-Film Couples of Al/Cu and Pb/Cu by Measurement of Contact Resistance
(a) Vol. **69**, No. 1, p. 139–149 (1982).
See also *1.5*, p. *38*; 9, p. 91; 21, p. 331; 21.1, p. 343.

H. GRUBER and E. KRAUTZ
*Magnetoresistance and Conductivity in the Binary System Titanum–Oxygen (I): Titanium Oxides with Metallic Conductivity
(a) Vol. **69**, No. 1, p. 287–295 (1982).
See also 8, p. 85; *14.3*, p. *203*; 15, p. 245; 18, p. 259; 22.6, p. 468.

H. R. PHILIPP
$AuAl_2$: Optical Properties and Consideration as a Transparent Electrode Material
(a) Vol. **69**, No. 2, p. 547–551 (1982).
See also *20.1*, p. *301*; 21, p. 331; 21.6, p. 374.

L. BROSSARD, H. BERNAS, L. THOMÉ, A. TRAVERSE, and P. NEDELLEC
*Influence of Disorder on Phonon Resistivity of Ion-Implanted Nickel Hydride
(a) Vol. **69**, No. 2, p. 727–732 (1982).
See also *11*, p. *119*; 21.1, p. 344.

E. M. KOGAN and V. V. USTINOV
*On the Theory of Longitudinal Magnetoresistance of Metallic Films and Wires
(b) Vol. **110**, No. 1, p. 245–255 (1982).
See also 21, p. 332.

M. S. RAHMAN KHAN
*In Situ Study of the Electrical Resistivity and Temperature Coefficient of Resistance of Bismuth Films
(b) Vol. **110**, No. 1, p. K41–K44 (1982).
See also 21.7, p. 379.

P. PUREUR, J. V. KUNZLER, W. H. SCHREINER, and D. E. BRANDÃO
*Electrical Resistivity of the Cobalt-Rich Co–Fe Alloys
(a) Vol. **70**, No. 1, p. 11–15 (1982).
See also 21.1, p. 345; 21.1.1, p. 360.

F. M. Mansy, N. K. Gobran, and G. Said
High-Temperature Creep Mechanism and Dissolution Process in Al–9.5 at% Zn Alloy
(a) Vol. **70**, No. 1, p. 57–61 (1982).
See also *10.1*, p. *97*; 21, p. 333; 21.1, p. 345.

W. Kleinn
*Restwiderstandsbestimmung aus Strom–Spannungs-Kurven kurzer freitragender Elektrotransportproben epitaktischer Goldschichten
(Determination of the Residual Resistivity from Current–Voltage Characteristics of Short Support-Free Samples of Epitaxial Gold Films)
(a) Vol. **70**, No. 1, p. 219–232 (1982).
See also *1.5*, p. *38*; 21.6, p. 375.

B. M. Gorelov, V. N. Laukhin, and M. K. Sheinkman
*Effect of Doping on the Metal–Dielectric Phase Transition in Tetrathiotetracene Iodide, TTT_2I_3
(b) Vol. **110**, No. 2, p. K145–K149 (1982).
See also *14.4*, p. *235*; 22.9, p. 504.

M. Matlak and J. Zieliński
Erratum to: "Electrical Resistivity and Insulator-to-Metal Transition for the Periodic Anderson Model"
(b) Vol. **110**, No. 2, p. K179 (1982).
See also *14.3*, p. *205*.

A. I. Lotkov, V. N. Grishkov, and V. V. Fadin
*The Influence of Palladium on the Martensitic Transformation of the Intermetallic Compound TiNi
(a) Vol. **70**, No. 2, p. 513–517 (1982).
See also *1.3*, p. *27*; 21.1, p. 347.

A. V. Andreev, A. V. Deryagin, and S. M. Zadvorkin
Anomalies in the Thermal Expansion and Resistivity of $NdCo_5$ at the Spin-Reorientation Phase Transition
(a) Vol. **70**, No. 2, p. K113–K115 (1982).
See also *18.2*, p. *269*; 21.1, p. 348.

B. Horvatić and V. Zlatić
Perturbation Expansion for the Asymmetric Anderson Hamiltonian (II): General Asymmetry
(b) Vol. **111**, No. 1, p. 65–75 (1982).
See also 8, p. 86; *13*, p. *142*; 15, p. 246; 21, p. 334.

A. B. Beznosov, G. S. Nikolskii, and A. Feher
*Band Structure, High Temperature Electrical Resistivity, and Exchange-Induced Magnetic Moment of the Itinerant Electrons in Gd–La Alloys
(b) Vol. **111**, No. 1, p. 89–93 (1982).
See also 18, p. 260; 21.4, p. 370.

E. Gratz, V. Sechovsky, V. Sima, Z. Smetana, and J. O. Ström-Olson
Magnetization, Electrical Resistivity, Thermopower, and Neutron Diffraction in $HoCu_2$
(b) Vol. **111**, No. 1, p. 195–201 (1982).
See also 15, p. 246; *18.4*, p. *280*; 21.1, p. 348; 21.4, p. 370.

J. L. Miane, G. Carmona, and P. Delhaes
Microwave Measurements of the Thermal Expansion of Organic Metals
(b) Vol. **111**, No. 1, p. 235–245 (1982).
See also 6, p. 73; *8*, p. *86*; 22.9, p. 505.

V. Christoph and A. L. Kuzemskii
*Electroconductivity of a Metallic System with a Nonspherical Fermi Surface
(b) Vol. **111**, No. 1, p. K1–K6 (1982).
See also 21, p. 335.

H. Matsui and M. Tamaki
*Electrical Resistivity of β-U_2N_3
(a) Vol. **71**, No. 1, p. K121–K125 (1982).
See also 18.2, p. 270; 21.5, p. 373.

A. R. Patel, N. C. Pandya, N. C. Chourasia, and G. K. Shivakumar
*Annealing Characteristics and Lattice Distortion Energy Spectra of Chromium–Copper Alloy Films
(a) Vol. **71**, No. 2, p. 323–327 (1982).
See also 10.1, p. 98; 21.1, p. 349.

V. Damodara Das and S. Vaidehi
Variation of Energy Gap and Resistivity Minimum Position with Thickness in Bismuth Thin Films
(a) Vol. **71**, No. 2, p. 351–356 (1982).
See also *14.3.1*, p. *214*; 21.7, p. 379.

S. K. Saha
*Temperature Dependent Electrical Resistivity of Evaporated Bismuth Film
(a) Vol. **71**, No. 2, p. K173–K175 (1982).
See also 21.7, p. 380.

A. P. Kopasov and I. N. Molkov
*On the Theory of Nonlinear Cyclotron Resonance
(b) Vol. **112**, No. 1, p. 339–350 (1982).
See also 21.7, p. 380.

I. I. Mazin, E. M. Savitskii, and Yu. A. Uspenskii
*A Simple Approach to Calculation of the Phonon-Limited Electrical Resistivity in Metals
(b) Vol. **112**, No. 1, p. K29–K33 (1982).
See also 21, p. 336.

H. Löffler and R. W. Gould
On the Influence of Small Magnesium Additions on the Decomposition Behaviour of an Al–Zn (6 at%) Mother Alloy
(a) Vol. **72**, No. 1, p. 147–152 (1982).
See also *10.1*, p. *99*; 21, p. 337; 21.1, p. 350.

K. S. Vahvaselkä
Total and Partial Interference Functions, Radial Atomic Distributions, and Electrical Resistivities for Liquid Indium and Liquid Gallium–Indium Alloys
(a) Vol. **72**, No. 1, p. 261–271 (1982).
See also *2*, p. *50*; 21, p. 337.

C. R. Brooks, P. J. Meschter, and T. G. Kollie
The Magnetic Heat Capacity of the Configurationally Disordered Ni–25 at% Fe Alloy
(a) Vol. **73**, No. 1, p. 189 – 198 (1982).
See also *8*, p. *88*; 12.1, p. 132; 18.2, p. 272; 21.1, p. 353; 21.1.1, p. 362.

A. M. Ghodgaonkar and K. Ramani
*Application of the Mayadas-Shatzkes Model for Electrical Conduction in Thin Films with Unlike Surfaces
(a) Vol. **73**, No. 1, p. K21 – K25 (1982).
See also 21, p. 339.

D. V. Gitsu, I. M. Golban, and V. G. Kantser
*Investigation of Transport Coefficients in Semimetals by the Variational Method (III): Commutational Effect of Magnetothermo-E.M.F. in Bismuth
(b) Vol. **113**, No. 2, p. 497 – 501 (1982).
See also 15, p. 248; 21.7, p. 380.

V. Kuckhermann, G. Thummes, and H. H. Mende
*Influence of the Real Fermi Surface on the Longitudinal Magnetoresistance Size Effect of Single Crystal Copper Whiskers
(a) Vol. **73**, No. 2, p. 439 – 445 (1982).
See also 18, p. 263; 21.1, p. 355.

J. P. Jernot and J. L. Chermant
Model to Predict the Tensile Strength of Sintered Materials
(a) Vol. **73**, No. 2, p. 579 – 586 (1982).
See also 7, p. 83; *12.1*, p. *133*; 21.1.1, p. 363.

H. Ratajczak and I. Gościańska
*Thermal Hall Hysteresis Loops in Thin Amorphous HoCo Films
(a) Vol. **73**, No. 2, p. 633 – 639 (1982).
See also 2, p. 53; *18.3*, p. *277*; 21.1, p. 355; 21.4, p. 371.

V. P. Pan, I. V. Zolotukhin, S. E. Litvin, B. G. Nikitin, and A. S. Shpigel
*Electrical Properties of Lanthanum–Aluminium Glass
(a) Vol. **73**, No. 2, p. K287 – K291 (1982).
See also *2*, p. *53*; 8, p. 89; 14.2, p. 201; 21, p. 341.

W. Schiller and J. Richter
*Temperature Dependence of the Thermoelectric Power of Disordered Alloys
(b) Vol. **114**, No. 1, p. 151 – 160 (1982).
See also 2, p. 54; *15*, p. *248*; 21, p. 341.

J. L. Bretonnet, J. G. Gasser, A. Bath, and R. Kleim
Structure Factors and Compressibilities of Liquid Metals
(b) Vol. **114**, No. 1, p. 243 – 253 (1982).
See also *2*, p. *54*; 21, p. 341.

L. K. Varga and T. Schmidt
*Resitivity Measurements on Amorphous Ni–P Alloys Prepared by Different Techniques
(a) Vol. **74**, No. 1, p. 279 – 283 (1982).
See also 2, p. 54; 21.1, p. 356.

14.2 Superconductivity. Superconducting Materials and Devices

Original Papers and Short Notes

L. V. PANINA, M. A. SAVCHENKO, and A. V. STEPHANOVICH
*The Theory of the Superconductive Phase in Rare Earth Metal Compounds
(b) Vol. **109**, No. 1, p. 37–48 (1982).
See also 21.4, p. 369.

L. B. DUBROVSKAYA, S. Z. NAZAROVA, and A. F. PREKUL
Superconducting and Normal Properties of Solid Solutions $Nb_{1-x}Ta_xC_y$
(a) Vol. **69**, No. 1, p. 167–172 (1982).
See also *1.3*, p. *25*; 14.3, p. 203; 18, p. 259; 21, p. 331.

W. ZAHN, P. B. WEBER, K. BLÜTHNER, and E.-B. KLEY
*Submicron High Critical Current Density Niobium-Lead Josephson Junctions
(a) Vol. **69**, No. 1, p. K25–K28 (1982).
See also 21, p. 331.

V. SHAMRAI, K. BOHMHAMMEL, and G. WOLF
The Heat Capacity of Nb_3SnH_x in Dependence on the Hydrogen Content
(b) Vol. **109**, No. 2, p. 511–517 (1982).
See also 6, p. 71; *8*, p. *85*; 21, p. 331.

M. C. L. ORLOWSKI
*Derivation of the Scattering Model for Josephson Effects from the Ginzburg-Landau Equations
(b) Vol. **109**, No. 2, p. 655–661 (1982).

J. J. WNUK
*Numerical Determination of Scaling Law Parameters in Type II Superconductors
(b) Vol. **109**, No. 2, p. K115–K118 (1982).

L. Y. VINNIKOV and A. O. GOLUBOK
Direct Observation of Magnetic Structure in Niobium Single Crystals with Hydride Precipitate Pinning Centres
(a) Vol. **69**, No. 2, p. 631–636 (1982).
See also *18*, p. *259*; 21, p. 332.

Y. KIMURA
*Superconducting Properties of the $PbMo_6S_8$ Conductor Produced by Solid State Reaction
(a) Vol. **69**, No. 2, p. K189–K192 (1982).

V. E. ARKHIPOV, V. I. VORONIN, A. E. KARKIN, and A. V. MIRMELSHTEIN
Radiation Disordering in V_3Si
(a) Vol. **70**, No. 1, p. 17–21 (1982).
See also *1.3*, p. *26*; *11*, p. *120*; 21.1, p. 345.

A. BECKMANN, E. KOLLEY, and W. KOLLEY
*On Superconductivity in Narrow-Band Alloys
(b) Vol. **110**, No. 2, p. 557–564 (1982).
See also 6, p. 73; 21.1, p. 347.

H.-G. Meyer and P. Seidel
*Tunneling into Proximity Electrodes with Magnetic Impurities
(b) Vol. **111**, No. 1, p. 271–279 (1982).

V. T. Kovachev and E. K. Nasarova
*Minimum in AC Losses of Low Bulk Pinning Nb_3Sn
(a) Vol. **71**, No. 2, p. 357–364 (1982).

A. Sikora and B. Makiej
*On the Occurrence of a Directional Asymmetry of the Critical Current in Type I Superconductor Containing Ferromagnetic Particles
(a) Vol. **71**, No. 2, p. K197–K200 (1982).
See also *18.2*, p. *270*; 21, p. 335.

A. Beckmann, E. Kolley, and W. Kolley
*Ginzburg-Landau Functional for Superconducting Alloys
(b) Vol. **112**, No. 1, p. K51–K55 (1982).
See also 21, p. 336.

W. Däumer, A. Kettschau, H. R. Khan, K. Lüders, Ch. J. Raub, H. Riesemeier, and G. Roth
*Superconductivity of Solid Solutions $V_{90}X_{10}$ (X = Cr, Mn, Nb, Mo, Pd, Ta, W, Re, Pt, Au)
(b) Vol. **112**, No. 1, p. K67–K71 (1982).
See also 21, p. 336; 21.1, p. 350; 21.6, p. 376.

S. Takács and J. Janetka
Influence of Resonance Tunneling on Resistivity in Semiconductors with Ionized Impurities and on Tunneling Current between Superconductors
(b) Vol. **113**, No. 1, p. 313–326 (1982).
See also *14.3.4*, p. *231*; 22, p. 389.

V. P. Iordatii and L. Z. Kon
*Influence of the Phonon Frequency Dispersion on the Critical Temperature of Superconductors
(b) Vol. **113**, No. 1, p. K67–K71 (1982).
See also 6, p. 79.

S. A. Ivanov, A. L. Kolesnikova, and A. E. Romanov
*Elastic Stresses and Elastic Energy of a Flux Line in a Half-Space
(a) Vol. **73**, No. 1, p. K31–K34 (1982).
See also *10.1*, p. *100*.

A. V. Verbilo, N. M. Kotov, A. P. Rusakov, and N. A. Kozlovskai
*Superconductive Properties of the Chevrel-Phase System $Cu_xMo_6S_8$ with $x = 3.2$ to 3.9
(a) Vol. **73**, No. 1, p. K73–K74 (1982).
See also 21, p. 339.

R. Heindl and R. Druilhé
Superconduction: On the Sintering of a "Chevrel Phase". – A New High Temperature Phase of $PbMo_6S_8$
(a) Vol. **73**, No. 1, p. K145–K146 (1982).

I. M. CHAPNIK
*On the Possibility of Prediction of Superconducting Properties in Transition Metal Compounds
(b) Vol. **113**, No. 2, p. K87–K92 (1982).
See also 21.1, p. 354.

E. KNAPEK and G. LEFRANC
*New Results on Superconducting Nb–Sn Sinter Material Applied for Electron Microscope Lenses
(a) Vol. **73**, No. 2, p. 447–453 (1982).

V. P. PAN, I. V. ZOLOTUKHIN, S. E. LITVIN, B. G. NIKITIN, and A. S. SHPIGEL
Electrical Properties of Lanthanum–Aluminium Glass
(a) Vol. **73**, No. 2, p. K287–K291 (1982).
See also *2*, p. *53*; 8, p. 89; *14.1*, p. *197*; 21, p. 341.

E. KOLLEY and W. KOLLEY
*On the Coexistence of Superconductivity and Itinerant Ferromagnetism
(b) Vol. **114**, No. 1, p. 135–144 (1982).
See also *18.2*, p. *273*.

G. KOZŁOWSKI and K. ROGACKI
*Magnetocaloric Effect and Spin Waves in Two-Sublattice Antiferromagnetic Superconductors
(b) Vol. **114**, No. 1, p. K5–K9 (1982).
See also *18.4*, p. *281*.

I. N. KUZMENKO, O. A. PALAGINA, and V. V. PUSTOVALOV
Study of Plastic Deformation of Aluminium near the Superconducting Transition
(a) Vol. **74**, No. 1, p. 51–57 (1982).
See also *10.1*, p. *101*; 21, p. 341.

B. I. BELEVTSEV, YU. F. KOMNIK, V. I. ODNOKOZOV, and A. V. FOMIN
*Discovery of Low-Temperature Metastable State in Amorphous Bismuth–Hydrogen Systems
(b) Vol. **114**, No. 2, p. 467–474 (1982).
See also *1.5*, p. *41*; 2, p. 55; 21.7, p. 381.

G. GIAQUINTA and C. DI MAURO
*Gapless Regime by Thermally Activated Vortex–Antivortex Plasma in 2D Superconductors
(b) Vol. **114**, No. 2, p. K183–K187 (1982).

S. TAKÁCZ
*The Summation of Pinning Forces in Superconductors with Small Defect Densities
(a) Vol. **74**, No. 2, p. 437–444 (1982).
See also 10.1, p. 102.

V. A. SIRENKO and V. S. FOMENKO
*Change of the Flow Stress at the Superconducting Transition in Lead–Indium Alloys
(a) Vol. **74**, No. 2, p. 459–465 (1982).
See also *10.1*, p. *102*; 21, p. 342.

H.-J. KÖHLER, T. HÄDRICH, P. SEIDEL, P. WEBER, K. BLÜTHNER, and K.-H. BERTHEL
*Aging Dependence of the Current Density Distribution in Nb–NbO_x–Pb Tunnel Junctions
(a) Vol. **74**, No. 2, p. K123–K125 (1982).

14.3 Semiconductors

Review Articles

Original Papers and Short Notes

L. B. DUBROVSKAYA, S. Z. NAZAROVA, and A. F. PREKUL
Superconducting and Normal Properties of Solid Solutions $Nb_{1-x}Ta_xC_y$
(a) Vol. **69**, No. 1, p. 167 – 172 (1982).
See also *1.3*, p. *25*; 14.2, p. 199; 18, p. 259; 21, p. 331.

H. GRUBER and E. KRAUTZ
*Magnetoresistance and Conductivity in the Binary System Titanium-Oxygen (I): Titanium Oxides with Metallic Conductivity
(a) Vol. **69**, No. 1, p. 287 – 295 (1982).
See also 8, p. 85; *14.1*, p. *193*; 15, p. 245; 18, p. 259; 22.6, p. 468.

V. A. M. BRABERS and J. C. J. M. TERHELL
*Electrical Conductivity and Cation Valencies in Nickel Manganite
(a) Vol. **69**, No. 1, p. 325 – 332 (1982).
See also *1*, p. *9*; 15, p. 245; 22.8.2, p. 499.

J. LECOMTE, J. P. LOUP, G. BOSSER, M. HERVIEU, and B. RAVEAU
Non-Stoichiometry and Electrical Conductivity of Strontium Niobates with Perovskite Structure (III): Thermodynamic Data for $Sr(Sr_{(1/3)+x}Nb_{(2/3)-x})O_{3-(3/2)x}$
(a) Vol. **69**, No. 1, p. 359 – 366 (1982).
See also *10.2*, p. *104*; *14.4*, p. *234*; 22.8.1, p. 492.

P. L. KUKK and H. A. AARNA
*Solubility of Ag in CdS Single Crystals
(a) Vol. **69**, No. 1, p. K109 – K112 (1982).
See also *10.2*, p. *104*; 22.4.1, p. 439.

K. TENNAKONE, W. G. D. DHARMARATNA and S. C. JAYEWARDENA
*Semiconduction in Ammonium Phosphomolybdate
(a) Vol. **69**, No. 1, p. K129 – K130 (1982).
See also 22.8.1, p. 492.

R. MACH, P. FLÖGEL, L. G. SUSLINA, A. G. ARESHKIN, J. MAEGE, and G. VOIGT
*The Influence of Compositional Disorder on Electrical and Optical Properties of ZnS_xSe_{1-x} Single Crystals
(b) Vol. **109**, No. 2, p. 607 – 615 (1982).
See also 13.5.1, p. 181; 20.1, p. 300; *20.3*, p. *319*; 22.4.4, p. 447.

H. T. LANGHAMMER, M. STORDEUR, H. SOBOTTA, and V. RIEDE
*Optical and Electrical Investigations of the Anisotropy of Sb_2Te_3 Single Crystals
(b) Vol. **109**, No. 2, p. 673 – 681 (1982).
See also 15, p. 245; *20.1*, p. *300*; 22.7, p. 479.

E. BARTA
*On the Optically Determined Relaxation Time of Heavily Doped p-Type Silicon
(b) Vol. **109**, No. 2, p. K85 – K88 (1982).
See also *20.1*, p. *300*; 22.1.2, p. 401.

A. SODEIKA, Z. SILEVIČIUS, Z. JANUŠKEVIČIUS, and A. SAKALAS
*The Influence of Intrinsic Defects on the Electrical Properties of Single Crystals of $CdSiP_2$ and $ZnGeP_2$
(a) Vol. **69**, No. 2, p. 491 – 495 (1982).
See also *13.4*, p. *164*; 22.8, p. 482.

C. LAFLAMME, M. AUBIN, R. PROVENCHER, and S. JANDL
*Resistivity Measurements of $ZrS_{3-x}Se_x$ Alloys
(a) Vol. **70**, No. 1, p. K67–K69 (1982).
See also 22.8, p. 483.

R. DURNÝ, P. MACKO, and R. D. TOMLINSON
*Electrical Properties of Neutron-Irradiated $CuInTe_2$
(a) Vol. **70**, No. 1, p. K89–K91 (1982).
See also *11*, p. *120*; 13.4, p. 166; 22.8, p. 483.

M. VAN DER MEER, R. SCHUCHARDT, and R. KEIPER
*Strong-Field Hopping in Disordered Semiconductors. A Problem of Directed Percolation
(b) Vol. **110**, No. 2, p. 571–580 (1982).
See also 2, p. 47; 22, p. 384.

D. VIGNAUD, J. L. FARVACQUE, and D. FERRE
*Anisotropy of the Free Carrier Piezoelectric Scattering by Dislocations in III–V Compounds
(b) Vol. **110**, No. 2, p. 601–609 (1982).
See also *10.2*, p. *107*; 22.2, p. 418.

YU. G. ARAPOV, A. B. DAVYDOV, I. M. TSIDILKOVSKII, and N. G. SHELUSHININA
*On the Anomalies in the Temperature Dependence of Conductivity in $Hg_{1-x}Cd_xTe$
(b) Vol. **110**, No. 2, p. 619–630 (1982).
See also 22.4.4, p. 447.

M. MATLAK and J. ZIELIŃSKI
*Erratum to: "Electrical Resistivity and Insulator-to-Metal Transition for the Periodic Anderson Model"
(b) Vol. **110**, No. 2, p. K179 (1982).
See also *14.1*, p. *194*.

V. I. KUZNETSOV and P. F. LUGAKOV
Study of Radiation Defect Clusters, Their Structure and Properties, in Proton-Irradiated Silicon
(a) Vol. **70**, No. 2, p. 387–395 (1982).
See also *11*, p. *121*; 22.1.2, p. 404.

E. ZIEGLER, A. HEINRICH, H. OPPERMANN, and G. STÖVER
*Growth and Electrical Properties of Non-Stoichiometric ZnO Single Crystals Doped with Co
(a) Vol. **70**, No. 2, p. 563–570 (1982).
See also 3, p. 57; 13.4, p. 167; 22.4, p. 437.

A. HAYDAR, A. CORET, and N. OUDJEHANE
*Dark Current–Voltage Characteristics of Cu_2O Crystals and Their Relation to Structural Defects
(a) Vol. **70**, No. 2, p. 683–690 (1982).
See also 22.6, p. 471.

T. G. ABDEL-MALIK, A. M. ABDEEN, and A. A. ALY
*Thermally Stimulated Current in β-Metal-Free Phthalocyanine Single Crystals
(a) Vol. **70**, No. 2, p. 703–712 (1982).
See also *13.4*, p. *167*; 22.9, p. 504.

J. Horák, L. Koudelka, J. Klikorka, and L. Šiška
*Transport Properties and Plasma Resonance of Bismuth Selenide Crystals Doped with Cadmium
(b) Vol. **111**, No. 2, p. 575–580 (1982).
See also *13.4*, p. *169*; 15, p. **247**; 22.7, p. 479.

I. C. da Cunha Lima, M. Fabbri, and A. Ferreira da Silva
*Diagonal and Off-Diagonal Disorder in Doped Semiconductors
(b) Vol. **111**, No. 2, p. K69–K74 (1982).
See also *8*, p. *87*; 22.1.2, p. 407.

R. S. Brazis and A. S. Mironas
*Investigation of Electric Conductivity and Avalanche Breakdown in InSb by Means of Nonlinear Magnetoplasma Resonance
(a) Vol. **71**, No. 2, p. 473–481 (1982).
See also 14.3.4, p. 229; 18, p. 261; 20, p. 295; 22.2.3, p. 428.

S. M. Wasim, G. S. Porras, and R. D. Tomlinson
*Some Electrical Characteristics of Cu- and In-Doped $CuInTe_2$
(a) Vol. **71**, No. 2, p. 523–530 (1982).
See also *13.4*, p. *170*; 22.8, p. 485.

T. Satow, O. Uemura, and Y. Sagara
The Neutron Diffraction Study of Liquid Antimony and Bismuth Chalcogenides
(a) Vol. **71**, No. 2, p. 555–561 (1982).
See also *2*, p. *49*; 22.7, p. 479.

E. Yu. Brailovskii, F. K. Karapetyan, I. G. Megela, and V. P. Tartachnik
*Radiation Defects in Electron-Irradiated InP Crystals
(a) Vol. **71**, No. 2, p. 563–568 (1982).
See also *11*, p. *122*; 22.2.2, p. 425.

M. Iliev, S. Angelov, I. Z. Kostadinov, V. Bojchev, and V. Hadjiev
*Critical Behaviour of Co_3O_4 Conductivity near the Antiferromagnetic Phase Transition
(a) Vol. **71**, No. 2, p. 627–632 (1982).
See also 18.4, p. 280; 22.8.2, p. 500.

G. Krabbes, E. I. Terukov und H. Oppermann
*Zeitabhängiges Verhalten des elektrischen Widerstandes beim α-Phasenübergang in stark dotiertem Cobalteisensulfid $Co_yFe_{1-y}S$
(Time Dependence of the Electrical Resistivity of α-Phase Transition in Highly Doped Cobalt Iron Sulfide ($Co_yFe_{1-y}S$)
(a) Vol. **71**, No. 2, p. K143–K146 (1982).
See also *1.2*, p. *18*; 22.8, p. 486.

E. V. Kučys, O. E. Panchuk, J. V. Kameneckas, and D. F. Junevičius
On the Kinetics of the Process of Decay of Supersaturated Solid Copper or Silver Solution in Cadmium Telluride
(a) Vol. **71**, No. 2, p. K157–K160 (1982).
See also 9, p. 92; *10.2*, p. *110*; 22.4.3, p. 446.

N. Kamegashira and K. Sato
*Anomaly in Isothermal Conductivity in the Holmium–Oxygen System
(a) Vol. **71**, No. 2, p. K165–K168 (1982).
See also 22.6, p. 472.

V. N. BRUDNYI, S. A. VOROBIEV, and A. A. TSOI
Positron Annihilation and Electrical Properties of Electron-Irradiated (≈ 2 MeV) InAs Crystals
(a) Vol. **72**, No. 2, p. 529–534 (1982).
See also *11*, p. *124*; *13.4*, p. *173*; 22.2.1, p. 422.

M. WOLF, E. I. TERUKOV, W. REICHELT, and H. OPPERMANN
Influence of ^{18}O Substitution on the High-Temperature Phase Transition of Cr-Doped V_2O_3 and Its Relation to Low-Temperature Phase Transition
(a) Vol. **72**, No. 2, p. 655–659 (1982).
See also *1.2*, p. *20*; 6, p. 78; 14.1, p. 196; 18, p. 262; 22.6, p. 473.

A. N. ANAGNOSTOPOULOS, C. MANOLIKAS, D. PAPADOPOULOS, and J. SPYRIDELIS
*Composition Faults in $ZnIn_2S_4$(III) Layered Crystals and Their Influence on the Anisotropic Conductivity of This Compound
(a) Vol. **72**, No. 2, p. 731–736 (1982).
See also 1.4, p. 35; 22.8, p. 488.

L. K. VODOPYANOV and S. P. KOZYREV
Infrared Reflection Spectroscopy of Ion-Implanted n-$Hg_{0.8}Cd_{0.2}Te$
(a) Vol. **72**, No. 2, p. 737–744 (1982).
See also 11, p. 125; *20.1*, p. *309*; 22.4.4, p. 448.

A. BLAISE, D. DAMIEN, and J. MULAK
*Electrical Resistivity of Some Neptunium Chalcogenides
(a) Vol. **72**, No. 2, p. K145–K148 (1982).
See also *14.1*, p. *196*.

E. M. LAWSON and S. J. PEARTON
*Hydrogen Passivation of Laser-Induced Acceptor Defects in p-Type Silicon
(a) Vol. **72**, No. 2, p. K155–K158 (1982).
See also *10.2*, p. *113*; 22.1.2, p. 411.

M. R. BROZEL, E. J. FOULKES, and B. TUCK
*The Electrical Properties of InP Diffused with Chromium or Iron
(a) Vol. **72**, No. 2, p. K159–K163 (1982).
See also *10.2*, p. 113; 22.2.2, p. 426.

V. E. BORISENKO and V. A. LABUNOV
Annealing of Antimony Implanted Silicon with Halogen Lamp Irradiation
(a) Vol. **72**, No. 2, p. K173–K176 (1982).
See also *11*, p. *125*; 22.1.2, p. 411.

D SENDOREK, M. LUBECKA, and A. WĘGRZYN
*Carrier Mobility in EuO
(b) Vol. **113**, No. 1, p. K19–K22 (1982).
See also *18.2*, p. *272*; 22.6, p. 473.

P. KUIVALAINEN, J. HELESKIVI, and M. LEPPIHALME
*Contribution of the Fermi Level Shift to the Low Temperature Magnetoresistance in αSi:H
(b) Vol. **113**, No. 1, p. K39–K42 (1982).
See also 18, p. 262; 22.1.2, p. 412.

V. L. BONCH-BRUEVICH and TO THI BICH NGUYET
*Low Temperature Conductivity and Optical Transitions of a Two-Dimensional Disordered Semiconductor under Weak Localization Conditions
(b) Vol. **114**, No. 1, p. 67–71 (1982).
See also 2, p. 53; *20.1*, p. *314*; 22, p. 391.

G. M. SHMELEV, G. I. TSURKAN, and NGUEN HONG SCHON
*Photo-Hall Effect in Longitudinal Magnetic Field
(b) Vol. **114**, No. 1, p. K41–K45 (1982).

J. AUTH, U. HENNIGER, and R. KEIPER
*On the Formulation of Transport Equations in Semiconductors with Smooth Macroscopic Inhomogeneity
(b) Vol. **114**, No. 1, p. K75–K78 (1982).
See also 22, p. 391.

E. HARTMANN and L. KOVÁCS
*Electrical Conductivity of Paratellurite (TeO_2) Crystals
(a) Vol. **74**, No. 1, p. 59–64 (1982).
See also 16, p. 253; 22.6, p. 475.

L. A. KAZAKEVICH and P. F. LUGAKOV
The Effect of Dislocations on the Radiation Defect Annealing Processes in Silicon
(a) Vol. **74**, No. 1, p. 113–122 (1982).
See also *10.2*, p. *116*; *11*, p. *127*; 13.4, p. 176; 22.1.2, p. 414.

L. G. KHVOSTANTSEV, V. A. SIDOROV, L. E. SHELIMOVA, and N. KH. ABRIKOSOV
*Phase Transitions in GeTe at Hydrostatic Pressure up to 9.3 GPa
(a) Vol. **74**, No. 1, p. 185–192 (1982).
See also *1.2*, p. *23*; 12.2, p. 138; 22.7, p. 480.

M. N. KHAN and E. E. KHAWAJA
*The Electrical and Optical Properties of Glasses of the Li_2O–GeO_2 and Na_2O–GeO_2 Systems
(a) Vol. **74**, No. 1, p. 273–277 (1982).
See also 2, p. 54; *20.1*, p. *314*; 22.6.1, p. 478.

D. MERGEL and R. LABUSCH
Variable Reconstruction of Dislocation Cores in Si
(b) Vol. **114**, No. 2, p. 545–551 (1982).
See also *10.2*, p. *117*; *13.4*, p. *177*; 16, p. 254; 22.1.2, p. 415.

M. I. AUSLENDER, N. G. BEBENIN, and V. P. KALASHNIKOV
The Asymptotics of the Electron Density of States in the s–f Exchange Model for Ferromagnetic Semiconductors near T_c
(b) Vol. **114**, No. 2, p. K147–K152 (1982).
See also *13*, p. *147*; *18.2*, p. *274*.

E. LITWIN-STASZEWSKA, L. KOŃCZEWICZ, R. PIOTRZKOWSKI, and W. SZYMAŃSKA
*On the Deep Impurity Level in n-InSb
(b) Vol. **114**, No. 2, p. K153–K156 (1982).
See also *13.4*, p. *177*; 22.2.3, p. 429.

A. OZA
*Electrical Resistivity of α-Cyclodextrin $KI–I_2 \cdot 4\, H_2O$ Crystals
(b) Vol. **114**, No. 2, p. K171–K176 (1982).
See also 22.9, p. 508.

14.3.1 Films

Review Articles

R. Mach and G. O. Müller
Physical Concepts of High-Field, Thin-Film Electroluminescence Devices
(a) Vol. **69**, No. 1, p. 11–66 (1982).
See also *14.3.3*, p. *219*; *20.3*, p. *318*; 22.4.1, p. 439; 22.4.2, p. 442.

Z. M. Jarzębski
*Preparation and Physical Properties of Transparent Conducting Oxide Films
(a) Vol. **71**, No. 1, p. 13–41 (1982).
See also 1.5, p. 38; 3, p. 57; 13.1, p. 148; *20.1*, p. *298*; 22.6, p. 468; 22.6.1, p. 477.

Original Papers and Short Notes

J. Szczyrbowski, A. Dietrich, and H. Hoffmann
Optical Properties of RF-Sputtered Indium Oxide Films
(a) Vol. **69**, No. 1, p. 217–226 (1982).
See also *20.1*, p. *299*; 22.6, p. 468.

S. Ahmed and D. Sigurd
*Electrical Properties of Glow Discharge a-Si:H Thin Films Prepared with a RF Plasma Etch Unit
(a) Vol. **69**, No. 1, p. 305–314 (1982).
See also 2, p. 45; 22.1.2, p. 400.

V. M. Rubinov and M. Tuychiev
*Investigation of the Current Characteristics of Thin-Film Structures Based on GaP under the Action of an Electron Beam
(a) Vol. **69**, No. 1, p. K113–K116 (1982).
See also *11*, p. *118*; 22.2.2, p. 424.

A. V. Dvurechenskii and I. A. Ryazantsev
*Substitutional Doping of Vacuum-Evaporated Amorphous Silicon by Ion Implantation
(a) Vol. **69**, No. 1, p. K117–K120 (1982).
See also 2, p. 46; *11*, p. *118*; 22.1.2, p. 401.

C. Gunkel, D. Schalch, and A. Scharmann
*Electron Beam Induced Leakage Currents in Silicon Nitride Thin Films
(a) Vol. **69**, No. 1, p. K131–K134 (1982).
See also *11*, p. *118*; 22, p. 383.

C. M. Vodenicharov, A. A. Stanchev, and S. G. Christov
*A New Method for Determination of the Metal–Insulator (Semiconductor) Work Function by Photoelectric Measurements
(a) Vol. **69**, No. 2, p. 541–546 (1982).
See also *14.3.3*, p. *219*; 16, p. 250; 22, p. 384.

J. Pichon, J. Salardenne, and A. S. Barriere
*Study of Electronic Localized States Responsible of the Polarization of FeF_3 Thin Films
(a) Vol. **69**, No. 2, p. 699–706 (1982).
See also 13.4, p. 165; *14.3.3*, p. *220*; 21.1.1, p. 359.

E. F. Krimmel, A. G. K. Lutsch, and E. Doering
*Contribution to Electron Beam Annealing of High-Dose Ion-Implanted Polysilicon
(a) Vol. **71**, No. 2, p. 451–456 (1982).
See also *11*, p. *122*; 22.1.2, p. 408.

A. V. Dvurechenskii, I. A. Ryazantsev, and V. A. Dravin
*Doping of Amorphous Silicon by Manganese Ion Implantation
(a) Vol. **71**, No. 2, p. K133–K136 (1982).
See also 2, p. 49; *11*, p. *123*; 13.4, p. 170; 22.1.2, p. 408.

C. R. Dutta and K. Barua
DC Electrical Properties of Vacuum Deposited Yb_2O_3 Films
(a) Vol. **71**, No. 2, p. K181–K186 (1982).
See also *13.4*, p. *170*; *14.3.4*, p. *229*; 22.6, p. 472.

R. Dimmich and F. Warkusz
*Optical Reflectance and Electrical Resistance of Polycrystalline Thin Metallic Films
(a) Vol. **72**, No. 1, p. 117–122 (1982).
See also *20.1*, p. *308*; 21.6, p. 377.

A. L. Dawar, A. O. Mohammad, Partap Kumar, O. P. Taneja, and P. C. Mathur
*Effect of Indium on the Electrical Transport Properties of Epitaxial Thin Films of SnTe
(a) Vol. **72**, No. 1, p. K83–K87 (1982).
See also 22, p. 388.

J. Tyczkowski, G. Czeremuszkin, and M. Kryszewski
*On Differentiation of Electrical Conduction Processes in Polymers
(a) Vol. **72**, No. 2, p. 751–762 (1982).
See also *14.3.4*, p. *230*; 22.9, p. 506.

M. Y. Nadeem and C. A. Hogarth
*DC Conduction in Thin Films of GeO_2/BaO Complexes prior to Electroforming
(a) Vol. **72**, No. 2, p. K203–K206 (1982).
See also 22.6.1, p. 477.

R. Dutta, P. K. Banerjee, and S. S. Mitra
*Optical and Electrical Properties of Hydrogenated Amorphous Silicon Carbide
(b) Vol. **113**, No. 1, p. 277–284 (1982).
See also 2, p. 51; *20.1*, p. *310*; 22.3, p. 434.

E. A. Silinsh, V. A. Kolesnikov, I. J. Muzikante, and D. R. Balode
On Charge Carrier Photogeneration Mechanisms in Organic Molecular Crystals
(b) Vol. **113**, No. 1, p. 379–393 (1982).
See also *16*, p. *253*; 22.9, p. 506.

A. L. Dawar, P. Kumar, S. K. Paradkar, T. D. Sadana, and P. C. Mathur
*Electrical Transport Properties of Epitaxial Films of $Pb_{1-x}Sn_xTe$
(a) Vol. **73**, No. 2, p. 503–507 (1982).
See also 13.1, p. 155; 22, p. 390.

V. I. Osinskii, S. A. Malyshev, and M. P. Ryzhkov
*Minority Carrier Transport and Radiative Recombination in $Al_xGa_{1-x}As$ Variable-Composition Structures
(a) Vol. **74**, No. 1, p. 43–50 (1982).
See also *20.3*, p. *328*; 22.2.4, p. 431.

14.3.2 Surfaces and Interfaces

Original Papers and Short Notes

E. Nebauer, E. Brink, P. Janietz, and T. Liecke
Are Au Profiles Well Measurable in n-GaAs by Electroetching?
(a) Vol. **69**, No. 1, p. K65–K68 (1982).
See also *9*, p. *91*; 22.2.1, p. 420.

A. M. Gabovich and A. I. Voitenko
*Influence of Semiconductor Dielectric Function Spatial Dispersion on Charge Electrostatic Energy near the Semiconductor/Vacuum Interface and Field Emission Current
(b) Vol. **110**, No. 2, p. 407–416 (1982).
See also 17, p. 255; 22.1.1, p. 396; 22.1.2, p. 404.

J. Marczewski, R. Nitecki, and M. Liberadzka
*The Influence of the Negative Bias-Temperature Test on the Properties of the Si–SiO_2 Interface
(a) Vol. **70**, No. 2, p. 555–561 (1982).
See also *13.3*, p. *159*; 20.1, p. 304; 22.1.2, p. 405.

W. Lorenz and M. Handschuh
*Dynamical Theory of Photocurrent–Voltage Curves and Photocurrent Efficiency on Semiconductor Electrodes
(a) Vol. **71**, No. 1, p. 127–134 (1982).
See also *16*, p. *251*; 22.2.2, p. 425.

M. Weidner and G. Ritter
*Spreading Resistance Profile Correction Scheme for Thin Layers on a Conducting Boundary
(a) Vol. **71**, No. 1, p. 153–158 (1982).
See also 22.1.2, p. 407.

W. J. Kordalski and B. M. Wilamowski
Method for Determination of Charge Density Distribution in Silicon Nitride of MNOS Structures
(a) Vol. **71**, No. 2, p. 387–398 (1982).
See also *14.3.3*, p. *222*; 22, p. 387.

K. Germanova, K. Marinova, and A. Trayanov
*On the Electronic Subbands in n-Inversion Layers on (111) Ge
(b) Vol. **112**, No. 2, p. 575–580 (1982).
See also *13.1*, p. *153*; 22.1.1, p. 397.

N. V. Vershinina and N. N. Gerasimenko
*Some Peculiarities of Radiation-Induced Changes in the Properties of Si–SiO_2 Structures Irradiated at Elevated Temperatures
(a) Vol. **72**, No. 2, p. 569–574 (1982).
See also *11*, p. *124*; 13.4, p. 173; 22.1.2, p. 411.

T. Suzuki, Y. Hirose, and H. Hirose
A Model of the Photodoping Mechanism in the Ag/As_2S_3 System
(a) Vol. **72**, No. 2, p. K165–K168 (1982).
See also *10.2*, p. *113*; *16*, p. *253*; 22, p. 389.

14.3.3 Devices. Junctions (Contact Problems See 14.3.4)

Review Articles

R. MACH and G. O. MÜLLER
*Physical Concepts of High-Field, Thin-Film Electroluminescence Devices
(a) Vol. **69**, No. 1, p. 11–66 (1982).
See also 14.3.1, p. **213**; *20.3*, p. *318*; 22.4.1, p. 439; 22.4.2, p. 442.

S. CHANDRA and R. K. PANDEY
*Semiconductor Photoelectrochemical Solar Cells
(a) Vol. **72**, No. 2, p. 415–454 (1982).
See also *16*, p. *249*; 22.2, p. 418; 22.4, p. 437.

R. J. J. ZIJLSTRA
Distribution Model of Flicker Noise in Semiconductors
(a) Vol. **73**, No. 1, p. 11–21 (1982).
See also *14.3*, p. *202*; 14.3.4, p. 227; 22.1, p. 393; 22.2, p. 418.

Original Papers and Short Notes

P. KRISPIN
Thermochemical Model Applied to a Deep-Level Defect in GaP
(a) Vol. **69**, No. 1, p. 193–200 (1982).
See also *13.4*, p. *163*; 22.2.2, p. 424.

G. OELGART and G. ARNOLD
*Avalanche Multiplication in GaP Studied by Means of the SEM
(a) Vol. **69**, No. 1, p. 207–216 (1982).
See also *11*, p. *118*; 14.3.4, p. **227**; 22.2.2, p. 424.

A. ZEHE, L. HESSE, and R. REYES S.
*Analysis of Ghost Peak Formation in the Emission Spectra of GaAs p–n Junctions
(a) Vol. **69**, No. 1, p. 281–285 (1982).
See also *20.3*, p. *318*; 22.2.1, p. 420.

K. SEN and R. S. SRIVASTAVA
*SIS Solar Cell Theory and Calculations
(a) Vol. **69**, No. 1, p. 413–418 (1982).
See also *16*, p. *249*; 22.1.2, p. 401.

S. K. KRAWCZYK
*Simple Equation Describing I–U Characteristics of MIS Solar Cells
(a) Vol. **69**, No. 1, p. K77–K81 (1982).
See also *16*, p. *250*; 22.1.2, p. 401.

C. M. VODENICHAROV, A. A. STANCHEV, and S. G. CHRISTOV
*A New Method for Determination of the Metal–Insulator (Semiconductor) Work Function by Photoelectric Measurements
(a) Vol. **69**, No. 2, p. 541–546 (1982).
See also *14.3.1*, p. *213*; 16, p. 250; 22, p. 384.

X. W. Fan and J. Woods
*Blue and Violet Electroluminescence Emission in $ZnSe_xS_{1-x}$ MIS Diodes
(a) Vol. **70**, No. 1, p. K27 – K30 (1982).
See also 13.5.1, p. 182; *20.3*, p. *321*; 22.4.4, p. 447.

V. Milanović and D. Tjapkin
Energy Band Calculation and Zero Energy Gap Conditions for Semiconductor Superlattices
(b) Vol. **110**, No. 2, p. 687 – 695 (1982).
See also *13.1*, p. *150*; 22.2.4, p. 430.

W. R. Fahrner, K. Heidemann, and P. Schöttle
*Results of Ion Implantation into Silicon in the 100 MeV Range (I): Oxygen and Boron Implantation
(a) Vol. **70**, No. 2, p. 463 – 472 (1982).
See also *11*, p. *121*; 22.1.2, p. 405.

V. P. Romanov
Stationary Distribution of Mobile Ions in a Dielectric with Regard to Their Elastic Interaction with the Medium
(a) Vol. **70**, No. 2, p. 525 – 532 (1982).
See also *14.4*, p. *235*; 22.6, p. 470.

Y. Hasegawa and Y. Abe
*Electrical and Optical Characteristics of a Schottky Barrier on a Cleaved Surface of the Layered Semiconductor InSe
(a) Vol. **70**, No. 2, p. 615 – 621 (1982).
See also *13.3*, p. *159*; 13.4, p. 167; 16, p. 250; 22.3, p. 433.

N. M. Ravindra, K. Sunil Kumar, and V. K. Srivastava
Temperature and Voltage Dependence of the Barrier Height in SnO_2/Si Solar Cells
(a) Vol. **70**, No. 2, p. 623 – 630 (1982).
See also *13.3*, p. *159*; 16, p. 251; 22.1.2, p. 405.

V. S. Lysenko, R. N. Litovskii, A. N. Nazarov, and V. P. Kulichkov
*Temperature Dependence of Capacitance–Voltage Characteristics in Implanted MOS Structures
(a) Vol. **70**, No. 2, p. 719 – 728 (1982).
See also 11, p. 121; 22.1.2, p. 405.

F. Schauer, M. Jedlička, and J. Polcer
*Amorphous–Crystalline Heterojunction with Built-in Interface Charge (I): Steady-State Transport
(a) Vol. **70**, No. 2, p. 755 – 762 (1982).
See also 2, p. 48; 13.3, p. 159; 22, p. 385; 22.4.2, p. 443.

A. I. Koifman, R. A. Muminov, and A. Ya. Yafasov
*Investigation of the Effect of Large Doses on the Si(Li) Detector Parameters
(a) Vol. **71**, No. 1, p. 59 – 64 (1982).
See also *11*, p. *121*; 22.1.2, p. 406.

W. R. Fahrner, K. Heidemann, and P. Schöttle
Results of Ion Implantation into Silicon in the 100 MeV Range (II): Electrical Properties
(a) Vol. **71**, No. 1, p. 121 – 125 (1982).
See also *11*, p. *122*; 22.1.2, p. 406.

K. Sen
*The Effect of Interface States on the Performance of MIS Solar Cells
(a) Vol. **71**, No. 2, p. 641 – 648 (1982).
See also 13.3, p. 159; *16*, p. *252*; 22.1.2, p. 408.

H. Nakashima and K. Hashimoto
*Barrier Height for Cobalt on Silicon
(a) Vol. **71**, No. 2, p. K201 – K203 (1982).
See also *14.3.4*, p. *230*; 21.1, p. 349; 22.1.2, p. 408.

W. R. Fahrner, D. Bräunig, and E. Borchert
*MOS and Thyristor Damage Parameters after 2.5 and 15 MeV High Temperature Electron Irradiation
(a) Vol. **72**, No. 1, p. 79 – 88 (1982).
See also *11*, p. *123*; 22.1.2, p. 409.

D. V. Morgan, K. Board, C. E. C. Wood, and L. F. Eastman
*Current Transport in Al/InAlAs/InGaAs Heterostructures
(a) Vol. **72**, No. 1, p. 251 – 260 (1982).
See also 22.2.4, p. 430.

A. van der Ziel and C. M. van Vliet
*Mobility-Fluctuation $1/f$ Noise in Nonuniform Nonlinear Samples and in Mesa Structures
(a) Vol. **72**, No. 1, p. K53 – K55 (1982).
See also *14.3.4*, p. *230*.

T. Rang
*Simple Relationship between the Breakdown Voltage, Concentration, and Junction Depth for Diffused pn Junction
(a) Vol. **72**, No. 1, p. K117 – K119 (1982).
See also 22.1.2, p. 410.

O. Hildebrand
Anomalous Impurity Diffusion in III–V Compounds: The Consequence of Self-Induced Field Effects
(a) Vol. **72**, No. 2, p. 575 – 584 (1982).
See also *9*, p. *92*; 22.2, p. 418.

E. Ziegler, W. Siegel, H. Blumtritt, and O. Breitenstein
*Electrical and EBIC Investigations of GaP Grain Boundaries
(a) Vol. **72**, No. 2, p. 593 – 605 (1982).
See also 1.4, p. 35; 22.2.2, p. 425.

H. Krause
*DC Characteristics of MOS Structures and Hopping Currents in Thermally Grown SiO_2 Films
(a) Vol. **72**, No. 2, p. 607 – 616 (1982).
See also *13.4*, p. *173*; 22.6, p. 472.

D. Seuret, O. Vigil, and F. Leccabue
*Electro-Optical Properties of $CuGa_{0.7}In_{0.3}Se_2ZnSe$ Heterojunction
(a) Vol. **72**, No. 2, p. 661 – 666 (1982).
See also *20.3*, p. *324*; 22.4.2, p. 443; 22.8, p. 488.

A. van der Ziel and G. Bosman
*Collision-Free Limit of Near-Thermal Noise in Short Solid-State Diodes
(a) Vol. **73**, No. 1, p. K93 – K96 (1982).

M. S. Gilenko, P. M. Karageorgy-Alkalaev, and A. Yu. Leiderman
*On the Noise of Semiconductor Diode Structures with Strong Carrier Accumulation
(a) Vol. **73**, No. 1, p. K147 – K151 (1982).
See also *14.3.4*, p. *231*; 22.1.2, p. 413.

T. X. Hoai, P. H. Binh, and N. D. Hung
*Study of the Nonlinear Recombination Process in Silicon by the Saturated Photovoltage Transient Method
(a) Vol. **73**, No. 2, p. K165 – K168 (1982).
See also *16*, p. *253*; 22.1.2, p. 414.

S. A. Asimov, D. A. Aronov, D. K. Isamukhamedova, and Yu. M. Yuabov
*Transient Processes of Current Reversal in MIS Structures with Majority and Minority Charge Carriers
(a) Vol. **73**, No. 2, p. K179 – K183 (1982).
See also 14.3.2, p. 218; 22.1.2, p. 414.

J. Pułtorak
*Influence of Surface Recombination-Generation in the Depletion Layer on the *I–U* Characteristic of a p–n Junction
(a) Vol. **73**, No. 2, p. K217 – K220 (1982).
See also 13.3, p. 160; 22.1.1, p. 398.

I. Zólomy
*Diffusion-Isolated MISS Devices
(a) Vol. **73**, No. 2, p. K249 – K251 (1982).
See also *14.3.4*, p. *232*; 22.1.2, p. 414.

D. Sengupta and Vikram Kumar
*Study of n-GaAs MOS Diodes with Spin-on SiO_2 Layer
(a) Vol. **73**, No. 2, p. K279 – K282 (1982).
See also 22.2.1, p. 423.

V. M. Bazovkin, G. L. Kurushev, and V. G. Polovinkin
*On the Theory of the MIS High-Frequency Capacitance. The Effect of Degeneration and Non-Parabolicity
(a) Vol. **74**, No. 1, p. 297 – 306 (1982).
See also 22, p. 391.

A. L. Dawar, Anil Kumar, O. P. Taneja, Partap Kumar, and P. C. Mathur
*Field Effect Studies in In_2Te_3 Metal–Insulator–Semiconductor Structure
(a) Vol. **74**, No. 2, p. 403 – 409 (1982).
See also 22, p. 392.

W. R. Fahrner, H. G. Bach, D. Bräunig, and M. Knoll
*Implantation Profiles of Buried Layers (MOS pn) from HF or LF *C*(*U*) Curves
(a) Vol. **74**, No. 2, p. 421 – 429 (1982).
See also *11*, p. *128*; 22.1.2, p. 415.

14.3.4 High-Field Phenomena, Space-Charge, Inhomogeneneities, Injected Carriers (Electroluminescence See 20.3; Junctions See 14.3.3)

Review Article

R. J. J. Zijlstra
Distribution Model of Flicker Noise in Semiconductors
(a) Vol. **73**, No. 1, p. 11–21 (1982).
See also *14.3*, p. *202*; 14.3.3, p. 219; 22.1, p. 393; 22.2, p. 418.

Original Papers and Short Notes

V. A. Kozlov and I. M. Nefedov
*Modified Self-Scattering Procedure for Monte-Carlo Simulation of Hot Electrons
(b) Vol. **109**, No. 1, p. 393–401 (1982).
See also *13*, p. *139*; 22, p. 382.

D. A. Aronov, R. Mamatkulov, and V. V. Rubinov
*The Effects of Field and Temperature Heating in Intrinsic Semiconductors with Non-Injecting Contacts
(a) Vol. **69**, No. 1, p. 159–165 (1982).
See also 16, p. 249; 22.1.1, p. 395.

G. Oelgart and G. Arnold
Avalanche Multiplication in GaP Studied by Means of the SEM
(a) Vol. **69**, No. 1, p. 207–216 (1982).
See also *11*, p. *118*; *14.3.3*, p. *219*; 22.2.2, p. 424.

H.-J. Fitting and J.-U. Friemann
*Monte-Carlo Studies of the Electron Mobility in SiO_2
(a) Vol. **69**, No. 1, p. 349–358 (1982).
See also 6, p. 71; *13*, p. *140*; 22.6, p. 468.

G. Bosman and R. J. J. Zijlstra
*Thermal Noise, Diffusion, and the Mobility of Hot Holes in Silicon
(b) Vol. **109**, No. 2, p. 683–691 (1982).
See also 22.1.2, p. 401.

H. Kostial and L. F. Kurtenok
*Temperature Dependence of the Multivalued Sasaki Effect
(b) Vol. **109**, No. 2, p. K109–K113 (1982).
See also *13*, p. *141*; 22.1.2, p. 402.

I. V. Ryzhikov, V. P. Sushkov, P. I. Tepaikin, and I. L. Kasatkin
Double Injection at p–n Junctions Based on $Al_xGa_{1-x}As$
(a) Vol. **69**, No. 2, p. 707–718 (1982).
See also *14.3.3*, p. *220*; *20.3*, p. *320*; 22.2.4, p. 430.

J. M. Noras
*Thermal Filling Effects on Constant Capacitance Transient Spectroscopy
(a) Vol. **69**, No. 2, p. K209–K213 (1982).

J. Sinkkonen
*Photoadmittance of an Inhomogeneous Semiconductor
(a) Vol. **69**, No. 2, p. K233–K237 (1982).
See also 16, p. 250; 22, p. 384.

Tran Chot
*Some New Experimental Results of a Study of Metal–n-Type Semiconductor Schottky Barrier Contacts
(a) Vol. **70**, No. 1, p. 311–316 (1982).
See also 22.1.2, p. 403; 22.2.1, p. 421; 22.2.2, p. 424; 22.4.1, p. 440; 22.4.2, p. 443; 22.4.3, p. 445.

W. G. Proctor, P. Lawaetz, Y. Marfaing, and R. Triboulet
*On the S-Shaped Characteristic of Semiconductors at Low Temperature
(b) Vol. **110**, No. 2, p. 637–641 (1982).
See also *13.4*, p. *167*; 22.4.3, p. 445.

A. A. Ignatov and V. I. Shashkin
*Current–Voltage Characteristics of the Semiconductor Superlattice Interacting with a Strong High-Frequency Field
(b) Vol. **110**, No. 2, p. K117–K119 (1982).
See also 22, p. 385.

R. A. Weeks, J. Narayan, and E. Sonder
*Electric Breakdown in MgO Crystals at Elevated Temperature
(a) Vol. **70**, No. 2, p. 631–639 (1982).
See also 10.2, p. 108; 19, p. 285; 22.6, p. 470.

J. Bracinı́k and J. Štelina
*Electric Field in the Acoustoelectric Stationary Domains of the Second Mode. Simple Model
(a) Vol. **70**, No. 2, p. K135–K139 (1982).
See also *7*, p. *82*; 22.2.3, p. 427.

J. Gazsó
*A Refinement upon the Method of Evaluating the Density of the Localized States from the Field Effect
(a) Vol. **70**, No. 2, p. K181–K183 (1982).
See also 2, p. 48; *13.4*, p. *168*; 14.3.1, p. 214; 22, p. 385.

A. R. Beattie
*Impact Ionisation and Quantum Efficiency in InSb
(b) Vol. **111**, No. 1, p. 141–153 (1982).
See also 22.2.3, p. 427.

T. M. Gasymov, A. A. Katanov, and F. K. Kocharli
Propagation of Strong Electromagnetic Waves in Kane Semiconductors under Conditions of Electron and Phonon Heating
(b) Vol. **111**, No. 1, p. 383–395 (1982).
See also *6*, p. *73*; *13*, p. *143*; 22, p. 386.

I. I. Fishchuk
*Theoretical Investigation of the Field Dependence of the Direct Current in Inhomogeneous Semiconductors
(b) Vol. **111**, No. 1, p. K17–K21 (1982).
See also 22, p. 386.

G. OELGART, H. HAEFNER, R. REULKE, and F. JOACHIM
*Degradation of Green Light-Emitting GaP:N Diodes
(a) Vol. **71**, No. 1, p. 89–98 (1982).
See also *20.3*, p. *323*; 22.2.2, p. 425.

M. M. HAFIZ, M. M. IBRAHIM, and A. S. AHMED
*Conduction and Switching in In_xSe_{1-x}
(a) Vol. **71**, No. 1, p. 259–263 (1982).
See also *14.3.1*, p. *214*; 22.3, p. 433.

J. BULLOT, O. GALLAIS, M. GAUTHIER, and J. LIVAGE
*Threshold Switching in V_2O_5 Layers Deposited from Gels
(a) Vol. **71**, No. 1, p. K1–K4 (1982).
See also *2*, p. *48*; 22.6, p. 471.

G. B. ABDULLAEV, O. B. TAGIEV, G. M. NIFTIEV, and T. KH. AZIZOV
*Injection and Thermoactivation Currents in $EuGa_2S_4$ Single Crystals
(a) Vol. **71**, No. 1, p. K45–K48 (1982).
See also *13.4*, p. *169*; 22.8, p. 484.

V. A. KISELEV
*Pseudo-Exciton Effect in the Schottky-Barrier Electroreflectance
(b) Vol. **111**, No. 2, p. 461–468 (1982).
See also 13.5.1, p. 183; *20.1*, p. *305*; 22.2.1, p. 421.

Y. SEKI, S. ENDO, and T. IRIE
*Negative Resistance in $CdIn_2S_4$ Single Crystals
(a) Vol. **71**, No. 2, p. 365–370 (1982).
See also 22.8, p. 485.

R. S. BRAZIS and A. S. MIRONAS
Investigation of Electric Conductivity and Avalanche Breakdown in InSb by Means of Nonlinear Magnetoplasma Resonance
(a) Vol. **71**, No. 2, p. 473–481 (1982).
See also *14.3*, p. *207*; 18, p. 261; 20, p. 295; 22.2.3, p. 428.

A. N. ANAGNOSTOPOULOS and A. KAROUTIS
*Electron Trap Distribution Determined by Means of Conductivity Measurements on Highly Anisotropic $ZnIn_2S_4$ Crystals
(a) Vol. **71**, No. 2, p. 543–548 (1982).
See also *13.4*, p. *170*; 22.8, p. 485.

M. G. KAPLUNOV, K. I. POKHODNIYA, and YU. G. BORODKO
Temperature Dependence of Reflection Spectra of Conductive TCNQ Salts
(a) Vol. **71**, No. 2, p. 649–656 (1982).
See also 6, p. 75; *20.1*, p. *307*; 22.9, p. 505.

C. R. DUTTA and K. BARUA
*DC Electrical Properties of Vacuum Deposited Yb_2O_3 Films
(a) Vol. **71**, No. 2, p. K181–K186 (1982).
See also *13.4*, p. *170*; 14.3.1, p. 215; 22.6, p. 472.

S. Takács and I. Janetka
*Influence of Resonance Tunneling on Resistivity in Semiconductors with Ionized Impurities and on Tunneling Current between Superconductors
(b) Vol. **113**, No. 1, p. 313–326 (1982).
See also **14.2, p. 200; 22, p. 389.**

A. Choudhury and D. Mukhopadhyay
*Optimum Second Harmonic Generation Efficiency in II–VI Compound Semiconductors at 77 K
(b) Vol. **113**, No. 1, p. K49–K51 (1982).
See also 22.4, p. 437.

E. G. Bagduev and M. Sh. Shikhsaidov
*The NDC and Low-Frequency Electric Instabilities in GaAs Crystals with Dislocations
(a) Vol. **73**, No. 1, p. 287–292 (1982).
See also 22.2.1, p. 422.

D. A. Aronov and V. V. Rubinov
*The Influence of Exclusion on Light Absorption by Free Carriers in "Pure" (almost Intrinsic) Semiconductors with Non-Injecting Contacts
(a) Vol. **73**, No. 1, p. K1–K5 (1982).
See also *20.1*, p. *311*; 22.1.1, p. 398.

R. S. Agrawal, Y. K. Sharma, V. K. Gautam, and C. S. Agrawal
*Exact J–U Characteristic of an Insulator with Traps Lying above the Fermi Level in Non-Constant Mobility Regime
(a) Vol. **73**, No. 1, p. K27–K30 (1982).
See also *14.3.3*, p. *224*; 22, p. 389.

S. J. Pearton and A. J. Tavendale
*The Motion of Deep Donor Centres in Reverse Biased n-GaAs Surface Barrier Diodes
(a) Vol. **73**, No. 1, p. K75–K78 (1982).
See also 13.4, p. 174; *14.3.3*, p. *224*; 22.1.1, p. 398; 22.2.1, p. 422.

V. Balynas and A. Krotkus
*On the Origin of S-Shaped Current–Voltage Characteristics in p-Type Tellurium
(a) Vol. **73**, No. 1, p. K101–K103 (1982).
See also 22.1.3, p. 417.

M. S. Gilenko, P. M. Karageorgy-Alkalaev, and A. Yu. Leiderman
*On the Noise of Semiconductor Diode Structures with Strong Carrier Accumulation
(a) Vol. **73**, No. 1, p. K147–K151 (1982).
See also *14.3.3*, p. *225*; 22.1.2, p. 413.

I. I. Fishchuk
*Effective Medium Theory for the DC Conductivity and Hall Effect of Inhomogeneous Semiconductors in High Electrical Fields
(b) Vol. **113**, No. 2, p. 549–557 (1982).
See also *2*, p. *52*; 22, p. 390.

V. I. Sokolov, T. P. Surkova, and V. V. Chernyaev
*Two Types of Excitons Bound with Ni in ZnSe and ZnS
(b) Vol. **114**, No. 2, p. K195–K199 (1982).
See also *13.5.1*, p. *186*; 20.1, p. 316; 22.4.1, p. 441; 22.4.2, p. 444.

U. Henniger and R. Keiper
*Diffusion Theory for Continuous Metal-Amorphous Semiconductor Contacts
(a) Vol. **74**, No. 2, p. 497–509 (1982).
See also *2*, p. *55*.

K.-H. Bäther, W. Brückner, W. Moldenhauer, and M. Wolf
*Homogeneity of ZnP Varistors. Distribution of Local Electric Parameters before and after Degradation
(a) Vol. **74**, No. 2, p. 559–564 (1982).
See also 22.4, p. 438.

Sh. G. Suleimanov
Investigation of Semi-Insulating GaAs S-Diodes in Magnetic Field
(a) Vol. **74**, No. 2, p. K97–K100 (1982).
See also *14.3.3*, p. *226*; 22.2.1, p. 423.

Yu. A. Vashpanov, V. V. Serdyuk, and V. A. Smyntyna
*Current Instabilities in Thin Cadmium Selenide Films
(a) Vol. **74**, No. 2, p. K131–K135 (1982).
See also 14.3.1, p. 216; 22.4.2, p. 444.

H. Krause
*Poole-Frenkel Currents in Thermally Grown SiO_2 Films
(a) Vol. **74**, No. 2, p. K151–K154 (1982).
See also *13.4*, p. *178*; 22.6, p. 476.

14.4 Dielectrics

Review Article

Original Papers and Short Notes

F. Callens, W. Maenhout-van der Vorst, and L. W. Ketellapper
*The Effect of the Solution pAg and pH on the Space Charge Characteristics of Silver Bromide Emulsion Grains
(a) Vol. **70**, No. 1, p. 189–195 (1982).
See also *10.2*, p. *106*; 22.5.1, p. 453.

A. D. Reddy, S. M. D. Rao, and G. S. Sastry
*Electrical Conductivity and StimulatedThermo-Current Studies on VO^{2+} Doped $Li(N_2H_5)SO_4$ Single Crystals
(a) Vol. **70**, No. 1, p. 269–275 (1982).
See also 14.4.1, p. 240; 22.8.1, p. 493.

K. B. Kadyrakunov, E. V. Nidaev, and L. S. Smirnov
*Behaviour of a Built-in Charge in a Dielectric at Pulsed Annealing of Si–SiO_2 Structures
(a) Vol. **70**, No. 1, p. K15–K17 (1982).
See also *13.3*, p. *158*; 22.1.2, p. 403; 22.6, p. 469.

S. A. Mazen, M. A. Ahmed, and B. A. Sabrah
Thermal Studies on the Electrical Conduction Mechanism of $CuFe_2O_4$
(a) Vol. **70**, No. 1, p. K71–K75 (1982).
See also 18.3, p. 276; 22.8.2, p. 499.

W. Tews and R. Zimmermann
*Electron-Energy-Loss Spectroscopy of Different Al_2O_3 Modifications (II): Lineshape-Analysis of Energy Derived Dielectric Function
(b) Vol. **110**, No. 2, p. 479–485 (1982).
See also 13.5.1, p. 182; *20.1*, p. *303*; 22.6, p. 470.

V. B. Ptashnik, T. Yu. Dunaeva, and Yu. M. Baikov
*The Electrical Conductivity of Solid Lithium Hydride Doped with Sulphur
(b) Vol. **110**, No. 2, p. K121–K124 (1982).
See also *10.2*, p. *107*; 22, p. 385.

P. E. van Camp, V. E. van Doren, and J. T. Devreese
*The Dielectric Matrix of Germanium in the Local Density Functional Formalism
(b) Vol. **110**, No. 2, p. K133–K136 (1982).
See also 22.1.1, p. 396.

B. M. Gorelov, V. N. Lauhkin, and M. K. Sheinkman
*Effect of Doping on the Metal–Dielectric Phase Transition in Tetrathiotetracene Iodide, TTT_2I_3
(b) Vol. **110**, No. 2, p. K145–K149 (1982).
See also *14.1*, p. *194*; 22.9, p. 504.

V. P. Romanov
*Stationary Distribution of Mobile Ions in a Dielectric with Regard to Their Elastic Interaction with the Medium
(a) Vol. **70**, No. 2, p. 525–532 (1982).
See also 14.3.3, p. 221; 22.6, p. 470.

J. Prakash
*On the Occurrence of Photo-Induced Polarization and Photo-Induced Depolarization
(a) Vol. **70**, No. 2, p. 747–753 (1982).
See also *10.2*, p. *108*; 22.5.2, p. 457.

J. DEL CERRO, S. RAMOS, and J. ISIDORO
Thermal Study of α-$LiIO_3$ Single Crystals below Room Temperature
(a) Vol. **72**, No. 1, p. K23–K26 (1982).
See also *8*, p. *87*; 22.8.1, p. 495.

T. M. BATIROV, V. M. FRIDKIN, R. NITSCHE, and K. A. VERKHOVSKAYA
Bulk Photovoltaic and Illumination Effects in the Argyrodite-Type Ionic Conductors Cu_6PS_5Br and Cu_6PS_5I
(a) Vol. **72**, No. 1, p. K105–K108 (1982).
See also *16*, p. *252*; 22.5, p. 451.

E. HEINER
The Dielectric Response of a Charge Density Wave (CDW) Distorted Electron–Phonon System in the Precritical Region (I): The Self-Consistent Method
(b) Vol. **112**, No. 2, p. 489–500 (1982).
See also 1.2, p. 19; *13*, p. *144*; 22, p. 388.

Z. OGORELEC
*Superionic Conductivity versus Ionicity
(b) Vol. **112**, No. 2, p. 621–625 (1982).
See also 22.5, p. 452.

M. SIU LI, M. DE SOUZA, and S. E. KAPPHAN
*Ionic Thermal Currents under Uniaxial Stress. A New Method to Determine Electric and Elastic Dipole Properties
(b) Vol. **112**, No. 2, p. 685–693 (1982).
See also *10.2*, p. *112*; 22.5.2, p. 459.

C. ANCE, J. P. FERRATON, J. M. BERGER, and F. DE CHELLE
Calculation Model for Valence Electrons and Hydrogen Concentration in a-Si:H
(b) Vol. **113**, No. 1, p. 105–111 (1982).
Erratum
(b) Vol. **114**, No. 2, p. 742 (1982).
See also *2*, p. *51*; *13.1*, p. *153*; 22.1.2, p. 412.

S. E. KAPPHAN
UV Absorption and Electro-Optical Alignment Behavior of Off-Center Ag^+ in RbI
(b) Vol. **113**, No. 1, p. 147–156 (1982).
See also *13.4*, p. *174*; *20.1*, p. *310*; 22.5.2, p. 460.

M. SAINT-PAUL and G. FROSSATI
*Dielectric Relaxation in Smoky Quartz at Very Low Temperatures
(b) Vol. **113**, No. 1, p. 373–377 (1982).
See also 22.6, p. 473.

M. GODLEWSKA, M. MASSALSKA-ARODŹ, and S. URBAN
Dielectric Relaxation in Liquid and Plastic Phases of 2,2-Dinitropropane
(a) Vol. **73**, No. 1, p. 65–69 (1982).
See also *1.2*, p. *20*; 22.9, p. 506.

J. K. Vij
*Pressure, Temperature, and Frequency Dependence of the Dielectric Properties of Strontium Barium Niobate
(a) Vol. **74**, No. 1, p. 225–232 (1982).
See also 12.2, p. 138; 22.8.1, p. 498.

V. N. Bondarev, A. B. Kuklov, and I. S. Rublyov
*Nonlinear Excitations in Superionic Crystals near the Nonuniform State Transition Temperature
(b) Vol. **114**, No. 2, p. 645–652 (1982).
See also *13.5*, p. *180*; 22, p. 392.

J. Shanker and R. Sundaraj
Dielectric and Anharmonic Behaviour of Alkali Halide Crystals
(b) Vol. **114**, No. 2, p. 659–665 (1982).
See also *4*, p. *65*; 12.2, p. 138; 22.5.2, p. 463.

V. Kumar Srivastava
Plasmon Approach to the Band Gap of Alkali Halides
(b) Vol. **114**, No. 2, p. 667–671 (1982).
See also *13.1*, p. *156*; 13.5.2, p. 187; 22.5.2, p. 463.

14.4.1 Ferroelectrics

Original Papers and Short Notes

D. Kostopoulos, S. Kouremenou, P. Varotsos, and S. Mourikis
*Thermocurrents in Ferroelectric Materials
(a) Vol. **70**, No. 1, p. K19—K21 (1982).
See also 22.8, p. 483.

N. R. Ivanov, J. Kroupa, J. Fousek, and B. Březina
*Birefringence of $PbHPO_4$-Family Ferroelectrics
(a) Vol. **70**, No. 1, p. K85—K87 (1982).
See also *20.1*, p. *303*; 22.8.1, p. 494.

F. Brehat, B. Wyncke, and M. El Sherif
Lattice Mode Anharmonicity in Sodium Nitrite
(b) Vol. **110**, No. 2, p. 499—506 (1982).
See also 1.2, p. 16; *6*, p. *73*; *20.1*, p. *303*; 22.8.1, p. 494.

J. A. Plascak and N. P. Silva
*A Double Chain Approximation to the Ising Model
(b) Vol. **110**, No. 2, p. 669—675 (1982).

V. S. Vikhnin and Yu. B. Borkovskaya
*EPR Line Critical Broadening Asymmetry and Shape Peculiarities Due to Dipole-Dipole Electric Interaction in Ferroelectrics
(b) Vol. **110**, No. 2, p. 717—727 (1982).
See also 19, p. 285.

S. V. Misul and I. A. Belobrova
Sequence of Phase Transitions in $(NH_4)_2ZnCl_4$
(a) Vol. **70**, No. 2, p. K167—K170 (1982).
See also *1.2*, p. *17*; 19, p. 286; 22.8, p. 484.

M. A. Hefni, N. M. Grecu, V. V. Grecu, and F. F. Popescu
Electron Paramagnetic Resonance Studies of Cu^{2+} Impurities in Ferroelectric $LiH_3(SeO_3)_2$
(b) Vol. **111**, No. 1, p. 407—416 (1982).
See also *19*, p. *286*; 22.8, p. 484.

K. H. Herrmann, G. A. Kalyuzhnaya, K.-P. Möllmann, and M. Wendt
*Photoferroelectric Behaviour in $Pb_{0.78}Sn_{0.22}Te$:In
(a) Vol. **71**, No. 1, p. K21—K24 (1982).
See also 22, p. 386.

K. Okada and I. Suzuki
*Comments on the Temperature Range of the Double Hysteresis Loop at the First-Order Ferroelectric Transition
(a) Vol. **71**, No. 1, p. K33—K35 (1982).

L. A. Shebanov, V. G. Osipyan, E. Zh. Freidenfeld, and E. H. Birks
*Ferroelectricity in $Bi_{10}Ti_3W_3O_{30}$
(a) Vol. **71**, No. 1, p. K61—K62 (1982).
See also 22.8, p. 484.

J. A. Plascak and S. R. Salinas
*Tunneling Model for Quasi-One-Dimensional Hydrogen-Bonded Crystals
(b) Vol. **113**, No. 1, p. 367–372 (1982).

A. D. Reddy, S. G. Sathyanarayan, and G. S. Sastry
*Proton Conduction in $(NH_4)HSO_4$ Single Crystals
(a) Vol. **73**, No. 1, p. K41–K45 (1982).
See also 22.8.1, p. 497.

G. Sorge, H. Beige, A. Almeida, and L. A. Shuvalov
*Elastic Nonlinearity of $KH_3(SeO_3)_2$ and $KD_3(SeO_3)_2$ Crystals
(a) Vol. **73**, No. 1, p. K63–K66 (1982).
See also *1.2*, p. *21*; 12.2, p. 136; 22.8, p. 489.

I. Henning, M. Mertig, R. Plath, G. Pompe, E. Hegenbarth, and R. Schalge
*Glass-Like Behaviour of $Sr_{1-x}Ba_xNb_2O_6$ (SBN) Single Crystals Demonstrated by Heat Capacity Measurements
(a) Vol. **73**, No. 1, p. K105–K109 (1982).
See also *8*, p. *89*; 22.8, p. 489.

N. S. Buinov, V. V. Mikhnevich, and S. R. Syrtsov
*Phase Transition in Narrow-Band Compensated Ferroelectric Semiconductors
(b) Vol. **113**, No. 2, p. K123–K126 (1982).
See also 13.4, p. 175; 22, p. 390.

J. Kroupa, J. Fousek, F. Smutný, and B. Březina
*Birefringence and Electrooptic Effect in K_2SeO_4
(b) Vol. **113**, No. 2, p. K153–K157 (1982).
See also *20.1*, p. *313*; 22.8.1, p. 497.

H. Amaral, R. Chaves, A. Levelut, and S. Ziolkiewicz
*Difffuse Phase Transition in Ferroelectric $SbSe_{0.40}S_{0.60}I$
(a) Vol. **73**, No. 2, p. 421–427 (1982).
See also 22.8, p. 489.

U. Straube, G. Sorge, and N. R. Ivanov
*Temperature-Dependent Electrostriction Coefficient and Pseudoproper Phase Transition in TSCC
(a) Vol. **73**, No. 2, p. 467–474 (1982).
See also 1.2, p. 22; *12.2*, p. *137*; 22.8, p. 490.

E. Bărbulescu, A. Bărbulescu, and D. Barb
*On the Ferroelectric Structural Phase Transitions in Solid Solutions
(a) Vol. **73**, No. 2, p. 603–610 (1982).
See also 22.8, p. 490.

G. Völkel
*The Importance of Anharmonic Effects in the Electron Spin–Lattice Relaxation at Ferroelectric Phase Transitions
(b) Vol. **114**, No. 1, p. K67–K69 (1982).
See also *19*, p. *290*; 22.9, p. 507.

15. Thermoelectric and Thermomagnetic Properties

Original Papers and Short Notes

O. Uemura, M. Itoh, and T. Satow
Electronic Properties of the Tl–$AgTlTe_2$ System in the Liquid State
(b) Vol. **109**, No. 1, p. 361 – 368 (1982).
See also *2*, p. *45*; *14.3*, p. *202*; 18, p. 259; 22.8, p. 481.

E. I. Terukov, N. Mattern, W. Reichelt, H. Oppermann, and M. Wolf
The Behaviour of Resistance and Thermopower and the Coexistence of Two Phases at Metal–Insulator Transitions
(a) Vol. **69**, No. 1, p. 97 – 101 (1982).
See also 1.2, p. 14; *14.1*, p. *193*; *14.3*, p. *202*; 22.6, p. 468.

H. Gruber and E. Krautz
Magnetoresistance and Conductivity in the Binary System Titanium–Oxygen (I): Titanium Oxides with Metallic Conductivity
(a) Vol. **69**, No. 1, p. 287 – 295 (1982).
See also 8, p. 85; *14.1*, p. *193*; *14.3*, p. *203*; 18, p. 259; 22.6, p. 468.

V. A. M. Brabers and J. C. J. M. Terhell
Electrical Conductivity and Cation Valencies in Nickel Manganite
(a) Vol. **69**, No. 1, p. 325 – 332 (1982).
See also *1*, p. *9*; *14.3*, p. *203*; 22.8.2, p. 499.

H. T. Langhammer, M. Stordeur, H. Sobotta, and V. Riede
Optical and Electrical Investigations of the Anisotropy of Sb_2Te_3 Single Crystals
(b) Vol. **109**, No. 2, p. 673 – 681 (1982).
See also *14.3*, p. *203*; *20.1*, p. *300*; 22.7, p. 479.

B. Gillot
*Thermopower–Composition Dependence in Iron–Chromium, Iron–Zinc, and Iron–Cobalt Ferrites
(a) Vol. **69**, No. 2, p. 719 – 725 (1982).
See also 22.8.2, p. 499.

S. Myhra
Acceptor Point Defects in n-Type α-Sn
(b) Vol. **110**, No. 1, p. 97 – 103 (1982).
See also *13.4*, p. *165*; *14.3*, p. *204*; 22.1, p. 393.

R. Bharati and R. A. Singh
*Electrical Transport in Chromium(III) Tungstate
(a) Vol. **70**, No. 1, p. K11 – K14 (1982).
See also *14.3*, p. *204*; 22.8.1, p. 493.

Z. Kletowski
Thermopower of Monocrystalline $RESn_3$ Compounds (RE = La, Ce, Pr, Nd, Sm, Eu, and Gd)
(a) Vol. **70**, No. 1, p. K51 – K53 (1982).
See also 21.4, p. 370; 21.7, p. 379.

R. J. Iwanowski
The Γ-Point Band Structure Parameters of $Cd_xHg_{1-x}Se$ ($x \leqq 0.1$) Mixed Crystals from Thermomagnetic Measurements
(b) Vol. **110**, No. 2, p. 461–470 (1982).
See also *13.1*, p. *150*; 22.4.4, p. 447.

B. Horvatić and V. Zlatić
Perturbation Expansion for the Asymmetric Anderson Hamiltonian (II): General Asymmetry
(b) Vol. **111**, No. 1, p. 65–75 (1982).
See also 8, p. 86; *13*, p. *142*; 14.1, p. 194; 21, p. 334.

D. Hauschildt, W. Fuhs, and H. Mell
Optically Induced Potential Fluctuations in a-Si:H Films
(b) Vol. **111**, No. 1, p. 171–176 (1982).
See also *14.3.1*, p. *214*; 22.1.2, p. 407.

E. Gratz, V. Sechovsky, V. Sima, Z. Smetana, and J. O. Ström-Olson
Magnetization, Electrical Resistivity, Thermopower, and Neutron Diffraction in $HoCu_2$
(b) Vol. **111**, No. 1, p. 195–201 (1982).
See also 14.1, p. 194; *18.4*, p. *280*; 21.1, p. 348; 21.4, p. 370.

R. Lenk
Integral Representation of the Incoherent Field in Disordered Materials (II): Hall Current and Thermoelectric Phenomena
(b) Vol. **111**, No. 1, p. 255–260 (1982).
See also *2*, p. *48*; 14, p. 191.

K. S. Dubey
Role of Boundary Scattering in Phonon Drag Thermoelectric Power and Lattice Thermal Conductivity of a Doped Semiconductor at Low Temperatures: Application to Phosphorus Doped Ge
(b) Vol. **111**, No. 1, p. K49–K54 (1982).
See also *6*, p. *74*; 8, p. 86; 22.1.1, p. 397.

L. G. Khvostantsev, A. I. Orlov, N. Kh. Abrikosov, T. E. Svechnikova, and S. N. Chizhevskaya
*Thermoelectric Properties and Phase Transitions in Bi_2Te_3 under Hydrostatic Pressure up to 9 GPa and Temperature up to 300 °C
(a) Vol. **71**, No. 1, p. 49–53 (1982).
See also *1.2*, p. *17*; 12.2, p. 135; 14.3, p. 206; 22.7, p. 479.

J. Rosenzweig and U. Birkholz
*Influence of Bipolar Photoconduction on the Thermoelectric Power of CdTe
(a) Vol. **71**, No. 1, p. K41–K44 (1982).
See also *16*, p. *251*; 22.4.3, p. 446.

A. Andreeff, Th. Frauenheim, E. A. Goremychkin, H. Griessmann, B. Lipold, W. Matz, O. D. Chistyakov, and E. M. Savitskii
Crystal Field in the Laves Phase Compound $PrNi_2$
(b) Vol. **111**, No. 2, p. 507–512 (1982).
See also *1*, p. *10*; 8, p. 87; *13.4*, p. *169*; 18.1, p. 266; 21.1, p. 349; 21.4, p. 370.

J. Horák, L. Koudelka, J. Klikorka, and L. Šiška
Transport Properties and Plasma Resonance of Bismuth Selenide Crystals Doped with Cadmium
(b) Vol. **111**, No. 2, p. 575–580 (1982).
See also *13.4*, p. *169*; *14.3*, p. *207*; 22.7, p. 479.

P. Becker, H. Siegert, and L. Bliek
*Investigation of Thermoelectric Effects in Silicon by X-Ray Interferometry
(a) Vol. **71**, No. 2, p. K139–K141 (1982).
See also 22.1.2, p. 408.

M. N. Avasthi
*Heat of Transport Study of the Superionic Conductor $Ag_7I_4VO_4$
(a) Vol. **71**, No. 2, p. K177–K180 (1982).
See also 22.8, p. 486.

D. Deschacht, A. Boyer, and E. Groubert
*The Thermoelectric Power of Polycrystalline Semimetal Films
(a) Vol. **71**, No. 2, p. K205–K209 (1982).
See also 21.7, p. 380.

A. H. Abou El Ela, N. Abdelmohsen, and H. H. A. Labib
Electrical Conductivity, Thermoelectric Power, and Thermal Conductivity of $AgTlSe_2$ in the Solid and Liquid State
(a) Vol. **71**, No. 2, p. K210–K224 (1982).
See also 2, p. 49; *8*, p. *87*; 14.3, p. 208; 22.8, p. 486.

T. Nakajima, O. Uemura, T. Satow, und T. Yagihashi
Electronic Properties of the Molton Tl–$AgTlSe_2$ System
(b) Vol. **112**, No. 1, p. 311–319 (1982).
See also *2*, p. *50*; *14.3*, p. *208*; 18, p. 261; 22.8, p. 486.

D. V. Gitsu, I. M. Golban, and V. G. Kantser
Investigation of Transport Coefficients in Semimetals by the Variational Method (I): General Solution of the Problem
(b) Vol. **112**, No. 2, p. 473–481 (1982).
See also *14.1*, p. *196*; 21.7, p. 380.

D. V. Gitsu, I. M. Golban, and V. G. Kantser
Investigation of Transport Coefficients in Semimetals by the Variational Method (II): Bismuth
(b) Vol. **113**, No. 1, p. 59–66 (1982).
See also *14.1*, p. *196*; 21.7, p. 380.

V. Ya. Prokhorenko, B. I. Sokolovskii, V. A. Alekseev, A. S. Basin, S. V. Stankus, and V. M. Sklyarchuk
*The Semiconductor–Metal Transition in Liquid Tellurium
(b) Vol. **113**, No. 2, p. 453–458 (1982).
See also 2, p. 52; *14.3*, p. *210*; 22.1.3, p. 417.

16. Photoconductivity. Photovoltaic Effects

Review Article

S. Chandra and R. K. Pandey
*Semiconductor Photoelectrochemical Solar Cells
(a) Vol. **72**, No. 2, p. 415 – 454 (1982),
See also *14.3.3*, p. *219*; 22.2, p. 418; 22.4, p. 437.

Original Papers and Short Notes

F. Henneberger, N. S. Averkiev, and R. J. Rasulov
*On a New Photogalvanic Effect Due to Free-Carrier Absorption
(b) Vol. **109**, No. 1, p. 343 – 351 (1982).
See also 6, p. 70; 13.1, p. 148; 22.1.3, p. 417.

G. M. Shmelev, G. I. Tsurkan, and E. M. Ephstein
*Photostimulated Radioelectrical Transverse Effect in Semiconductors
(b) Vol. **109**, No. 1, p. K53 – K58 (1982).
See also 22, p. 383.

A. Witek
The Effect of Illumination on Anomalous Ultrasonic Attenuation in Doped Single Crystals of $Bi_{12}GeO_{20}$
(a) Vol. **69**, No. 1, p. 67 – 70 (1982).
See also *7*, p. *82*; 22.6.1, p. 477.

A. P. Galushka, G. E. Davidyuk, and N. S. Bogdanyuk
Interaction of 1.2 MeV Electron Irradiation Produced Defects with Cu Impurity Atoms in CdS Single Crystals
(a) Vol. **69**, No. 1, p. 71 – 76 (1982).
See also *11*, p. *118*; *14.3*, p. *202*; 20.1, p. 299; 20.3, p. 318; 22.4.1, p. 439.

D. Mergel and R. Labusch
Optical Excitations of Dislocation States in Silicon
(a) Vol. **69**, No. 1, p. 151 – 158 (1982).
See also *10.2*, p. *103*; *13.4*, p. *163*; 22.1.2, p. 400.

D. A. Aronov, R. Mamatkulov, and V. V. Rubinov
The Effects of Field and Temperature Heating in Intrinsic Semiconductors with Non-Injecting Contacts
(a) Vol. **69**, No. 1, p. 159 – 165 (1982).
See also *14.3.4*, p. *227*; 22.1.1, p. 395.

K. Sen and R. S. Srivastava
*SIS Solar Cell Theory and Calculations
(a) Vol. **69**, No. 1, p. 413 – 418 (1982).
See also *14.3.3*, p. *219*; 22.1.2, p. 401.

S. G. Abdullaeva and V. A. Aliev
*Special Features of Negative Photoconductivity in $TlGaSe_2$
(a) Vol. **69**, No. 1, p. K33 – K36 (1982).
See also 13.4, p. 163; 22.8, p. 481.

N. M. Ravindra, K. Sunil Kumar, and V. K. Srivastava
Temperature and Voltage Dependence of the Barrier Height in SnO_2/Si Solar Cells
(a) Vol. **70**, No. 2, p. 623–630 (1982).
See also *13.3*, p. *159*; 14.3.3, p. 221; 22.1.2, p. 405.

J. Rosenzweig and U. Birkholz
*Photo-Hall Effect and Imperfection Analysis in CdTe in the Case of Bipolar Photoconductivity
(a) Vol. **70**, No. 2, p. K171–K174 (1982).
See also 13.4, p. 168; 22.4.3, p. 446.

W. Lorenz and M. Handschuh
*Dynamical Theory of Photocurrent–Voltage Curves and Photocurrent Efficiency on Semiconductor Electrodes
(a) Vol. **71**, No. 1, p. 127–134 (1982).
See also *14.3.2*, p. *217*; 22.2.2, p. 425.

A. A. Berezin and F. L. Weichman
*Cuprous Oxide–Copper Photovoltaic Cells Formed by High-Temperature Reduction
(a) Vol. **71**, No. 1, p. 265–274 (1982).
See also 20.1, p. 305; 22.6, p. 471.

G. A. Stanciu, I. M. Popescu, and C. M. Stoichiţa
*Evolution of the Photocurrent in a Region with Inversion Layer of the pnp Transistor
(a) Vol. **71**, No. 1, p. K17–K19 (1982).
See also *14.3.3*, p. *222*; 22.1.2, p. 407.

G. K. Padam, B. Singh, S. K. Gupta, and G. D. Sootha
Kinetic Behaviour of Oxygen Radicals in CdS
(a) Vol. **71**, No. 1, p. K37–K40 (1982).
See also 14.3, p. 206; *19*, p. *287*; 22.4.1, p. 440.

J. Rosenzweig and U. Birkholz
*Influence of Bipolar Photoconduction on the Thermoelectric Power of CdTe
(a) Vol. **71**, No. 1, p. K41–K44 (1982).
See also *15*, p. *246*; 22.4.3, p. 446.

B. G. Tagiev, E. F. Bagirzade, and G. M. Mamedov
*Interference Oscillations in the Temperature Dependence of Photocurrent and Transmission of GaSe, GaTe, and $GaSe_xTe_{1-x}$ Single Crystals
(a) Vol. **71**, No. 1, p. K49–K51 (1982).
See also *20.1*, p. *305*; 22.3, p. 433; 22.3.1, p. 436.

C. Viger and G. Fleury
*Spectral Response of Amorphous Se at 233 K
(a) Vol. **71**, No. 1, p. K103–K105 (1982).
See also *2*, p. *49*; 22.1.3, p. 417.

L. P. Kazakova, L. Tóth, and M. A. Taguyrdzhanov
*Reversible Photoinduced Change of Hole Drift Mobility in Amorphous AsSe Thin Films
(a) Vol. **71**, No. 1, p. K107–K109 (1982).
See also *2*, p. *49*; 22, p. 386.

V. J. RAO, W. A. ANDERSON, and G. RAJESWARAN
*The Role of Hydrogen in Improving Photovoltaic Response of Solar Cells on Polycrystalline Silicon
(a) Vol. **72**, No. 2, p. 745–749 (1982).
See also *14.3.3*, p. *224*; 20.1, p. 309; 22.1.2, p. 411.

S. K. KRAWCZYK and A. JAKUBOWSKI
*Investigation of Lateral Non-Uniformities in MOS Structures by the Surface Photovoltage Technique
(a) Vol. **72**, No. 2, p. K149–K153 (1982).
See also 13.3, p. 160; *14.3.3*, p. *224*; 22.1.2, p. 411.

T. SUZUKI, Y. HIROSE, and H. HIROSE
*A Model of the Photodoping Mechanism in the Ag/As_2S_3 System
(a) Vol. **72**, No. 2, p. K165–K168 (1982).
See also *10.2*, p. *113*; 14.3.2, p. 217; 22, p. 389.

H. G. FESTL, P. HERTEL, E. KRÄTZIG, and R. VON BALTZ
*Investigations of the Photovoltaic Tensor in Doped $LiNbO_3$
(b) Vol. **113**, No. 1, p. 157–164 (1982).
See also *14.4.1*, p. *242*; 22.8.1, p. 496.

E. A. SILINSH, V. A. KOLESNIKOV, I. J. MUZIKANTE, and D. R. BALODE
*On Charge Carrier Photogeneration Mechanisms in Organic Molecular Crystals
(b) Vol. **113**, No. 1, p. 379–393 (1982).
See also 14.3.1, p. 215; 22.9, p. 506.

J. DAUNAY, JAC. DAUNAY, and P. BUGNET
*A Possible Origin of the Internal Electric Field Responsible of the Anomalous Photovoltage in Twinned ZnSe Single Crystals
(a) Vol. **73**, No. 1, p. 199–203 (1982).
See also 22.4.2, p. 443.

A. CASTALDINI, A. CAVALLINI, P. GONDI, and E. BONETTI
Generation and Recombination Images of Dislocations in Si by Scanning Microscopy
(a) Vol. **73**, No. 2, p. 617–624 (1982).
See also *10.2*, p. *115*; 13.4, p. 175; 22.1.2, p. 414.

T. X. HOAI, P. H. BINH, and N. D. HUNG
*Study of the Nonlinear Recombination Process in Silicon by the Saturated Photovoltage Transient Method
(a) Vol. **73**, No. 2, p. K165–K168 (1982).
See also *14.3.3*, p. *225*; 22.1.2, p. 414.

J. JIMENEZ, M. A. GONZALEZ, L. F. SANZ, and J. BONNAFE
*On the Photoconductivity Response of GaAs Doped with Chromium and Oxygen
(a) Vol. **73**, No. 2, p. K189–K192 (1982).
See also 13.4, p. 176; 22.2.1, p. 423.

E. HARTMANN and L. KOVÁCS
Electrical Conductivity of Paratellurite (TeO_2) Crystals
(a) Vol. **74**, No. 1, p. 59–64 (1982).
See also *14.3*, p. *211*; 22.6, p. 475.

17. Emission of Electrons and Ions

Review Article

F. BECHSTEDT
Electronic Relaxation Effects in the Core Level Spectra of Solids
(b) Vol. **112**, No. 1, p. 9–49 (1982).
See also 4, p. 61; *13.1*, p. *148*; 20, p. 293; 21, p. 330; 22, p. 382; 23, p. 509.

Original Papers and Short Notes

G. GERGELY and H.-J. MÜSSIG
*Notes on Quantitative AES of the Fe–O System
(a) Vol. **69**, No. 1, p. K69–K72 (1982).
See also *1.6*, p. *42*; 22.6, p. 468.

A. J. MCEVOY and W. GISSLER
*ESCA Spectra and Electronic Properties of Some Ruthenium Compounds
(a) Vol. **69**, No. 1, p. K91–K96 (1982).
See also 13, p. 140; 21.4, p. 369; 22.6, p. 469.

V. CHÁB, J. ZEMEK, S. KOC, and P. MIKUŠÍK
*Photoemission Study of Oxygen Adsorbed on Sputtered a-Si(H) and a-Si
(b) Vol. **109**, No. 2, p. 495–501 (1982).
See also 2, p. 46; *13.3*, p. *158*; 22.1.2, p. 401.

G. I. ROSENMAN, E. I. BOIKOVA, and YU. L. CHEPELOV
*Emission Study of Surface States in Ferroelectrics
(a) Vol. **69**, No. 2, p. K173–K177 (1982).
See also *13.3*, p. *158*; 14.4.1, p. 240; 22.8.1, p. 493.

A. F. AKKERMAN, V. A. BOTVIN, M. YA. GRUDSKII, V. V. SMIRNOV, G. YA. CHERNOV, and M. V. SHILENKOVA
*Secondary Electron Radiation from Different Targets Exposed to Photons with Energy 0.1 to 3 MeV
(b) Vol. **110**, No. 1, p. 285–297 (1982).
See also *11*, p. *120*; 21, p. 332.

A. M. GABOVICH and A. I. VOITENKO
Influence of Semiconductor Dielectric Function Spatial Dispersion on Charge Electrostatic Energy near the Semiconductor/Vacuum Interface and Field Emission Current
(b) Vol. **110**, No. 2, p. 407–416 (1982).
See also *14.3.2*, p. *217*; 22.1.1, p. 396; 22.1.2, p. 404.

R. BLESSING and H. PAGNIA
*Electron Emission from Gold Island Films
(b) Vol. **110**, No. 2, p. 537–542 (1982).
See also 21.6, p. 375.

H. EICHENMÜLLER
*New Investigations of the Chemical Nature of the Centres of the Exoelectron Emission on Metal Surfaces
(a) Vol. **70**, No. 2, p. 483–487 (1982).
See also 1.6, p. 42; 21.1, p. 347.

YU. D. DEKHTYAR and G. L. SAGALOVICH
*Some Aspects of Activation Energy Calculations for Thermostimulated Exoelectronic Emission
(a) Vol. **73**, No. 1, p. K35–K39 (1982).
See also 22.1.2, p. **412**.

G. RUDLOF and H. GLAEFEKE
*Comparison of Isochronal and Time-Linear Thermal Annealing Curves and Conclusions about the Correlation Glow-Peak Nature of Traps and the Mechanism of TSEE for α-Al_2O_3
(a) Vol. **73**, No. 2, p. K253–K258 (1982).
See also *13.4*, p. *176*; 22.6, p. 475.

V. KORTOV, V. ISAKOV, H. GLAEFEKE, and H.-J. FITTING
*Energy Distribution of Exoelectrons from NaCl (111) and (100) Surfaces
(a) Vol. **73**, No. 2, p. K275–K278 (1982).
See also 13.3, p. 160; 22.5.2, p. 462.

J. SZUBER
Photoemission Yield Investigations of the Electronic Properties of UHV Annealed NiO Single Crystal Surfaces
(a) Vol. **74**, No. 1, p. K83–K87 (1982).
See also *13.3*, p. *161*; 22.6, p. 476.

M. ELANGO, A. KIKAS, E. NÕMMISTE, J. PRUULMANN, and A. SAAR
*The Quantum Yield Spectra of Electron Emission of Solids in XUV Region. Alkali Chlorides at 100 to 250 eV
(b) Vol. **114**, No. 2, p. 487–493 (1982).
See also 22.5.2, p. 463.

V. DRCHAL and J. KUDRNOVSKÝ
Theory of Auger Spectra from Disordered Alloys. The Effect of Electron Correlations in Filled Narrow Bands
(b) Vol. **114**, No. 2, p. 627–635 (1982).
See also *13.4*, p. *177*; 21, p. 342.

A. FUJIMORI and N. TSUDA
*XPS and UPS Spectra of Non-Stoichiometric CeH_x
(b) Vol. **114**, No. 2, p. K139–K142 (1982).
See also *13.1*, p. *156*; 21.4, p. 372.

I. TALE, V. KORTOV, and V. POPOV
Trap Spectroscopy in Alkali Halides at Synchronous Measurements of Exoelectron Emission and Luminescence by Fractional Glow Technique
(a) Vol. **74**, No. 2, p. 395–401 (1982).
See also *10.2*, p. *117*; 20.3, p. 328; 22.5.2, p. 464.

17.1 Field Emission Microscope Investigations

Original Papers and Short Notes

J.-M. Derochette
*Adsorption of H_2 and N_2 on Ir(100) Studied by Field Emission
(a) Vol. **71**, No. 1, p. K99–K102 (1982).
See also *1.6*, p. *43*; 21, p. 335.

H. C. Eaton, A. C. Chou, and J. C. Carter
Topographic Analysis of Grain Boundary Dislocations in Cold-Worked Tungsten Using Field Ion Microscopy
(a) Vol. **72**, No. 2, p. 817–824 (1982).
See also *1.6*, p. *43*; *10.1*, p. *100*; 21, p. 338.

18. Magnetic Properties

Original Papers and Short Notes

A. M. Sandauer and P. Byszewski
Evidence of the Conduction–Valence Band Overlap Induced by a Magnetic Field in Zero-Gap $Hg_{1-x}Mn_xTe$
(b) Vol. **109**, No. 1, p. 167–171 (1982).
See also *13.1*, p. *148*; *14.3*, p. *202*; 22.8, p. 481.

M. Błaszak
*Spin–Orbit Interaction in Superexchange with Orbital Degeneracy (I): Coupling Coefficients and Operator Algebra for the Chain $O^* \subset SU2$
(b) Vol. **109**, No. 1, p. 195–210 (1982).
See also *13*, p. *139*.

J. Metzdorf and F. R. Kessler
*Quantitative Analysis of the Nonresonant Interband Faraday Rotation in Ge and Si
(b) Vol. **109**, No. 1, p. 283–294 (1982).
See also *20*, p. *293*; 22.1.1, p. 395; 22.1.2, p. 400.

O. Uemura, M. Itoh, and T. Satow
Electronic Properties of the Tl–$AgTlTe_2$ System in the Liquid State
(b) Vol. **109**, No. 1, p. 361–368 (1982).
See also *2*, p. *45*; *14.3*, p. *202*; 15, p. 245; 22.8, p. 481.

L. B. Dubrovskaya, S. Z. Nazarova, and A. F. Prekul
Superconducting and Normal Properties of Solid Solutions $Nb_{1-x}Ta_xC_y$
(a) Vol. **69**, No. 1, p. 167–172 (1982).
See also *1.3*, p. *25*; 14.2, p. 199; 14.3, p. 203; 21, p. 331.

H. Gruber and E. Krautz
Magnetoresistance and Conductivity in the Binary System Titanium–Oxygen (I): Titanium Oxides with Metallic Conductivity
(a) Vol. **69**, No. 1, p. 287–295 (1982).
See also 8, p. 85; *14.1*, p. *193*; *14.3*, p. *203*; 15, p. 245; 22.6, p. 468.

M. Błaszak
*Spin–Orbit Interaction in Superexchange with Orbital Degeneracy (II): Effective Tensor Hamiltonian
(b) Vol. **109**, No. 2, p. 519–524 (1982).
See also *13*, p. *140*.

S. Nara and G. Heber
*Dynamics of Two-Spin Cluster in Spin Glass (II): The Magnetization Process at $T = 0$
(b) Vol. **109**, No. 2, p. 573–581 (1982).
See also *13*, p. *140*; 22.8, p. 482.

L. Y. Vinnikov and A. O. Golubok
*Direct Observation of Magnetic Structure in Niobium Single Crystals with Hydride Precipitate Pinning Centres
(a) Vol. **69**, No. 2, p. 631–636 (1982).
See also 14.2, p. 199; 21, p. 332.

S. ROBASZKIEWICZ and M. BARTKOWIAK
Ground State Phase Diagram of the Extended Hubbard Model with Ising-Like Exchange Interactions
(b) Vol. **111**, No. 2, p. K107–K111 (1982).
See also *1.2*, p. *18*; *13*, p. *143*; 14, p. 191.

I. BOSE and S. CHATTERJEE
*A Migdal-Kadanoff Type of Recursion Scheme for Two-Dimensional Square Lattice
(b) Vol. **111**, No. 2, p. K137–K140 (1982).
See also *13*, p. *143*.

X.-Z. DONG, B. GRÖGER, T. JENDRYSIK, and H. KRONMÜLLER
*Annealing Effect of Domain Patterns of the Nearly Non-Magnetostrictive Amorphous Alloy $Co_{58}Ni_{10}Fe_5Si_{11}B_{16}$
(a) Vol. **71**, No. 2, p. 441–449 (1982).
See also 2, p. 49; *20*, p. *295*; 22.8, p. 485.

R. S. BRAZIS and A. S. MIRONAS
Investigation of Electric Conductivity and Avalanche Breakdown in InSb by Means of Nonlinear Magnetoplasma Resonance
(a) Vol. **71**, No. 2, p. 473–481 (1982).
See also *14.3*, p. *207*; 14.3.4, p. 229; 20, p. 295; 22.2.3, p. 428.

A. N. CHAKRAVARTI, K. P. GHATAK, G. B. RAO, and K. K. GHOSH
Carrier Statistics in Zero-Gap Semiconductors in the Presence of a Quantizing Magnetic Field
(b) Vol. **112**, No. 1, p. 75–81 (1982).
See also *13.1*, p. *152*; 22.4.3, p. 446.

E. GERLACH and B. HARBECKE
*Electron Transport in a Magnetic Field
(b) Vol. **112**, No. 1, p. 143–150 (1982).
See also *14*, p. *191*.

A. NATH and U. S. GHOSH
*Ligand Field Theory of the Magnetic Behaviours of Neodymium Ethyl-sulphate Nonahydrate
(b) Vol. **112**, No. 1, p. 187–192 (1982).
See also *13*, p. *144*; 22.8.1, p. 495.

T. NAKAJIMA, O. UEMURA, T. SATOW, and T. YAGIHASHI
Electronic Properties of the Molton Tl–$AgTlS_2$ System
(b) Vol. **112**, No. 1, p. 311–319 (1982).
See also *2*, p. *50*; *14.3*, p. *208*; 15, p. 247; 22.8, p. 486.

P. COLOMBET, M. TREMBLET, and M. DANOT
*Magnetic Properties of the Diluted B-Spinels $Cu_yCr_y^{III}M_{2-y}^{VI}S_4$ (M^{VI} = Sn, Ti)
(a) Vol. **72**, No. 1, p. 105–116 (1982).
See also *1.3*, p. *29*; 5, p. 67; 19, p. 288; 22.8.2, p. 500.

K. ZÁVĚTA, J. SCHNEIDER, A. HANDSTEIN, and Z. KALVA
*The Influence of Annealing of the Magnetization of Amorphous $(Ni_{0.93}Fe_{0.07})_{80}P_{10}B_{10}$ Alloy
(a) Vol. **72**, No. 1, p. K79–K82 (1982).
See also 2, p. 50; 22.8, p. 487.

T. TOMOV, D. KLISSURSKI, and I. MITOV
Mössbauer Study of the Formation of Solid Solutions in the α-Fe_2O_3–Al_2O_3 System
(a) Vol. **73**, No. 1, p. 249–254 (1982).
See also *5*, p. *68*; 22.6, p. 474.

N. P. GAVALESHKO, A. N. LYAKHOVICH, P. P. VATAMANYUK, and A. I. SAVCHUK
*Faraday Effect in III–VI Layer Compounds
(b) Vol. **113**, No. 2, p. 447–451 (1982).
See also 13.5.1, p. 185; *20.1*, p. *311*; 22.3, p. 434.

K. I. GROZDEV and S. I. TZENOV
*Indirect Polarization of Spins in Anisotropic Double Ising Chain
(b) Vol. **113**, No. 2, p. K93–K96 (1982).
See also *13*, p. *146*.

V. KUCKHERMANN, G. THUMMES, and H. H. MENDE
Influence of the Real Fermi Surface on the Longitudinal Magnetoresistance Size Effect of Single Crystal Copper Whiskers
(a) Vol. **73**, No. 2, p. 439–445 (1982).
See also *14.1*, p. *197*; 21.1, p. 355.

G. ALBANESE, M. CARBUCICCHIO, L. PARETI, and S. RINALDI
*Cation Distribution and Magnetic Properties of $BaZn_2Fe_{16-x}In_xO_{27}$ and $BaZn_2Fe_{16-x}Sc_xO_{27}$ Hexagonal Ferrites
(a) Vol. **73**, No. 2, p. K193–K197 (1982).
See also *5*, p. *68*; 22.8.2, p. 501.

M. MATLAK, A. RAMAKANTH, and K. SKROBIS
Thermodynamic Properties of the s–f Model (Spin 7/2, Atomic Limit)
(b) Vol. **114**, No. 1, p. 145–150 (1982).
See also *13*, p. *146*.

E. Z. IMAMOV and V. D. KREVCHIK
*On the Diamagnetism of Deep Impurities in Quantized Semiconductor Films
(b) Vol. **114**, No. 1, p. 201–207 (1982).
See also 13.4, p. 176; 22, p. 391

J. SIVARDIÈRE
*Magnetism in $S = 3/2$ Ising Systems: A Simple Model for a Tetracritical Point
(b) Vol. **114**, No. 1, p. K15–K19 (1982).

A. S. T. PIRES, N. P. SILVA, and B. J. O. FRANCO
*A Linear Chain Approximation to the Ising Model with Competing Interactions
(b) Vol. **114**, No. 1, p. K63–K66 (1982).

C. RUDOWICZ
*Comment on Energy of Magnetic Anisotropy
(b) Vol. **114**, No. 1, p. K79–K81 (1982).
See also *13*, p. *146*.

G. A. PETRAKOVSKII, A. G. KLIMENKO, F. S. RAKHMENKULOV, and A. V. BARANOV
*Specific Heat and Low Field Magnetic Properties of $Co(S_{1-x}Se_x)_2$
(a) Vol. **74**, No. 1, p. K31–K35 (1982).
See also *8*, p. *89*; 21.1, p. 357.

18.1 Paramagnetic Properties

Original Papers and Short Notes

M. V. MEDVEDEV and M. V. SADOVSKII
*Random Bond Ising Model in Self-Avoiding Walk Approximation
(b) Vol. **109**, No. 1, p. 49–57 (1982).
See also 18.2, p. 267; 18.4, p. 279.

S. K. MISRA and P. MIKOLAJCZAK
*EPR Linewidth Studies of Gd^{3+}-Doped $Yb_xY_{1-x}Cl_3 \cdot 6\,H_2O$ Single Crystals. Yb^{3+} Spin–Lattice Relaxation Times
(b) Vol. **109**, No. 1, p. 59–66 (1982).
See also *19*, p. *282*; 22.8, p. 481.

K. ELK
*On the Susceptibility of the Periodic Anderson Model
(b) Vol. **109**, No. 1, p. 217–222 (1982).
See also 13, p. 139; 21, p. 330.

S. FAKIOĞLU
Renormalization Group and Critical Exponent for Two-Dimensional Ising Ferromagnet with Long-Range Interaction
(b) Vol. **109**, No. 1, p. K63–K68 (1982).
See also *18.2*, p. *267*.

A. APOSTOLOV, H. HRISTOV, T. MIDLAG, M. MIKHOV, and V. SKUMRIEV
*The Bulk Amorphous Alloy Tb_4Co_3 – An Example for a System with a Transition to a "Spin-Glass Like" State
(a) Vol. **69**, No. 1, p. K7–K10 (1982).
See also *2*, p. *46*; 18.2, p. 267; 21.2, p. 344 21.4, p. 369.

M. V. MEDVEDEV and M. V. SADOVSKII
*Random Site Ising Model in Self-Avoiding Walk Approximation
(b) Vol. **109**, No. 2, p. 449–456 (1982).
See also 18.2, p. 267; 18.4, p. 279.

A. R. MCGURN and R. A. TAHIR-KHELI
*Baxter-Wu Model with Random Bonds
(b) Vol. **109**, No. 2, p. K103–K107 (1982).
See also 18.2, p. 268; 18.4, p. 279.

E. KOLLEY and W. KOLLEY
*On Random Itinerant Ferromagnets in the Percolation Limit
(b) Vol. **110**, No. 1, p. K63–K67 (1982).
See also *18.2*, p. *269*; 21, p. 333.

K. N. SHRIVASTAVA
Second- and Third-Order Processes in the Theory of Spin–Lattice Relaxation
(b) Vol. **111**, No. 1, p. 77–82 (1982).
See also *6*, p. *73*; *13*, p. *142*; 19, p. 286.

18.2 Ferromagnetic Properties

Original Papers and Short Notes

M. V. MEDVEDEV and M. V. SADOVSKII
Random Bond Ising Model in Self-Avoiding Walk Approximation
(b) Vol. **109**, No. 1, p. 49 – 57 (1982).
See also *18.1*, p. *265*; 18.4, p. 279.

A. BOBÁK
*Effect of Fluctuations of the Coordination Number on Some Properties of Amorphous Ferromagnets
(b) Vol. **109**, No. 1, p. 161 – 166 (1982).
See also *2*, p. *45*.

B. S. LEE
*Infinite Range Ising Models with a Tricritical Point
(b) Vol. **109**, No. 1, p. 423 – 428 (1982).

A. URBANIAK-KUCHARCZYK
*Spin Correlation Function in Nickel
(b) Vol. **109**, No. 1, p. K59 – K61 (1982).
See also 21.1, p. 343.

S. FAKIOĞLU
*Renormalization Group and Critical Exponent for Two-Dimensional Ising Ferromagnet with Long-Range Interaction
(b) Vol. **109**, No. 1, p. K63 – K68 (1982).
See also 18.1, p. 265.

H. J. BLYTHE, F. WALZ, and H. KRONMÜLLER
*Investigation of the Magnetic Aftereffect in Dilute Fe–Ni Alloys after Low-Temperature Neutron Irradiation
(a) Vol. **69**, No. 1, p. 237 – 248 (1982).
See also *10.1*, p. *96*; 11, p. 118; 21.1, p. 344; 21.1.1, p. 359.

A. APOSTOLOV, H. HRISTOV, T. MIDLAG, M. MIKHOV, and V. SKUMRIEV
The Bulk Amorphous Alloy Tb_4Co_3 – An Example for a System with a Transition to a "Spin-Glass Like" State
(a) Vol. **69**, No. 1, p. K7 – K10 (1982).
See also *2*, p. *46*; *18.1*, p. *265*; 21.1, p. 344; 21.4, p. 369.

H. J. BLYTHE, F. DWORSCHAK, I. M. RICHARDSON, and F. WALZ
*Magnetic Relaxation Spectra of Some Dilute Binary Iron-Based Alloys after Low-Temperature Electron Irradiation
(a) Vol. **69**, No. 1, p. K97 – K101 (1982).
See also *10.1*, p. *96*; 11, p. 118; 21.1.1, p. 359.

M. V. MEDVEDEV and M. V. SADOVSKII
Random Site Ising Model in Self-Avoiding Walk Approximation
(b) Vol. **109**, No. 2, p. 449 – 456 (1982).
See also *18.1*, p. *265*; 18.4, p. 279.

R. Meyer and H. Kronmüller
*Micromagnetic Theory of Phase Transitions in Inhomogeneous Ferromagnets (V): Magnetic Phase Transition of Amorphous $Fe_{10}Ni_{70}B_{16}P_4$
(b) Vol. **109**, No. 2, p. 693–703 (1982).
Erratum
(b) Vol. **114**, No. 2, p. 741 (1982).
See also 21.1, p. 344.

M. V. Medvedev and A. V. Zaborov
*Magnetic States of a Quenched Random Site Heisenberg Magnet with Competitive Exchange Interactions (I): High-Temperature Magnetic States in Simple and Body-Centered Cubic Lattice
(b) Vol. **109**, No. 2, p. 773–784 (1982).
See also *18.4*, p. *279*.

H. Cofta and Z. Onyszkiewicz
*A Simple Spin Model of Spontaneous Metamagnetism
(b) Vol. **109**, No. 2, p. K93–K96 (1982).
See also *18.4*, p. *279*.

A. R. McGurn and R. A. Tahir-Kheli
Baxter-Wu Model with Random Bonds
(b) Vol. **109**, No. 2, p. K103–K107 (1982).
See also *18.1*, p. *265*; 18.4, p. 279.

A. Bandyopadhyay, S. B. Patil, D. K. Chakrabarty, and C. Radhakrishnamurty
*Magnetic Clusters in Some Barium Lanthanide Cobaltates
(a) Vol. **69**, No. 2, p. 441–445 (1982).
See also 22.8.1, p. 492.

H. J. Blythe, F. Walz, I. M. Richardson, and F. Dworschak
*Low Temperature Magnetic Relaxation in a Dilute Iron–Silicon Alloy after Electron Irradiation at 44 K
(a) Vol. **69**, No. 2, p. 563–567 (1982).
See also *11*, p. *119*; 21.1.1, p. 359.

T. Kanaizuka
*Interpretation of Mössbauer Spectra of β-FeB and Its Low Temperature Modification, α-FeB
(a) Vol. **69**, No. 2, p. 739–744 (1982).
See also 1.2, p. 15; *5*, p. *66*; 21.1.1, p. 359.

H. Figiel
The Y Transferred Hyperfine Fields in $Y_2(Co_{1-x}Mn_x)_{17}$ Compounds
(a) Vol. **69**, No. 2, p. K219–K221 (1982).
See also *19*, p. *284*; 21, p. 332; 21.1, p. 345.

L. Palermo and X. A. da Silva
*Induced Magnetism in a System Consisting of van Vleck Ions Coupled to an Electron Gas in the Narrow Band Limit
(b) Vol. **110**, No. 1, p. 307–311 (1982).
See also *13.4*, p. *165*; 21.4, p. 370.

M. I. AUSLENDER, N. G. BEBENIN, A. B. ZOLOTOVSKII, and V. P. KALASHNIKOV
On the Density of States of a Broad-Band Ferromagnetic Semiconductor at $T \gtreqqless T_c$
(b) Vol. **110**, No. 1, p. 369–377 (1982).
See also *13*, p. *141*; 22, p. 384.

E. KOLLEY and W. KOLLEY
*On Random Itinerant Ferromagnets in the Percolation Limit
(b) Vol. **110**, No. 1, K63–K67 (1982).
See also *18.1*, p. *265*; 21, p. 333.

S. PARVIANEN and M. LEHTINEN
*Magnetic Contribution to the Thermal Expansion in Some $SiFe_xCo_{1-x}$ Alloys
(a) Vol. **70**, No. 1, p. 79–85 (1982).
See also *8*, p. *86*; 21.1, p. 346; 21.1.1, p. 360.

H. KIDO, M. SHIMADA, and M. KOIZUMI
*Synthesis and Magnetic Properties of GdCoSi and GdMnSi
(a) Vol. **70**, No. 1, p. K23–K26 (1982).
See also *1.3*, p. *27*; 21.1, p. 346; 21.4, p. 370.

M. V. MEDVEDEV and A. V. ZABOROV
*Magnetic States of a Quenched Random Site Heisenberg Magnet with Competitive Exchange Interactions (II): Magnetic States with Partial Spin-Glass Freezing in Simple and Body-Centered Cubic Lattices
(b) Vol. **110**, No. 2, p. 387–397 (1982).
See also *18.4*, p. *279*.

Y. J. WANG and H. KRONMÜLLER
*The Influence of the Surface Conditions on the Magnetic Properties in Amorphous Alloys $Fe_{40}Ni_{40}B_{20}$ and $Co_{58}Ni_{10}Fe_5Si_{11}B_{16}$
(a) Vol. **70**, No. 2, p. 415–421 (1982).
See also 1.6, p. 42; 21.1, p. 347; 21.1.1, p. 360.

X. Z. DONG and H. KRONMÜLLER
*Magnetic Domain Structure of Some Nearly Non-Magnetostrictive Amorphous Alloys under Elastic Stresses
(a) Vol. **70**, No. 2, p. 451–462 (1982).
See also 21.1, p. 347; 21.1.1, p. 360.

J. D. LIVINGSTON
*Magnetomechanical Properties of Amorphous Metals
(a) Vol. **70**, No. 2, p. 591–596 (1982).
See also 2, p. 47; *12.1*, p. *131*; 21, p. 334.

O. N. AWASTHI and B. K. BASU
On Magnetoelastic Dispersion in a Nickel Single Crystal
(a) Vol. **70**, No. 2, p. 713–717 (1982).
See also *7*, p. *82*; 21.1, p. 347.

A. V. ANDREEV, A. V. DERYAGIN, and S. M. ZADVORKIN
*Anomalies in the Thermal Expansion and Resistivity of $NdCo_5$ at the Spin-Reorientation Phase Transition
(a) Vol. **70**, No. 2, p. K113–K115 (1982).
See also 14.1, p. 194; 21.1, p. 348.

A. V. Andreev, E. N. Tarasov, A. V. Deryagin, and S. M. Zadvorkin
*Influence of Symmetry of the Crystalline Structure on the Magnetic Anisotropy of Tb_2Co_7
(a) Vol. **71**, No. 2, p. K245–K247 (1982).
See also *1.2*, p. *19*; 1.3, p. 29; 21.1, p. 349; 21.4, p. 371.

Yu. A. Izyumov and V. M. Laptev
Soliton Magnetoelastic Excitation in the Heisenberg Ferromagnetic Chain
(b) Vol. **112**, No. 1, p. 155–159 (1982).
See also *13.5.4*, p. *189*.

S. K. Ghatak
*Cluster Variational Treatment of an Anisotropic Heisenberg Magnet
(b) Vol. **112**, No. 1, p. K35–K38 (1982).

B. Kuhlow
Magnetic Ordering in $CrCl_3$ at the Phase Transition
(a) Vol. **72**, No. 1, p. 161–168 (1982).
See also *18.4*, p. *280*; *20*, p. *296*; 22.5, p. 451.

H. Mehrer, G. Flik, J. Hrováth, and H. Kronmüller
*Crystallization and Aging Effects in Some Amorphous Ferromagnets
(a) Vol. **72**, No. 1, p. 215–223 (1982).
See also *2*, p. *50*; 21.1, p. 350; 21.1.1, p. 361.

A. del Moral, J. F. D. Montenegro, J. S. Abell, and M. R. Ibarra
On the Magnetization and Magnetostriction of ErZn
(a) Vol. **72**, No. 1, p. 353–360 (1982).
See also 12.1, p. 132; *18.1*, p. *266*; 21.1, p. 351; 21.4, p. 371.

S. Parviainen
*Magnetic Contribution to the Thermal Expansion of $TiFe_{0.5}Co_{0.5}$
(a) Vol. **72**, No. 2, p. K141–K144 (1982).
See also *8*, p. *88*; 21.1, p. 352; 21.1.1, p. 361.

O. S. Kolotov, V. A. Pogozhev, R. V. Telesnin, G. V. Smirnov, Yu. V. Shvydko, S. Kadeĉkova, M. Kotrbova, and J. Novak
*Pulse Magnetization Reversal in $FeBO_3$ Crystals
(a) Vol. **72**, No. 2, p. K197–K201 (1982).
See also 3, p. 58; 22.8.1, p. 496.

V. V. Dyakin and B. V. Karpenko
*On the Theory of a Heisenberg Impurity Ferromagnet
(b) Vol. **113**, No. 1, p. 51–58 (1982).

P. R. Silva and F. C. Sá Barreto
*Transition Temperature for Effective Field Ising Systems
(b) Vol. **113**, No. 1, p. 67–72 (1982).

A. I. Mitsek, D. I. Sirota, and N. P. Kolmakova
On the Theory of Acoustic Properties of Many-Domain Ferromagnets
(b) Vol. **113**, No. 1, p. 79–86 (1982).
See also *7*, p. *83*; 18.4, p. 280.

H. J. Blythe
*Magnetic Disaccommodation in Ordered Ni_3Fe Following Either Cold-Work or Low-Temperature Electron Irradiation
(a) Vol. **73**, No. 2, p. K311–K315 (1982).
See also *11*, p. *127*; 21.1, p. 356; 21.1.1, p. 363.

J. R. Faleiro Ferreira and N. P. Silva
*Extended Linear Chain Approximation to the Anisotropic Ferromagnetic Ising Model
(b) Vol. **114**, No. 1, p. 47–51 (1982).

H. Cofta and P. Pawlicki
*The Model of Metamagnetic Phase Transition in Mn_3GaC
(b) Vol. **114**, No. 1, p. 53–60 (1982).
See also *18.4*, p. *281*; 22.8, p. 490.

A. Z. Maksymowicz
*Curie Temperature and Short-Range Atomic Order in Amorphous Ferromagnets
(b) Vol. **114**, No. 1, p. 125–133 (1982).
See also *2*, p. *54*.

E. Kolley and W. Kolley
*On the Coexistence of Superconductivity and Itinerant Ferromagnetism
(b) Vol. **114**, No. 1, p. 135–144 (1982).
See also *14.2*, p. *201*.

P. R. Silva and F. C. Sá Barreto
*Internal Energy and Short-Range Order in Ising Models
(b) Vol. **114**, No. 1, p. 227–233 (1982).

L. Iannarella, A. P. Guimarães, and X. A. da Silva
*A Simple Model Approach to Localized-Itinerant Magnetism. Application to Rare-Earth Intermetallics
(b) Vol. **114**, No. 1, p. 255–263 (1982).
See also 21.4, p. 372.

H. J. Blythe, F. Walz, and F. Dworschak
*Magnetic Relaxation in Dilute Iron–Nickel Alloys Following Low-Temperature Electron Irradiation
(a) Vol. **74**, No. 1, p. 181–184 (1982).
See also *10.1*, p. *101*; 11, p. 128; 21.1, p. 356; 21.1.1, p. 363.

J. Degauque and B. Astie
*Evolution of the Configuration of Magnetic Domains Interacting with Structural Defects in High Purity Iron
(a) Vol. **74**, No. 1, p. 201–210 (1982).
See also *10.1*, p. *101*; 21.1.1, p. 363.

R. A. Jago, P. E. Clark, and P. L. Rossiter
The Santa Catharina Meteorite and the Equilibrium State of Fe–Ni Alloys
(a) Vol. **74**, No. 1, p. 247–254 (1982).
See also *1.3*, p. *31*; *5*, p. *68*; 21.1, p. 356; 21.1.1, p. 363.

18.2.1 Ferromagnetic Films

Original Papers and Short Notes

J. M. Alameda and F. López
*Transverse Susceptibility and Structure Constant in Cobalt Thin Films
(a) Vol. **69**, No. 2, p. 757–765 (1982).
See also 21.1, p. 344.

A. Duda
*Resonance Spectrum of Elliptically Polarized Spin Waves in Thin Magnetic Films
(b) Vol. **111**, No. 1, p. 303–310 (1982).
See also *19*, p. *286*.

J. A. Misiak
*Low-Angle Electron Deflection Studies of Domain Wall Structure in Thin NiFe Films
(a) Vol. **71**, No. 1, p. 43–48 (1982).
See also 21.1, p. 348; 21.1.1, p. 361.

J. Strzeszewski and K. Świderczak
*Electron Microscopic Studies of Crystallization and Magnetic Domain Structure of Mn–Ga Thin Films
(a) Vol. **73**, No. 1, p. 31–36 (1982).
See also *1.2*, p. *20*; 3, p. 59; 21.1, p. 353.

V. I. Malyutin, V. E. Osukhovskii, Yu. D. Vorobiev, and A. S. Abramov
*The Structure Constant and Micromagnetic Structure of Etched $Ni_{80}Fe_{20}$ Films
(a) Vol. **73**, No. 1, p. K111–K114 (1982).
See also *1.3*, p. *30*; 21.1, p. 354; 21.1.1, p. 362.

V. Seshu Bai and K. V. S. Rama Rao
*Ferromagnetic Resonance Investigations in $MnSb_{1-x}In_x$ Films
(a) Vol. **73**, No. 2, p. K303–K305 (1982).
See also *19*, p. *290*; 21, p. 341.

W. Korneta and Z. Pytel
*The Autocorrelation Time in Thin Ferromagnetic Film above T_C
(b) Vol. **114**, No. 1, p. K83–K85 (1982).

P. Görnert, M. Nevřiva, J. Šimšová, W. Andrä, W. Schüppel, P. Šumšái, and R. Bubáková
*Co Containing Garnet Films with Low Magnetization
(a) Vol. **74**, No. 1, p. 107–112 (1982).
See also *1.5*, p. *41*; 22.8.2, p. 502.

18.3 Ferrimagnetic Properties

Original Papers and Short Notes

T. Duczmal, S. P. Kunsevich, V. P. Palekhin, and J. Pietrzak
*Temperature Dependence of Magnetocrystalline Anisotropy Constant of $BaCo_2Ti_2Fe_8O_{19}$ Single Crystals
(a) Vol. **69**, No. 1, p. K125–K127 (1982).
See also 19, p. 283; 22.8.2, p. 499.

A. V. Muromtsev and E. A. Raitman
*Relaxation of Complex Susceptibility of Ferrites in AC Magnetic Field
(a) Vol. **70**, No. 1, p. 29–34 (1982).
See also 22.8.2, p. 499.

S. A. Mazen, M. A. Ahmed, and B. A. Sabrah
Thermal Studies on the Electrical Conduction Mechanism of $CuFe_2O_4$
(a) Vol. **70**, No. 1, p. K71–K75 (1982).
See also *14.4*, p. *235*; 22.8.2, p. 499.

F. Walz, V. A. M. Brabers, S. Chikazumi, H. Kronmüller, and M. O. Rigo
*Magnetic After-Effects in Single- and Poly-Crystalline Magnetite
(b) Vol. **110**, No. 2, p. 471–478 (1982).
See also 22.6, p. 470.

V. G. Kleparskii, I. Pintér, and G. Serfözö
*Temperature Dependence of the Damping Parameter on Moving Domain Walls in Ferrite Garnet Film
(a) Vol. **70**, No. 2, p. K101–K106 (1982).
See also 19, p. 286; 22.8.2, p. 500.

K. P. Koval, E. V. Babkin, and V. G. Pynko
*Growth-Induced Anisotropy in Epitaxial Crystals of γ-Fe_2O_3
(a) Vol. **70**, No. 2, p. K117–K120 (1982).
See also 22.6, p. 471.

Ph. Bauer and J. M. Genin
Dipolar Contribution and Quadrupole Effect in $GdFe_2$
(a) Vol. **71**, No. 2, p. K149–K152 (1982).
See also *5*, p. *67*; 21.1.1, p. 361.

J. Janicki, J. Pietrzak, A. Porębska, and J. Suwalski
*Mössbauer Study of Copper Ferrites
(a) Vol. **72**, No. 1, p. 95–98 (1982).
See also *5*, p. *67*; 22.8.2, p. 500.

M. Wurlitzer
*Photomagnetic Effect in YIG(Fe^{2+})
(a) Vol. **72**, No. 1, p. 169–176 (1982).
See also *20*, p. *296*; 22.8.2, p. 500.

A. S. PARSHIN and N. S. CHISTYAKOV
*Spin Wave Resonance in Thin Monocrystalline Iron-Manganese Ferrite Films
(b) Vol. **112**, No. 2, p. K89–K92 (1982).
See also 22.8.2, p. 501.

Z. DRZAZGA and T. MYDLARZ
*Magnetic Properties of the 3d Metal-Rich $HoCo_5$ Compounds
(a) Vol. **72**, No. 2, p. 771–776 (1982).
See also 21.1, p. 352; 21.4, p. 371.

E. BURZO, D. SEITABLA, and M. CHIPARĂ
*Magnetic Properties and Resonance Study of $Gd(Co_xAl_{1-x})_3$ Compounds
(b) Vol. **113**, No. 1, p. 87–92 (1982).
See also *18.1*, p. *266*; 19, p. 289; 21, p. 338; 21.1, p. 352; 21.4, p. 371.

K. TAKADA and K. GOTO
*A Numerical Explanation for a Relation of the Domain Width to the Thickness in Barium Ferrite Crystals
(a) Vol. **73**, No. 1, p. 209–213 (1982).
See also 22.8.2, p. 501.

H. RATAJCZAK and I. GOŚCIAŃSKA
*Thermal Hall Hysteresis Loops in Thin Amorphous HoCo Films
(a) Vol. **73**, No. 2, p. 633–639 (1982).
See also 2, p. 53; *14.1*, p. *197*; 21.1, p. 355; 21.4, p. 371.

I. ONYSZKIEWICZ and J. PIETRZAK
*Magnetocrystalline Anisotropy of Tetragonal Copper Ferrite
(a) Vol. **73**, No. 2, p. 641–650 (1982).
See also 19, p. 290; 22.8.2, p. 501.

R. Z. SADIKHOV, I. G. KERIMOV, L. M. VALIEV, and A. D. NAMAZOV
*Temperature Dependence of Spontaneous Magnetization of the $FeCr_{2-x}In_xS_4$ ($0 \leqq x \leqq 1$) System
(a) Vol. **73**, No. 2, p. K209–K212 (1982).
See also 22.8.2, p. 501.

A. E. GELYASIN and E. A. VASILEV
*Influence of Temperature on Structure and Properties of $Mn_{1+x}Sb_{1-x}Sn_x$ Alloys
(a) Vol. **73**, No. 2, p. K221–K224 (1982).
See also 5, p. 68; *18.2*, p. *272*; 21, p. 341.

I. ONYSZKIEWICZ, N. T. MALAFAEV, A. A. MURAKHOVSKII, and J. PIETRZAK
Magnetic and FMR Investigation of the Structural Phase Transition in Cu-Ferrite with Cooperative Jahn-Teller Effect
(a) Vol. **73**, No. 2, p. K243–K247 (1982).
See also *1.2*, p. *22*; 19, p. 290; 22.8.2, p. 501 .

S. A. MAZEN and M. A. AHMED
The Deficiency Effect of Iron on the Conduction Mechanism of $Co_{0.5}Zn_{0.5}Fe_2O_4$
(a) Vol. **73**, No. 2, p. K307–K310 (1982).
See also *14.4*, p. *238*; 22.8.2, p. 502.

18.4 Antiferromagnetic Properties

Original Papers and Short Notes

M. V. Medvedev and M. V. Sadovskii
Random Bond Ising Model in Self-Avoiding Walk Approximation
(b) Vol. **109**, No. 1, p. 49–57 (1982).
See also *18.1*, p. *265*; 18.2, p. **267**.

U. Lindner
*Solitary Excitations in 1-D Antiferromagnetic Systems with Finite Band Width
(b) Vol. **109**, No. 1, p. K41–K46 (1982).
See also *13.5*, p. *179*.

W. Bażela, I. Szott, and A. Szytuła
*Magnetic Properties of the $Co_xNi_{1-x}Mn$ Ge System in the Hexagonal Phase
(a) Vol. **69**, No. 1, p. K49–K54 (1982).
Erratum
(a) Vol. **71**, No. 1, p. K131 (1982).
See also *1.1*, p. *11*; 22.8, p. **482**.

M. V. Medvedev and M. V. Sadovskii
Random Site Ising Model in Self-Avoiding Walk Approximation
(b) Vol. **109**, No. 2, p. 449–456 (1982).
See also *18.1*, p. *265*; 18.2, p. **267**.

M. V. Medvedev and A. V. Zaborov
*Magnetic States of a Quenched Random Site Heisenberg Magnet with Competitive Exchange Interactions (I): High-Temperature Magnetic States in Simple and Body-Centered Cubic Lattices
(b) Vol. **109**, No. 2, p. 773–784 (1982).
See also *18.2*, p. *268*.

H. Cofta and Z. Onyszkiewicz
*A Simple Spin Model of Spontaneous Metamagnetism
(b) Vol. **109**, No. 2, p. K93–K96 (1982).
See also *18.2*, p. *268*.

A. R. McGurn and R. A. Tahir-Kheli
*Baxter-Wu Model with Random Bonds
(b) Vol. **109**, No. 2, p. K103–K107 (1982).
See also *18.1*, p. *265*; 18.2, p. **268**.

M. V. Medvedev and A. V. Zaborov
*Magnetic States of a Quenched Random Site Heisenberg Magnet with Competitive Exchange Interactions (II): Magnetic States with Partial Spin-Glass Freezing in Simple and Body-Centered Cubic Lattices
(b) Vol. **110**, No. 2, p. 387–397 (1982).
See also *18.2*, p. *269*.

S. V. Tarasenko and G. K. Chepurnykh
*Study of Relaxation Processes in Antiferromagnets Using Sound Absorption Measurements
(b) Vol. **110**, No. 2, p. 507–511 (1982).
See also 7, p. **82**; *13.5.4*, p. *189*.

T. P. Arsenio, Z. Arguello, P. H. Domingues, N. C. Furtado, and C. A. Taft
*Temperature Dependence of Hyperfine Field and Resonance Fraction in Single Crystal $RbFeS_2$
(b) Vol. **110**, No. 2, p. K129–K131 (1982).
See also *5*, p. *66*; 22.8.2, p. 500.

S. Katsura, S. Fujiki, T. Suenaga, and I. Nagahara
*Spin Glass and Superantiferromagnet in the Site Diluted Ising Model on the Frustrated Square and F.C.C. Lattices with First- and Second-Neighbor Interactions
(b) Vol. **111**, No. 1, p. 83–88 (1982).
See also *18.2*, p. *270*; 22.4.4, p. 447.

E. Gratz, V. Sechovsky, V. Sima, Z. Smetana, and J. O. Ström-Olson
*Magnetization, Electrical Resistivity, Thermopower, and Neutron Diffraction in $HoCu_2$
(b) Vol. **111**, No. 1, p. 195–201 (1982).
See also 14.1, p. 194; 15, p. 246; 21.1, p. 348; 21.4, p. 370.

M. M. Kotlyarskii and E. A. Popov
*Magnetic Phase Diagram of $NaMnCl_3$
(b) Vol. **111**, No. 1, p. K13–K16 (1982).
See also *20*, p. *295*; 22.8, p. 484.

Diep-The-Hung and J. L. Motchane
*Magnetic Properties of Oxygen Monolayers Adsorbed on Graphite
(b) Vol. **111**, No. 2, p. 639–648 (1982).
See also 22.1, p. 393.

M. Iliev, S. Angelov, I. Z. Kostadinov, V. Bojchev, and V. Hadjiev
Critical Behaviour of Co_3O_4 Conductivity near the Antiferromagnetic Phase Transition
(a) Vol. **71**, No. 2, p. 627–632 (1982).
See also *14.3*, p. *207*; 22.8.2, p. 500.

A. Blaise, J. M. Fournier, and D. Damien
*Magnetic Properties of Plutonium Diantimonides
(a) Vol. **71**, No. 2, p. K147–K148 (1982).
See also 18.1, p. 266; 21.5, p. 373; 21.7, p. 380.

M. D. Kaplan
*Metamagnetoelastic and Metamagnetoelectric Behaviour of the Jahn-Teller Antiferroelectrics
(b) Vol. **112**, No. 1, p. 351–356 (1982).
See also *1.2*, p. *19*; 22.8, p. 487.

B. Kuhlow
*Magnetic Ordering in $CrCl_3$ at the Phase Transition
(a) Vol. **72**, No. 1, p. 161–168 (1982).
See also 18.2, p. 271; *20*, p. *296*; 22.5, p. 451.

A. I. Mitsek, D. I. Sirota, and N. P. Kolamkova
On the Theory of Acoustic Properties of Many-Domain Ferromagnets
(b) Vol. **113**, No. 1, p. 79–86 (1982).
See also *7*, p. *83*; 18.2, p. 271.

P. DAY, K. TURNER, D. VISSER, and T. E. WOOD
*Neutron Diffraction and Magneto-Optical Study of the Effect of Chloride Doping on the Incommensurate Magnetic Phase of $NiBr_2$
(b) Vol. **113**, No. 2, p. 623–634 (1982).
See also *18.2*, p. *272*; 20.1, p. 312; 22.5, p. 452.

D. DĂDÂRLAT, T. PETRIŞOR, and I. POP
*The Magnetic Behaviour of Diluted CrGe Alloys
(b) Vol. **113**, No. 2, p. K111–K114 (1982).
See also 21.1, p. 355.

A. CASTETS, D. GIGNOUX, J. C. GOMEZ-SAL, F. RODRIGUEZ-GONZALEZ, and E. ROUDAUT
Magnetic Properties and Structure of Tb_2Pt
(a) Vol. **73**, No. 2, p. 475–481 (1982).
See also *1*, p. *10*; *18.2*, p. *272*; 21.4, p. 371; 21.6, p. 378.

H. COFTA and P. PAWLICKI
*The Model of Metamagnetic Phase Transition in Mn_3GaC
(b) Vol. **114**, No. 1, p. 53–60 (1982).
See also *18.2*, p. *273*; 22.8, p. 490.

G. KOZŁOWSKI and K. ROGACKI
*Magnetocaloric Effect and Spin Waves in Two-Sublattice Antiferromagnetic Superconductors
(b) Vol. **114**, No. 1, p. K5–K9 (1982).
See also *14.2*, p. *201*.

W. SCHUMACHER and U. LINDNER
*Antiferromagnetic and Incommensurable Magnetic Solutions of the Hubbard Model
(b) Vol. **114**, No. 1, p. K55–K57 (1982).

Y. SOMEYA
*Phase Diagram of Random Mixtures with Competing Spin Anisotropies: $Fe_{1-x}Co_xCl_2$ and $K_2Mn_{1-x}Fe_xF_4$
(a) Vol. **74**, No. 1, p. 167–172 (1982).
See also 21.1, p. 356; 21.1.1, p. 363.

C. MIYAKE, H. TAKEUCHI, H. OHYA-NISHIGUCHI, and S. IMOTO
Electron Spin Resonance and Magnetic Susceptibility of M–U–O Ternary Mixed Oxides (I): Anomalies of Electron Spin Resonance and Magnetic Susceptibility of $Li_{2n-5}UO_n$ ($n = 3$, 4, and 6)
(a) Vol. **74**, No. 1, p. 173–180 (1982).
See also *18.1*, p. *266*; *19*, p. *290*; 22.6.1, p. 477.

J. A. TUSZYŃSKI and H. COFTA
*On Thermodynamics of Spontaneous Metamagnetism
(b) Vol. **114**, No. 2, p. 359–364 (1982).
See also *18.2*, p. *274*; 22.8, p. 491.

P. H. DOMINGUES, T. P. ARSENIO, N. C. FURTADO, and C. A. TAFT
Quadrupolar and Magnetic Hyperfine Anisotropy in Cesium Dithioferrate
(b) Vol. **114**, No. 2, p. K161–K163 (1982).
See also *5*, p. *69*; 22.8.2, p. 502.

19. Magnetic Resonance

Review Article

M. Prasad
*Quantum-Statistical Theories of the Cyclotron Resonance Lineshape in Parabolic Semiconductors
(b) Vol. **109**, No. 1, p. 11–35 (1982).
See also 13.4, p. 162; 22, p. 382.

Original Papers and Short Notes

S. K. Misra and P. Mikolajczak
*EPR Linewidth Studies of Gd^{3+}-Doped $Yb_xY_{1-x}Cl_3 \cdot 6\,H_2O$ Single Crystals. Yb^{3+} Spin-Lattice Relaxation Times
(b) Vol. **109**, No. 1, p. 59–66 (1982).
See also *18.1*, p. *265*; 22.8, p. 481.

E. G. Sieverts, S. H. Muller, C. A. J. Ammerlaan, R. L. Kleinhenz, and J. W. Corbett
*EPR Observation of an Au–Fe Complex in Silicon (II): Electronic Structure
(b) Vol. **109**, No. 1, p. 83–94 (1982).
See also *13.4*, p. *162*; 22.1.2, p. 400.

D. M. Daraselia and A. Bräuer
*Investigation of Superhyperfine Interactions of Mn^{2+} in α-$LiIO_3$ and Evaluation of Lithium Quadrupolar Interactions
(b) Vol. **109**, No. 1, p. 223–231 (1982).
See also *13.4*, p. *162*; 22.8.1, p. 492.

U. Schmid and P. Reineker
*Coherent Motion of Triplet Excitons in Molecular Crystals with Two Differently Oriented Molecules per Unit Cell. Application to the Charge Transfer Crystal Anthracene/Tetracyanobenzene
(b) Vol. **109**, No. 1, p. 265–273 (1982).
See also *13.5.1*, p. *181*; 22.9, p. 503.

A. A. Alybakov, N. Toichiev, and Sh. Akchalov
*Investigation of EPR and Thermoluminescence Glow-Curves of X-Irradiated $KCl:CrO_4^{2-}$, Ca^{2+} Crystals
(b) Vol. **109**, No. 1, p. 295–301 (1982).
See also *20.3*, p. *318*; 22.5.2, p. 454.

B. V. R. Chowdari, Y. Ravisekhar and C. Muralikrishna
Radiation Damage of Ni^{2+}-Doped $NH_4MgPO_4 \cdot 6\,H_2O$
(b) Vol. **109**, No. 1, p. 315–324 (1982).
See also *13.4*, p. *163*; 20.1, p. 298; 22.8.1, p. 492.

D. Ya. Osokin
*Coherent Quasi-Steady States in Nitrogen-14 NQR Multipulse Experiments
(b) Vol. **109**, No. 1, p. K7–K10 (1982).

I. MERTIG, E. MROSAN, R. ZELLER, and P. H. DEDERICHS
*Anisotropic Cyclotron Masses and Electron–Phonon Interaction in Silver
(b) Vol. **109**, No. 1, p. K17–K21 (1982).
See also *13.2*, p. *157*; 21.6, p. 374.

M. ZAŁUŻNY
*On the Cyclotron Resonance in Size-Quantized Films
(b) Vol. **109**, No. 1, p. K69–K73 (1982).
See also *13.1*, p. *148*.

P. J. ALONSO, V. M. ORERA, F. PALACIO, and R. ALCALA
*Optical and EPR Studies of NH_4MnF_3 above 80 K
(b) Vol. **109**, No. 1, p. K81–K83 (1982).
See also *20.1*, p. *299*; 22.5, p. 450.

G. E. ARCHANGELSKII, F. KAREL, J. MAREŠ, S. PAČESOVÁ, and J. PASTRŇÁK
*The Luminescence and EPR Spectra of Manganese Activated AlN
(a) Vol. **69**, No. 1, p. 173–183 (1982).
See also *20.3*, p. *318*; 22.2, p. 418.

A. LEBLÉ, J. J. ROUSSEAU, J. C. FAYET, J. PANNETIER, J. L. FOURQUET, and R. DE PAPE
*Order–Disorder Transition of NH_4AlF_4 through EPR Investigations
(a) Vol. **69**, No. 1, p. 249–256 (1982).
See also 1.2, p. 14; 22.8, p. 481.

T. DUCZMAL, S. P. KUNSEVICH, V. P. PALEKHIN and J. PIETRZAK
Temperature Dependence of Magnetocrystalline Anisotropy Constant of $BaCO_2Ti_2Fe_8O_{19}$ Single Crystals
(a) Vol. **69**, No. 1, p. K125–K127 (1982).
See also *18.3*, p. *276*; 22.8.2, p. 499.

ZH. EGEMBERDIEV, S. ZAZUBOVICH, V. SEEMAN, and V. NAGIRNYI
Luminescence Accompanying the Recombination of Electrons with V_F Centres in KCl Crystals
(b) Vol. **109**, No. 2, p. 473–482 (1982).
See also *10.2*, p. *104*; *20.3*, p. *319*; 22.5.2, p. 455.

M. HÖHNE
*Electron Paramagnetic Resonance of Gold in Silicon (II): Cluster Centres
(b) Vol. **109**, No. 2, p. 525–534 (1982).
See also 13.4, p. 164; 22.1.2, p. 401.

H. NAGASAWA
Mixed Valence State and Metal–Non-Metal Transition of KCP
(b) Vol. **109**, No. 2, p. 749–759 (1982).
See also *13*, p. *140*; *14*, p. *190*; 22.8, p. 482.

M. ONODA, T. TAKAHASHI, and H. NAGASAWA
*Nuclear Quadrupole Effect and Phase Transition in the Quasi-One-Dimensional Conductor β-$Na_{0.33}V_2O_5$
(b) Vol. **109**, No. 2, p. 793–797 (1982).
See also *1.2*, p. *15*; 22.8, p. 482.

V. V. Ustinov and D. Z. Khusainov
*On the Lineshape of the Conduction Electron Spin Resonance in Thick Metallic Samples under the Anomalous Skin-Effect Conditions
(b) Vol. **110**, No. 1, p. 363–368 (1982).
See also 21, p. 333.

M. N. Aliev, M. M. Rahman, and M. Z. Asadov
*Influence of Hyperfine Interaction on Magnetic Resonance Line-Shape on the Nuclei of Paramagnetic Impurities in Magnetically Diluted Crystals
(b) Vol. **110**, No. 1, p. K57–K61 (1982).

E. G. Sieverts and S. H. Muller
*Heat–Treatment Centers in Silicon and Their g-Values
(b) Vol. **110**, No. 1, p. K89–K92 (1982).
See also *13.4*, p. *166*; 22.1.2, p. 403.

D. Backs
*EPR of Mn^{2+} in Faulted Cubic and Mixed Cubic–Hexagonal Crystals of ZnS
(b) Vol. **110**, No. 1, p. K93–K96 (1982).
See also 22.4.1, p. 440.

M. Stojić
*The Origin of Iron Interstitial in Quenched Silicon Single Crystals
(a) Vol. **70**, No. 1, p. K39–K42 (1982).
See also 10.2, p. 106; 22.1.2, p. 404.

H. Overhof
*The Temperature Dependence of the Spin Resonance in Boron Doped a-Si:H
(b) Vol. **110**, No. 2, p. 521–530 (1982).
See also 2, p. 47; *13.4*, p. *166*; 22.1.2, p. 404.

H. Haberlandt
*The E_1' Centre in SiO_2. MO Calculation of the ^{17}O, ^{29}Si, and ^{1}H Hyperfine Splittings
(b) Vol. **110**, No. 2, p. 531–535 (1982).
See also *13.4*, p. *167*; 22.6, p. 470.

M. Mladenova, A. Malinovski, and M. Georgiev
Conversion of Impurity–Vacancy Complexes under γ-Irradiation and Subsequent F Band Bleaching in $KCl:Eu^{2+}$
(b) Vol. **110**, No. 2, p. 643–652 (1982).
See also *10.2*, p. *107*; 22.5.2, p. 456.

V. S. Vikhnin and Yu. B. Borkovskaya
EPR Line Critical Broadening Asymmetry and Shape Peculiarities Due to Dipole–Dipole Electric Interaction in Ferroelectrics
(b) Vol. **110**, No. 2, p. 717–727 (1982).
See also *14.4.1*, p. *241*.

R. A. Weeks, J. Narayan, and E. Sonder
Electric Breakdown in MgO Crystals at Elevated Temperature
(a) Vol. **70**, No. 2, p. 631–639 (1982).
See also 10.2, p. 108; *14.3.4*, p. *228*; 22.6, p. 470.

V. G. Kleparskii, I. Pintér, and G. Serfözö
Temperature Dependence of the Damping Parameter on Moving Domain Walls in Ferrite Garnet Film
(a) Vol. **70**, No. 2, p. K101–K106 (1982).
See also *18.3*, p. *276*; 22.8.2, p. 500.

H. Ishida, R. Ikeda, and D. Nakamura
*Pre-Melting State of Methylammonium Iodide as Revealed by Proton Magnetic Resonance
(a) Vol. **70**, No. 2, p. K151–K154 (1982).
See also *1.2*, p. *17*; 22.9, p. 504.

S. V. Misul and I. A. Belobrova
Sequence of Phase Transitions in $(NH_4)_2ZnCl_4$
(a) Vol. **70**, No. 2, p. K167–K170 (1982).
See also *1.2*, p. *17*; 14.4.1, p. 241; 22.8, p. 484.

I. Bakonyi, P. Panissod, and K. Tompa
*Correction of the Knight Shift for Demagnetizing Effects
(b) Vol. **111**, No. 1, p. 59–64 (1982).

K. N. Shrivastava
Second- and Third-Order Processes in the Theory of Spin–Lattice Relaxation
(b) Vol. **111**, No. 1, p. 77–82 (1982).
See also *6*, p. *73*; *13*, p. *142*; 18.1, p. 265.

D. Backs and H. J. Freyer
*Photosensitive EPR of Ge^{3+} and Sn^{3+} in Twinned Cubic Zinc Sulfide
(b) Vol. **111**, No. 1, p. 117–124 (1982).
See also *13.4*, p. *168*; 22.4.1, p. 440.

S. K. Misra, P. Mikolajczak, and M. Jalochowski
*Host-Lattice Effects in the EPR Spectra of Gd^{3+} Doped Hydrated Single Crystals of Rare-Earth Chlorides, Nitrates, and Sulphates
(b) Vol. **111**, No. 1, p. 227–234 (1982).
See also *13.4*, p. *168*; 18.1, p. 266; 22.8, p. 484.

A. Duda
*Resonance Spectrum of Elliptically Polarized Spin Waves in Thin Magnetic Films
(b) Vol. **111**, No. 1, p. 303–310 (1982).
See also *18.2.1*, p. *275*.

D. Ball
*The Paramagnetic Resonance of Nd^{3+} and Yb^{3+} in Zircon Structure Silicates
(b) Vol. **111**, No. 1, p. 311–320 (1982).
See also *13.4*, p. *168*; 22.8.1, p. 494.

M. A. Hefni, N. M. Grecu, V. V. Grecu, and F. F. Popescu
*Electron Paramagnetic Resonance Studies of Cu^{2+} Impurities in Ferroelectric $LiH_3(SeO_3)_2$
(b) Vol. **111**, No. 1, p. 407–416 (1982).
See also 14.4.1, p. 241; 22.8, p. 484.

G. J. Troup, A. Edgar, D. R. Hutton, and P. P. Phakey
*8 mm Wavelength EPR Spectrum of Cr^{3+} in Laser-Quality Alexandrite
(a) Vol. **71**, No. 1, p. K29 – K31 (1982).
See also 22.8, p. 484.

G. K. Padam, B. Singh, S. K. Gupta, and G. D. Sootha
*Kinetic Behaviour of Oxygen Radicals in CdS
(a) Vol. **71**, No. 1, p. K37 – K40 (1982).
See also 14.3, p. 206; 16, p. 251; 22.4.1, p. 440.

W. Korczak and M. Subotowicz
*Analysis of Temperature Variation of EPR Linewidths for Gd^{3+} Ion in CeF_3 Single Crystals
(b) Vol. **111**, No. 2, p. 455 – 460 (1982).
See also 13.4, p. 169; 22.5, p. 450.

M. Ikeya
*ESR Studies of Manganese Centres in LiH X-Rayed at Low Temperature
(b) Vol. **111**, No. 2, p. 525 – 530 (1982).
See also *13.4*, p. *169*; 22.5.2, p. 458.

V. V. Nemoshkalenko, A. V. Zhalko-Titarenko, V. N. Antonov, Vl. N. Antonov, and E. Mrosan
*Anisotropy of Cylotron Masses in Gold
(b) Vol. **111**, No. 2, p. K81 – K83 (1982).
See also *13.2*, p. *157*; 21.6, p. 376.

J. Rosa, K. Polák and J. Kubátová
ESR and Optical Studies of Impurity Centres in γ- and X-Irradiated $LiNbO_3$
(b) Vol. **111**, No. 2, p. K85 – K87 (1982).
See also *13.4*, p. *170*; 20.1, p. 306; 22.8.1, p. 495.

E. I. Neimark and B. E. Vugmeister
*Nuclear Spin–Lattice Relaxation Due to Quadrupole Interaction with Current Carriers in Quantizing Magnetic Fields
(b) Vol. **111**, No. 2, p. K113 – K115 (1982).

W. Pastusiak and J. Kuriata
*A New Model for the Interpretation of Spin–Lattice Coefficients for S-State ($4f^7$) Ions
(b) Vol. **111**, No. 2, p. K147 – K151 (1982).
See also *13.4*, p. *170*; 22.5.3, p. 465.

G. K. Padam, B. Singh, S. K. Gupta, and G. D. Sootha
*Effect of Impurities on the Oxygen Radicals and F^+ Centres in CdS
(a) Vol. **71**, No. 2, p. K169 – K172 (1982).
See also *10.2*, p. *110*; 22.4.1, p. 441.

Z. Kruczyński and J. Pietrzak
*EPR Linewidth Anomaly of Cr^{3+} in γ-Alum $NaAl(SO_4)_2 \cdot 12\ H_2O$
(a) Vol. **71**, No. 2, p. K231 – K234 (1982).
See also 22.8, p. 486.

Y. M. M. Hornos, H. Panepucci, C. J. Magon, and M. C. Terrile
*Discrete Saturation Effects of the ESR Spectrum of H^0 in CaF_2
(b) Vol. **112**, No. 1, p. 91–96 (1982).
See also *10.2*, p. *110*; 22.5.3, p. 465.

J. Goebbels, H. R. Khan, K. Lüders, and Ch. J. Raub
*NMR on Metastable Pt–P and Ni–Pt–P
(b) Vol. **112**, No. 1, p. 115–118 (1982).
See also 2, p. 49; 21.1, p. 350.

D. L. Sastry and K. V. S. Rama Rao
*ESR and Optical Absorption Studies of ClO_3 in $Ba(ClO_4)_2 \cdot 3\,H_2O$ Single Crystals
(b) Vol. **112**, No. 1, p. 133–142 (1982).
See also 11, p. 123; *20.1*, p. *307*; 22.8.1, p. 495.

P. Colombet, M. Tremblet, and M. Danot
Magnetic Properties of the Diluted B-Spinels $Cu_yCr_y^{III}M_{2-y}^{IV}S_4$ (M^{IV} = Sn, Ti)
(a) Vol. **72**, No. 1, p. 105–116 (1982).
See also *1.3*, p. *29*; 5, p. 67; *18*, p. *261*; 22.8.2, p. 500.

A. E. Moskalev, I. A. Belobrova, L. I. Zherebtsova, and I. P. Aleksandrova
Ferroelectric Phase Transition in $(NH_4)_2ZnBr_4$
(a) Vol. **72**, No. 1, p. K19–K21 (1982).
See also *1.2*, p. *19*; *14.4.1*, p. *242*; 22.8, p. 487.

K. G. Nikiforov, A. G. Gurevich, S. I. Radautsan, L. M. Emiryan, and V. E. Tezlevan
*Effect of Nonstochiometry on Ferromagnetic Resonance in $CdCr_2S_4$ Crystals
(a) Vol. **72**, No. 1, p. K37–K39 (1982).
See also 10.2, p. 111; 22.8, p. 487.

H. Rosenberger, G. Scheler, and Yu. N. Moskvich
*^1H Chemical Shifts in Polycrystalline $RbHSeO_4$ and NH_4HSeO_4
(a) Vol. **72**, No. 1, p. K49–K51 (1982).
See also 22.8, p. 487.

E. Yu. Brailovskii, I. G. Megela, N. H. Pambuchchyan, and V. V. Teslenko
*EPR Study of Electron-Irradiated InP:Fe
(a) Vol. **72**, No. 1, p. K109–K111 (1982).
See also *11*, p. *124*; 22.2.2, p. 425.

M. Korkmaz, Ö. Korkmaz, and S. Kospançali
*EPR of Copper-Doped $(NH_4)_2C_2O_4 \cdot H_2O$ Single Crystals
(b) Vol. **112**, No. 2, p. 423–427 (1982).
See also *13.4*, p. *172*; 22.8.1, p. 496.

H. Opyrchał, K. D. Nierzewski, and B. Macalik
Effect of γ-Irradiation on Eu^{2+} Ions in KCl Crystals
(b) Vol. **112**, No. 2, p. 429–434 (1982).
See also *10.2*, p. *111*; 11, p. 124; 20.1, p. 309; 20.3, p. 324; 22.5.2, p. 459.

G. G. Fedoruk and I. Z. Rutkovskii
*Free Induction of E_1'-Centres in Crystalline Quartz
(b) Vol. **112**, No. 2, p. 453–456 (1982).
See also 13.4, p. 172; 22.6, p. 472.

M. Wójcik, H. Le Gall, and J. M. Desvignes
*Two Type of Tetrahedral Sites in the Mixed {Ca_2Na} [Mg_2] (V_3)O_{12} Garnet
(b) Vol. **112**, No. 2, p. 483–488 (1982).
See also 22.8.2, p. 501.

J. Kreissl and W. Gehlhoff
*EPR Investigation of Manganese–Boron Pairs in Silicon
(b) Vol. **112**, No. 2, p. 695–704 (1982).
See also *13.4*, p. *172*; 22.1.2, p. 410.

W. van Puymbroeck, N. Schrijvers, A. Bouwen, and D. Schoemaker
*Electron Spin Resonance Study of H_D-Type $BrCl^-$ Defects in KCl Doped with Divalent Cation Impurities
(b) Vol. **112**, No. 2, p. 725–733 (1982).
See also *10.2*, p. *112*; 22.5.2, p. 459.

R. Göring, K. Kneipp, and H. Nass
Investigation of Glasses in the BaO–B_2O_3–GeO_2 System with Low BaO Content by NMR and Raman Spectroscopy
(a) Vol. **72**, No. 2, p. 623–630 (1982).
See also *2*, p. *51*; 6, p. 78; 20.1, p. 309; 22.6, p. 473.

S. Jerzak, S. Waplak, and L. A. Shuvalov
Influence of an External Electric Field and Uniaxial Stress on the Domain Structure of $RbH_3(SeO_3)_2$ Crystals Studied by EPR Spectra of the Cr^{3+} Ion
(a) Vol. **72**, No. 2, p. 783–787 (1982).
See also *14.4.1*, p. *242*; 22.8.1, p. 496.

E. Burzo, D. Seitabla, and M. Chipară
Magnetic Properties and Resonance Study of $Gd(Co_xAl_{1-x})_3$ Compounds
(b) Vol. **113**, No. 1, p. 87–92 (1982).
See also *18.1*, p. *266*; *18.3*, p. *277*; 21, p. 338; 21.1, p. 352; 21.4, p. 371.

B. Henderson and A. H. Garrison
Ion Implantation in Alkaline Earth Oxides
(b) Vol. **113**, No. 1, p. 267–276 (1982).
See also *11*, p. *125*; 20.1, p. 310; 20.3, p. 325; 22.6, p. 473.

T. J. Bastow
Temperature and Pressure Dependence of the Electric Field Gradient at the Br Sites in $PbBr_2$ and $SnBr_2$
(b) Vol. **113**, No. 1, p. 361–365 (1982).
See also *13.4*, p. *174*; 22.5, p. 452.

A. S. Litovchenko, A. V. Brodovoi, and A. A. Melnikov
*Study of the Temperature Dependences of Magnetic Susceptibilities and ^7Li NMR Spectra of Ferriferrous Micas
(a) Vol. **73**, No. 1, p. K79–K82 (1982).
See also 18.1, p. 266; 22.8.1, p. 497.

M. Maryško, Z. Frait, L. Půst, M. Polcarová, J. Brádler, and P. Šumšál
*FMR in Misoriented (111) Garnet Films
(a) Vol. **74**, No. 1, p. K49–K53 (1982).
See also *18.3*, p. *278*; 22.8.2, p. 502.

J. Pirnat, J. Lužnik, and Z. Trontelj
*Temperature Dependence of ^{81}Br NQR in Rb_3ZnBr_5
(a) Vol. **74**, No. 1, p. K55–K57 (1982).
See also 22.8, p. 490.

H. Bross
Conduction Spin Density and Knight Shift of Copper
(b) Vol. **114**, No. 2, p. 529–535 (1982).
See also *13.2*, p. *157*; 21.1, p. 357.

I. Bakonyi, I. Kovács, and I. Pócsik
*On the Field-Dependent Broadening of NMR Lines in Paramagnets
(b) Vol. **114**, No. 2, p. 609–614 (1982).

G. B. Demidovich, S. N. Karyagin, and V. F. Kiselev
Spin Centres on Clean Germanium and Silicon Surfaces and Their Interaction with Adsorbed Molecules
(b) Vol. **114**, No. 2, p. 705–713 (1982).
See also *13.3*, p. *161*; 22.1.1, p. 398; 22.1.2, p. 415.

N. Ishii, S. Oozora, M. Kumeda, and T. Shimizu
*Calculations of g-Value and Linewidth of the ESR Signal in Amorphous Silicon Nitride
(b) Vol. **114**, No. 2, p. K111–K114 (1982).
See also *2*, p. *55*; 22, p. 392.

V. M. Polyanovskii
*The Influence of Finite Conduction Bandwidth on Gigantic Acoustomagnetic Oscillations in Semiconductors
(b) Vol. **114**, No. 2, p. K115–K117 (1982).
See also *7*, p. *83*.

V. K. Jain and S. K. Yadav
*EPR of Mn^{2+} in $K_2Zn(XO_4)_2 \cdot 6\,H_2O$ (X = S, Se) Single Crystals
(b) Vol. **114**, No. 2, p. K131–K134 (1982).
See also 22.8, p. 491.

L. Půst and Z. Frait
*FMR and FMAR Linewidths of Fe–3 wt% Si Single Crystals between 3.5 and 300 K
(b) Vol. **114**, No. 2, p. K157–K160 (1982).
See also 21.1.1, p. 364.

L. Komorowski, A. Chyla, and R. Kowal
Phase Transitions in n-Alkyltriphenylphosphonium Tetracyanoquinodimethanides
(a) Vol. **74**, No. 2, p. 453–457 (1982).
See also *1.2*, p. *23*; 8, p. 90; 14.3, p. 212; 22.9, p. 508.

20. Optical Properties

Review Articles

V. V. Nemoshkalenko, V. N. Antonov, Vl. N. Antonov, W. John, H. Wonn, and P. Ziesche
*Electronic Structure and Soft X-Ray Emission Spectra of 5d Transition Metals
(b) Vol. **111**, No. 1, p. 11–52 (1982).
Erratum
(b) Vol. **114**, No. 2, p. 741 (1982).
See also *13.1*, p. *148*; **21**, p. 330; 21.1, p. 343; 21.6, p. 374.

F. Bechstedt
Electronic Relaxation Effects in the Core Level Spectra of Solids
(b) Vol. **112**, No. 1, p. 9–49 (1982).
See also 4, p. 61; *13.1*, p. *148*; 17, p. 255; 21, p. 330; 22, p. 382; 23, p. 509.

Original Papers and Short Notes

A. Ślebarski and J. Auleytner
*Soft X-Ray Emission Spectra of Aluminium and Dysprosium from $DyCo_2$–$DyAl_2$ Intermetallic Compounds
(b) Vol. **109**, No. 1, p. 125–129 (1982).
See also 21, p. 330; 21.1, p. 343; 21.4, p. 369.

V. I. Litvinov, K. A. Valuev, and K. D. Tovtsyuk
Redistribution of Impurities in the Field of a Strong Electromagnetic Waves
(b) Vol. **109**, No. 1, p. 180 104 (1082).
See also *13.4*, p. *162*; 14.3, p. 202; 22, p. 382.

J. Metzdorf and F. R. Kessler
*Quantitative Analysis of the Nonresonant Interband Faraday Rotation in Ge and Si
(b) Vol. **109**, No. 1, p. 283–294 (1982).
See also *18*, p. *259*; 22.1.1, p. 395; 22.1.2, p. 400.

H. Sørum, O. M. Weng, and J. Bremer
*The CrK Series X-Ray Emission Spectra
(b) Vol. **109**, No. 1, p. 335–342 (1982).
See also 13.1, p. 148; 21.1, p. 343.

W. Radloff, E. Below, H. Wagner, and P. Kleinert
*Laser Absorption Measurement of OH Ion Distribution Profiles in Silica Glasses
(a) Vol. **69**, No. 1, p. K21–K24 (1982).
See also *2*, p. *46*; 22.6.1, p. 477.

E. Gaubas, K. Jarašiūnas, and J. Vaitkus
*Light Induced Transient Grating Decay in Si and Some $A^{II}B^{VI}$ Compounds
(a) Vol. **69**, No. 1, p. K87–K90 (1982).
See also 22.4.1, p. 439; 22.4.2, p. 442; 22.4.3, p. 445.

I. I. Gegusin, V. N. Datsyuk, and R. V. Vedrinskii
*Electron Structure and X-Ray Absorption Spectra of Potassium Chloride (I): Band Structure Approach
(b) Vol. **109**, No. 2, p. 563–572 (1982).
See also *13.1*, p. *149*; 22.5.2, p. 455.

E. I. SHTYRKOV, V. S. LOBKOV, S. A. MOISEEV, and N. G. YARMUKHAMETOV
*Forced Deformation of the Reversed Three-Pulsed Photon-Echo Signal on the Ruby R_1-Line
(a) Vol. 70, No. 2, p. K93–K95 (1982).
See also 22.6, p. 471.

V. KUMAR, A. R. CHETAL, and K. S. SRIVASTAVA
*Calculation of the Density of Compounds from the Chemical Shift of the X-Ray Absorption Edge
(a) Vol. 70, No. 2, p. K107–K111 (1982).
See also *13*, p. *142*; 21.1, p. 348; 22, p. 385.

M. M. KOTLYARSKII and E. A. POPOV
*Magnetic Phase Diagram of $NaMnCl_3$
(b) Vol. 111, No. 1, p. K13–K16 (1982).
See also *18.4*, p. *280*; 22.8, p. 484.

R. V. VEDRINSKII, I. I. GEGUSIN, V. N. DATSYUK, A. A. NOVAKOVICH, and V. L. KRAIZMAN
*Electron Structure and X-Ray Absorption Spectra of Potassium Chloride (II): Effect of a Core Hole in the Cluster Approach
(b) Vol. 111, No. 2, p. 433–442 (1982).
See also *13.1*, p. *151*; 22.5.2, p. 457.

I. M. DYKMAN and L. A. PROKHNITSKII
Influence of Electron Heating on Reflection and Attenuation Coefficients of Electromagnetic Waves in Semiconductors with Superlattice
(b) Vol. 111, No. 2, p. 449–454 (1982).
See also *13*, p. *143*; *14.3*, p. *206*; 22, p. 387.

I. KAMOHARA and Y. H. OHTSUKI
*Light Scattering by the Surface Wake
(b) Vol. 111, No. 2, p. 619–624 (1982).
See also *13.3*, p. *159*.

X.-Z. DONG, B. GRÖGER, T. JENDRYSIK, and H. KRONMÜLLER
*Annealing Effect of Domain Patterns of the Nearly Non-Magnetostrictive Amorphous Alloy $Co_{58}Ni_{10}Fe_5Si_{11}B_{16}$
(a) Vol. 71, No. 2, p. 441–449 (1982).
See also 2, p. 49; *18*, p. *261*; 22.8, p. 485.

R. S. BRAZIS and A. S. MIRONAS
Investigation of Electric Conductivity and Avalanche Breakdown in InSb by Means of Nonlinear Magnetoplasma Resonance
(a) Vol. 71, No. 2, p. 473–481 (1982).
See also *14.3*, p. *207*; 14.3.4, p. 229; 18, p. 261; 22.2.3, p. 428.

S. GHOSH and V. K. AGRAWAL
*Effect of Magnetostatic Field on the Stimulated Raman Scattering of Transversely Polarized Electromagnetic Waves and Raman Instability in Semiconductor Plasmas
(b) Vol. 112, No. 1, p. 119–131 (1982).
See also 6, p. 76; *13*, p. *144*; 22.2.3, p. 428.

B. Kuhlow
*Magnetic Ordering in $CrCl_3$ at the Phase Transition
(a) Vol. **72**, No. 1, p. 161–168 (1982).
See also 18.2, p. 271; *18.4*, p. *280*; 22.5, p. 451.

M. Wurlitzer
*Photomagnetic Effect in YIG(Fe^{2+})
(a) Vol. **72**, No. 1, p. 169–176 (1982).
See also *18.3*, p. *276*; 22.8.2, p. 500.

A. N. Gusatinskii, G. I. Alperovich, and A. V. Soldatov
X-Ray Spectroscopic Investigation and Calculation of the Energy Band Structure of Rare Earth Monophosphide and Monosulphide
(b) Vol. **112**, No. 2, p. 599–607 (1982).
See also *13.1*, p. *153*; 22, p. 388.

O. P. Burmistrova and V. A. Kosobukin
Radiative Decay of Surface Collective Excitations in Polarizable Cylinders
(b) Vol. **112**, No. 2, p. 675–683 (1982).
See also 6, p. 77; *13.3*, p. *160*; *13.5.2*, p. *187*.

Yu. E. Perlin, A. A. Kaminskii, M. G. Blazha, and V. N. Enakii
*Multiphonon Nonradiative Transitions of TR^{3+} Ions in $Y_3Al_5O_{12}$ and $Lu_3Al_5O_{12}$ Garnets
(b) Vol. **112**, No. 2, p. K125–K130 (1982).
See also 6, p. 77; *13.4*, p. *172*; 22.8.2, p. 501.

A. G. Revesz and S. H. Wemple
The Optical Properties of Noncrystalline Silicon and $Si_{1-x}H_x$ Films
(a) Vol. **72**, No. 2, p. 721–729 (1982).
See also *1.5*, p. *40*; 2, p. 51; *4*, p. *63*; 22.1.2, p. 411.

H. Sørum
*Intensity Distributions of the K Emission Spectra of Iron
(b) Vol. **113**, No. 1, p. 197–202 (1982).
See also 13.1, p. 154; 21.1.1, p. 362.

T. K. Boletskaya, B. E. Egorushkin, and V. P. Fadin
Electron Structure and Optical Conductivity of NiTi
(b) Vol. **113**, No. 1, p. 307–311 (1982).
See also *13.1*, p. *154*; 14.1, p. 196; 21.1, p. 353.

G. Dräger, F. Werfel, and O. Brümmer
*Polarized OKα Valence Band Spectra of TiO_2 (Rutile)
(b) Vol. **113**, No. 1, p. K15–K17 (1982).
See also *13.1*, p. *154*; 22.6, p. 473.

V. V. Hizhnyakov and A. V. Sherman
Self-Trapping of Polaritons
(b) Vol. **113**, No. 2, p. 471–479 (1982).
See also *13.5*, p. *179*; *13.5.1*, p. *185*; 22.5.2, p. 461.

T. A. Sasaki, T. Soga, and H. Adachi
Electronic Structure of Molybdenum Dioxide Calculated by the Xα Method
(b) Vol. **113**, No. 2, p. 647 – 655 (1982).
See also *4*, p. *64*; *13*, p. *146*; 22.6, p. 474.

B. Jagannatha Rao and A. R. Chetal
*Study of CuK Absorption Edge in Cu(I) Compounds
(b) Vol. **113**, No. 2, p. 727 – 730 (1982).
See also *13.1*, p. *155*; 22.5, p. 452.

N. R. Lokhande
*A Study of EXAFS Associated with the K Absorption Discontinuity of Nickel in Pure Metals and in Ni–S Amorphous Alloys
(b) Vol. **114**, No. 1, p. K35 – K39 (1982).
See also 2, p. 54; 21.1, p. 356.

Yu. F. Solomonov and V. K. Subashiev
*The Study of the Third Order Nonlinear Susceptibility in GaSe
(a) Vol. **74**, No. 1, p. 75 – 78 (1982).
See also 22.3, p. 434.

A. A. Dakhel
*X-Ray Study of Nd–Si and Tb–Si Compounds
(a) Vol. **74**, No. 1, p. K73 – K77 (1982).
See also 21.4, p. 372; 22.1.2, p. 415.

20.1 Spectra. Optical Constants (X-Ray Spectra See 20)

Review Article

Original Papers and Short Notes

J. P. LASCARYA, J. DIOURI, J. P. NADAI, and J. FERRE
*Infrared Absorption Spectra of Eu-Rich EuO. Excitations of the Bound Magnetic Polaron
(b) Vol. **109**, No. 1, p. 417–422 (1982).
See also 13.5.3, p. 188; 22.6, p. 468.

P. P. REDDY and Y. P. REDDY
*Absorption Spectrum of Fe^{3+} in Strontium Tartrate Tetrahydrate Single Crystals
(b) Vol. **109**, No. 1, p. K27–K29 (1982).
See also 22.9, p. 503.

I. V. BODNAR, A. P. BOLOGA, and B. V. KORZUN
*Composition Dependence of the Band Gap of $CuGa_xIn_{1-x}Se_2$ Solid Solutions
(b) Vol. **109**, No. 1, p. K31–K33 (1982).
See also *13.1*, p. *148*; 22.8, p. 481.

G. JEANDEL, A. WARIN, G. VILLERMAIN-LECOLIER et G. MORLOT
*Absorption induite dans l'infrarouge lointain par l'ion Y^{3+} dans BaF_2
(Long-Wavelength Infrared Absorption Due to Y^{3+} Ion in BaF_2)
(b) Vol. **109**, No. 1, p. K39–K40 (1982).
See also *6*, p. *70*; 22.5.3, p. 465.

P. J. ALONSO, V. M. ORERA, F. PALACIO, and R. ALCALA
*Optical and EPR Studies of NH_4MnF_3 above 80 K
(b) Vol. **109**, No. 1, p. K81–K83 (1982).
See also *19*, p. *283*; 22.5, p. 450.

A. P. GALUSHKA, G. E. DAVIDYUK, and N. S. BOGDANYUK
Interaction of 1.2 MeV Electron Irradiation Produced Defects with Cu Impurity Atoms in CdS Single Crystals
(a) Vol. **69**, No. 1, p. 71–76 (1982).
See also *11*, p. *118*; *14.3*, p. *202*; 16, p. 249; 20.3, p. 318; 22.4.1, p. 439.

N. SAITO and T. YAMAGUCHI
*Optical and Compositional Properties of Amorphous $Si_xC_yF_z$ Films Prepared by Reactive Sputtering
(a) Vol. **69**, No. 1, p. 133–138 (1982).
See also 1.3, p. 25; *2*, p. *45*; 22.1.2, p. 400.

N. R. S. TAIT and D. W. L. TOLFREE
*Optical Absorption Measurements on Carbon Foils
(a) Vol. **69**, No. 1, p. 201–206 (1982).
See also 22.1, p. 393.

J. SZCZYRBOWSKI, A. DIETRICH, and H. HOFFMANN
*Optical Properties of RF-Sputtered Indium Oxide Films
(a) Vol. **69**, No. 1, p. 217–226 (1982).
See also 14.3.1, p. 213; 22.6, p. 468.

V. A. TREPAKOV, L. N. AMITIN†, A. I. KUZNETSOV, and A. L. KHOLKIN
*Vacuum Ultraviolet Reflection Spectra of Pb-Containing ABO_3 Perovskites. Application to Electrooptic Properties
(a) Vol. **69**, No. 1, p. 425–429 (1982).
See also *13.1*, p. *149*; 22.8.1, p. 492.

NGUYEN BA AN, NGUYEN VAN HIEU, NGUYEN TOAN THANG, and NGUYEN AI VIET
*On the Role of the Coulomb Interaction of Charge Carriers in the Electronic Raman Scattering on Donor Levels in Direct Band Gap Semiconductors
(b) Vol. **109**, No. 2, p. 463–472 (1982).
See also *13.4*, p. *164*; 13.5.1, p. 181; 22, p. 383.

M. YOSHIDA, N. OHNO, K. NAKAMURA, and Y. NAKAI
Exciton Spectra of Orthorhombic Indium Bromide
(b) Vol. **109**, No. 2, p. 503–509 (1982).
See also *13.5.1*, p. *181*; 22.5, p. 450.

R. MACH, P. FLÖGEL, L. G. SUSLINA, A. G. ARESHKIN, J. MAEGE, and G. VOIGT
The Influence of Compositional Disorder on Electrical and Optical Properties of ZnS_xSe_{1-x} Single Crystals
(b) Vol. **109**, No. 2, p. 607–615 (1982).
See also 13.5.1, p. 181; *14.3*, p. *203*; *20.3*, p. *319*; 22.4.4, p. 447.

T. SKETTRUP
*Experimental Check of Different Additional Boundary Conditions in Case of Spatial Dispersion
(b) Vol. **109**, No. 2, p. 663–672 (1982).
See also *13.5.1*, p. *181*; 22.4, p. 437.

H. T. LANGHAMMER, M. STORDEUR, H. SOBOTTA, and V. RIEDE
*Optical and Electrical Investigations of the Anisotropy of Sb_2Te_3 Single Crystals
(b) Vol. **109**, No. 2, p. 673–681 (1982).
See also *14.3*, p. *203*; 15, p. 245; 22.7, p. 479.

E. BARTA
*On the Optically Determined Relaxation Time of Heavily Doped p-Type Silicon
(b) Vol. **109**, No. 2, p. K85–K88 (1982).
See also *14.3*, p. *203*; 22.1.2, p. 401.

V. D. KREVCHIK and A. YA. YAFASOV
*Two Photon Interband Absorption in Quantized Semiconductor Films with Participation of Deep Impurity Centres
(b) Vol. **109**, No. 2, p. K97–K101 (1982).
See also 13.1, p. 149; *13.4*, p. *164*; 22, p. 384.

T. G. KERIMOVA, SH. S. MAMEDOV, and E. YU. SALAEV
*Special Features of the Electronic Spectrum of $CdGa_2S_4$ in the Region of the Fundamental Absorption Edge
(b) Vol. **109**, No. 2, p. K119–K122 (1982).
See also *13.1*, p. *149*; 22.8, p. 482.

F. HENNEBERGER and V. MAY
*Non-Linear Optical Response and Optical Bistability Due to Excitonic Molecules
(b) Vol. **109**, No. 2, p. K139–K143 (1982).
See also *13.5.1*, p. *181*; 22.5, p. 450.

B. S. SUNDERSHESHU and T. KENDELEWICZ
*Temperature Dependence of the Absorption Edge of $Cd_{1-x}Mn_xTe$ Mixed Crystals
(a) Vol. **69**, No. 2, p. 467–474 (1982).
See also 13.1, p. 149; 22.4.4, p. 447.

A. N. GEORGOBIANI, S. I. RADAUTSAN, and I. M. TIGINYANU
Elcctroabsorption and Non-Equilibrium Carrier Recombination in $CdGa_2S_4$ Single Crystals
(a) Vol. **69**, No. 2, p. 513–520 (1982).
See also *13.4*, p. *165*; 16, p. 250; *20.3*, p. *320*; 22.8, p. 483.

H. R. PHILIPP
*$AuAl_2$: Optical Properties and Consideration as a Transparent Electrode Material
(a) Vol. **69**, No. 2, p. 547–551 (1982).
See also 14.1, p. 193; 21, p. 331; 21.6, p. 374.

E. GLASER, G. GÖTZ, N. SOBOLEV, and W. WESCH
Investigations of Radiation Damage Production in Ion Implanted Silicon
(a) Vol. **69**, No. 2, p. 603–614 (1982).
See also *11*, p. *119*; 14.3, p. 204; 19, p. 284; 22.1.2, p. 402.

YU. S. BULYSHEV, I. M. KASHIRSKII, V. V. SINITSKII, G. F. MYATCHINA, T. G. ERMAKOVA, and V. A. LOPYREV
Photoinjection of Charge Carriers in Poly-1-Vinyl-1,2,4-Triazole
(a) Vol. **69**, No. 2, p. 637–641 (1982).
See also *14.3.3*, p. *220*; 16, p. 250; 22.9, p. 503.

B. S. ACHARYA and M. L. MUKHERJEE
Role of Tin Centres on the Thermoluminescence of KCl:$SnCl_2$ Single Crystals
(a) Vol. **69**, No. 2, p. 649–655 (1982).
See also *10.2*, p. *105*; *20.3*, p. *320*; 22.5.2, p. 455.

M. ALDISSI, M. ROLLAND, and F. SCHUE
*Middle Infrared Study of the Cis Trans Isomerisation of $(CH)_x$ Induced by Oxidation
(a) Vol. **69**, No. 2, p. 733–737 (1982).
See also 14.3, p. 204; 22.9, p. 503.

A. H. ABOU EL ELA and S. MAHMOUD
*Optical Absorption of Thin Zinc Films
(a) Vol. **69**, No. 2, p. K193–K197 (1982).
See also 13.1, p. 149; 21.1, p. 345.

A. GRAJA, R. SWIETLIK, J. PETZELT, and L. DOBIÁŠOVÁ
Phase Transition in $MTPP(TCNQ)_2$ Single Crystals by Infrared Reflection Measurements
(a) Vol. **69**, No. 2, p. K205–K208 (1982).
See also *1.2*, p. *15*; 6, p. 71; 22.9, p. 503.

S. A. KULAGIN and I. S. OSADKO
Relaxation and Optical Processes in Impurity Centres with Two-Well Adiabatic Potential
(b) Vol. **110**, No. 1, p. 57–68 (1982).
See also *13.4*, p. *165*; 20.3, p. 320.

O. GOEDE and W. HEIMBRODT
Isoelectronic Impurity Te in $CdS_{1-x}Se_x$ Mixed Crystals
(b) Vol. **110**, No. 1, p. 175–182 (1982).
See also *13.5.1*, p. *182*; *20.3*, p. *320*; 22.4.4, p. 447.

K. N. Reddy, M. L. Rao, and V. H. Babu
Thermoluminescence and Optical Absorption Studies of Z_1-Centres in NaCl Crystals Doped with Samarium
(a) Vol. **70**, No. 1, p. 335–341 (1982).
See also *10.2*, p. *106*; *20.3*, p. *321*; 22.5.2, p. 456.

N. R. Ivanov, J. Kroupa, J. Fousek, and B. Březina
*Birefringence of $PbHPO_4$-Family Ferroelectrics
(a) Vol. **70**, No. 1, p. K85–K87 (1982).
See also *14.4.1*, p. *241*; 22.8.1, p. 494.

S. Bahn
Luminescence and Optical Investigation of NaF:C^{3+} Single Crystals
(b) Vol. **110**, No. 2, p. 399–405 (1982).
See also *10.2*, p. *107*; *20.3*, p. *321*; 22.5.2, p. 456.

B. Z. Malkin, A. A. Kaminskii, N. R. Agamalyan, L. A. Bumagina, and T. I. Butaeva
Spectra of Rare-Earth Ions in the Crystal Fields of Double Tungstates and Molybdates
(b) Vol. **110**, No. 2, p. 417–422 (1982).
See also *13.4*, p. *166*; 20.3, p. 321; 22.8.1, p. 494.

L. Merten and A. Lehmeyer
*Reflectivity of Uniaxial Antiferromagnetic and Ferrimagnetic Crystals
(b) Vol. **110**, No. 2, p. 423–430 (1982).
See also *13.5.4*, p. *189*; 18, p. 260.

Fat Duen Ho
The F^+ Center in Lithium Tantalate
(b) Vol. **110**, No. 2, p. 437–443 (1982).
See also *10.2*, p. *107*; *13.4*, p. *166*; 22.8.1, p. 494.

W. Tews and R. Zimmermann
*Electron-Energy-Loss Spectroscopy of Different Al_2O_3 Modifications (II): Lineshape–Analysis of Energy Derived Dielectric Function
(b) Vol. **110**, No. 2, p. 479–485 (1982).
See also 13.5.1, p. 182; *14.4*, p. *235*; 22.6, p. 470.

F. Brehat, B. Wyncke, and M. El-Sherif
*Lattice Mode Anharmonicity in Sodium Nitrite
(b) Vol. **110**, No. 2, p. 499–506 (1982).
See also 1.2, p. 16; *6*, p. *73*; 14.4.1, p. 241; 22.8.1, p. 494.

I. A. Parfianovich and V. N. Salomatov
The F_A(II) Center Relaxed Excited State and Models of F_A(II)-Like Defects in Alkali Halide Crystals
(b) Vol. **110**, No. 2, p. 513–520 (1982).
See also *10.2*, p. *107*; 22.5.2, p. 456.

S. Bednarek and U. Rössler
*Method of Invariants Applied to Indirect Gap Absorption
(b) Vol. **110**, No. 2, p. 565–570 (1982).
See also 6, p. 73; 22.1.2, p. 404; 22.2.2, p. 424.

YU. V. KRYUCHENKO and V. I. SUGAKOV
*Electrodynamics of a Spatially Dispersive Medium with Spherical Cavity
(b) Vol. **111**, No. 1, p. 177–185 (1982).
See also *13.5.1*, p. *183*; 22.4.1, p. 440.

R. BALTRAMIEYUNAS and E. KUOKSHTIS
Collective Phenomena in Edge Emission of ZnTe and ZnSe Crystals
(b) Vol. **111**, No. 1, p. 187–193 (1982).
See also 13.5.1, p. 183; *20.3*, p. *322*; 22.4.2, p. 443; 22.4.3, p. 446.

D. FRÖHLICH and R. KENKLIES
*Polarization Dependence of Two-Photon Magnetoabsorption of the 1S Exciton in Cu_2O
(b) Vol. **111**, No. 1, p. 247–253 (1982).
See also *13.5.1*, p. *183*; 22.6, p. 471.

I. YA. FUGOL, O. N. GRIGORASHCHENKO, and E. V. SAVCHENKO
Polariton and Spatial Dispersion Effects in Exciton Spectra of Xenon Cryocrystals
(b) Vol. **111**, No. 1, p. 397–406 (1982).
See also *6*, p. *74*; *13.5.1*, p. *183*; 20.3, p. 322; 23, p. 509.

R. J. BARTLETT and C. G. OLSON
*Absorption Spectrum of Ni from 40 to 220 eV
(b) Vol. **111**, No. 1, p. K33–K36 (1982).
See also 13.1, p. 150; 21.1, p. 348.

E. E. MATYAS and A. G. KAROZA
*Plasma Edge Reflectance Measurements in InAs–AlAs Alloys
(b) Vol. **111**, No. 1, p. K45–K48 (1982).
See also *14.3*, p. *206*; 22.2.4, p. 430.

A. A. BEREZIN and F. L. WEICHMAN
Cuprous Oxide–Copper Photovoltaic Cells Formed by High-Temperature Reduction
(a) Vol. **71**, No. 1, p. 265–274 (1982).
See also *16*, p. *251*; 22.6, p. 471.

B. G. TAGIEV, E. F. BAGIRZADE, and G. M. MAMEDOV
*Interference Oscillations in the Temperature Dependence of Photocurrent and Transmission of GaSe, GaTe, and $GaSe_xTe_{1-x}$ Single Crystals
(a) Vol. **71**, No. 1, p. K49–K51 (1982).
See also *16*, p. *251*; 22.3, p. 433; 22.3.1, p. 436.

V. A. KISELEV
*Pseudo-Exciton Effect in the Schottky-Barrier Electroreflectance
(b) Vol. **111**, No. 2, p. 461–468 (1982).
See also 13.5.1, p. 183; *14.3.4*, p. *229*; 22.2.1, p. 421.

T. IKARI, S. SHIGETOMI, and K. HASHIMOTO
*Crystal Structure and Raman Spectra of InSe
(b) Vol. **111**, No. 2, p. 477–481 (1982).
See also 1.3, p. 28; *6*, p. *74*; 22.3, p. 433.

V. P. Izvekov, A. Watterich, T. Györgyi, and E. Pungor
*Infrared Properties of Doped and Irradiated NaCl Crystals
(b) Vol. **111**, No. 2, p. K153–K156 (1982).
See also *6*, p. *75*; 22.5.2, p. 458.

A. A. Kaminskii, A. G. Petrosyan, G. A. Denisenko, T. I. Butaeva, V. A. Fedorov, and S. E. Sarkisov
Spectroscopic Properties and 3 μm Stimulated Emission of Er^{3+} Ions in the $(Y_{1-x}Er_x)_3Al_5O_{12}$ and $(Lu_{1-x}Er_x)_3Al_5O_{12}$ Garnet Crystal Systems
(a) Vol. **71**, No. 2, p. 291–312 (1982).
See also *20.2*, p. *317*; *20.3*, p. *323*; 22.8, p. 485.

H. Arashi and M. Ishigame
Raman Spectroscopic Studies of the Polymorphism in ZrO_2 at High Pressures
(a) Vol. **71**, No. 2, p. 313–321 (1982).
See also *1.2*, p. *18*; 6, p. 75; 22.6, p. 472.

K. Yasuda, M. Takeda, H. Masuda, and A. Yoshida
Changes of the Band Gap of SiC Crystals by High Energy Electron Irradiation and Annealing
(a) Vol. **71**, No. 2, p. 549–553 (1982).
See also *11*, p. *122*; *13.1*, p. *151*; 22.3, p. 434.

M. G. Kaplunov, K. I. Pokhodniya, and Yu. G. Borodko
*Temperature Dependence of Reflection Spectra of Conductive TCNQ Salts
(a) Vol. **71**, No. 2, p. 649–656 (1982).
See also 6, p. 75; 14.3.4, p. 229; 22.9, p. 505.

T. G. Kerimova, F. R. Adyalova, A. Sh. Khidirov, and E. Yu. Salaev
*Optical Properties of α- and β-$CdIn_2Se_4$ in the Region of the Fundamental Absorption Edge
(a) Vol. **71**, No. 2, p. K211–K213 (1982).
See also 13.1, p. 151; 22.8, p. 486.

B. S. Umarov, V. S. Gorelik, M. M. Sushinskii, and J. F. Vetelino
*Raman Scattering from Nonfundamental States in Ammonium Chloride Crystals at Low Temperatures
(b) Vol. **112**, No. 1, p. 69–73 (1982).
See also *6*, p. *75*; 22.5, p. 451.

D. L. Sastry and K. V. S. Rama Rao
*ESR and Optical Absorption Studies of ClO_3 in $Ba(ClO_4)_2 \cdot 3\,H_2O$ Single Crystals
(b) Vol. **112**, No. 1, p. 133–142 (1982).
See also 11, p. 123; *19*, p. *288*; 22.8.1, p. 495.

F. W. Fluck, H. P. Geserich, and C. Politis
*Optical Investigations on HfC_xN_{1-x} Compounds
(b) Vol. **112**, No. 1, p. 193–200 (1982).
See also 22.8, p. 486.

M. Yamawaki and C. Hamaguchi
Effect of Electric Fields on Luminescence in GaAs
(b) Vol. **112**, No. 1, p. 201–210 (1982).
See also 13.5.1, p. 184; *20.3*, p. *323*; 22.2.1, p. 422.

J. Kikas and M. Rätsep
*Correlation of Transition Frequencies in Impurity Spectra of Inhomogeneous Solids
(b) Vol. **112**, No. 2, p. 409–415 (1982).
See also *10.2*, p. *111*; 22.5.2, p. 459.

H. Opyrchał, K. D. Nierzewski, and B. Macalik
Effect of γ-Irradiation on Eu^{2+} Ions in KCl Crystals
(b) Vol. **112**, No. 2, p. 429–434 (1982).
See also *10.2*, p. *111*; 11, p. 124; 19, p. 288; 20.3, p. 324; 22.5.2, p. 459.

K. Bange, A. Liebegall, and B. Lorentz
*Interband Magnetoreflectivity at $Pb_{1-x}Sn_xSe$ Single Crystals
(b) Vol. **112**, No. 2, p. 543–547 (1982).
See also 13.1, p. 153; *18*, p. *262*; 22.8, p. 487.

H. D. Lutz and G. Kliche
*Far-Infrared Reflection Spectra, Optical and Dielectric Constants, Effective Charges, and Lattice Dynamics of the Skutterudites CoP_3, $CoAs_3$, and $CoSb_3$
(b) Vol. **112**, No. 2, p. 549–557 (1982).
See also *6*, p. *77*; 13.5.2, p. 187; 22, p. 388.

B. Pfalzgraf, O. Roder, and J.-M. Spaeth
*Optical Investigation of Ag^+ Centers in Cesium Halides
(b) Vol. **112**, No. 2, p. 565–574 (1982).
See also *10.2*, p. *112*; 13.5.1, p. 184; 22.5.2, p. 459.

K. R. Allakhverdiev, M. M. Godzhaev, A. I. Nadzhafov, and R. M. Sardarly
Light Scattering Spectra of the Solid Solutions $TlGa_xIn_{1-x}Se_{2x}S_{2(1-x)}$
(b) Vol. **112**, No. 2, p. K93–K97 (1982).
See also *6*, p. *77*; 22.8, p. 488.

G. M. Shmelev, G. I. Tsurkan, and V. M. Polanovskii
*Absorption of Frequency-Modulated Electromagnetic Wave in Semiconductors
(b) Vol. **112**, No. 2, p. K121–K123 (1982).
See also 22, p. 388.

R. Göring, K. Kneipp, and H. Nass
Investigation of Glasses in the BaO–B_2O_3–GeO_2 System with Low BaO Content by NMR and Raman Spectroscopy
(a) Vol. **72**, No. 2, p. 623–630 (1982).
See also *2*, p. *51*; 6, p. 78; 19, p. 289; 22.6, p. 473.

L. K. Vodopyanov and S. P. Kozyrev
*Infrared Reflection Spectroscopy of Ion-Implanted n-$Hg_{0.8}Cd_{0.2}Te$
(a) Vol. **72**, No. 2, p. 737–744 (1982).
See also 11, p. 125; 14.3, p. 209; 22.4.4, p. 448.

V. J. Rao, W. A. Anderson, and G. Rajeswaran
The Role of Hydrogen in Improving Photovoltaic Response of Solar Cells on Polycrystalline Silicon
(a) Vol. **72**, No. 2, p. 745–749 (1982).
See also *14.3.3*, p. *224*; *16*, p. *253*; 22.1.2, p. 411.

K. R. ALLAKHVERDIEV, T. G. MAMEDOV, E. YU. SALAEV, and I. K. EFENDIEVA
*The Fundamental Absorption Edge of $TlInSe_2$
(b) Vol. **113**, No. 1, p. K43–K47 (1982).
See also *13.1*, p. *154*; 22.8, p. 488.

R. CASES, R. ALCALÁ, and V. M. ORERA
*F'-Centers in CsF
(b) Vol. **113**, No. 1, p. K61–K65 (1982).
See also *10.2*, p. *113*; 20.3, p. 325; 22.5.2, p. 460.

M. SEKITA and T. KOBAYASHI
*Oscillator Strengths of Transitions between Stark Levels of $Nd_xGd_{1-x}P_5O_{14}$
(a) Vol. **73**, No. 1, p. 61–64 (1982).
See also 13.4, p. 174; 22.8.1, p. 496.

D. A. ARONOV and V. V. RUBINOV
*The Influence of Exclusion on Light Absorption by Free Carriers in "Pure"(almost Intrinsic) Semiconductors with Non-Injecting Contacts
(a) Vol. **73**, No. 1, p. K1–K5 (1982).
See also *14.3.4*, p. *231*; 22.1.1, p. 398.

A. I. BOBRYSHEVA, I. I. JERU, and S. A. MOSKALENKO
*The Isotopic Shift of the Excitonic Absorption Lines in Cu_2O Crystals
(b) Vol. **113**, No. 2, p. 439–445 (1982).
See also 6, p. 79; *13.5.1*, p. *184*; 22.6, p. 474.

N. P. GAVALESHKO, A. N. LYAKHOVICH, P. P. VATAMANYUK, and A. I. SAVCHUK
*Faraday Effect in III–VI Layer Compounds
(b) Vol. **113**, No. 2, p. 447–451 (1982).
See also 13.5.1, p. 185; *18*, p. *263*; 22.3, p. 434.

B. A. ORLOWSKI
The Initial and Final States for UPS Optical Transitions in $Cd_{0.2}Hg_{0.8}Te$ Crystal
(b) Vol. **113**, No. 2, p. 487–492 (1982).
See also *13.1*, p. *154*; 22.4.4, p. 448.

E. I. GRANCHAROVA, J. P. LASCARAY, J. DIOURI, and J. ALLÈGRE
*Comments on the Behaviour of the Absorption Edge of $Cd_{1-x}Mn_xTe$ for $x > 0.4$
(b) Vol. **113**, No. 2, p. 503–508 (1982).
See also 13.4, p. 174; 22.4.4, p. 448.

K. SCHMITT
*Absorption and Stress-Induced Linear Dichroism of Pb^{++} Centres in KCl and RbCl
(b) Vol. **113**, No. 2, p. 559–568 (1982).
See also 13.4, p. 175; 22.5.2, p. 461.

S. PERMOGOROV, A. REZNITSKII, S. VERBIN, G. O. MÜLLER, P. FLÖGEL, and M. NIKIFOROVA
Localized Excitons in $CdS_{1-x}Se_x$ Solid Solutions
(b) Vol. **113**, Vol. 2, p. 589–600 (1982).
See also *13.5.1*, p. *185*; *20.3*, p. *326*; 22.4.4, p. 448.

J. Kroupa, J. Fousek, F. Smutný, and B. Březina
*Birefringence and Electrooptic Effect in K_2SeO_4
(b) Vol. **113**, No. 2, p. K153–K157 (1982).
See also *14.4.1*, p. *243*; 22.8.1, p. 497.

M. M. El Oker, E. A. Abou-Saif, S. A. El Sahhar, and A. A. Mohamed
*Optical Properties and Structure of Thin Indium Films
(a) Vol. **73**, No. 2, p. 389–394 (1982).
See also *1.5*, p. *41*; 21, p. 340.

P. Bussemer and U. Dietrich
*Conditions for Extremal Values of the Optical Dielectric Constant in Binary Mixtures
(a) Vol. **73**, No. 2, p. 429–437 (1982).
See also *4*, p. *64*; 22, p. 390.

P. Veeresham, U. V. Subba Rao, and V. Hari Babu
Thermoluminescence of Z_1-Centres in $KCl_xBr_{1-x}:Ca^{2+}$ Crystals
(a) Vol. **73**, No. 2, p. 565–572 (1982).
See also *10.2*, p. *115*; *20.3*, p. *326*; 22.5.2, p. 461.

T. Kala
*Study of the Influence of Manganese and Lanthanum Oxides on the Band Structure of $Pb(Zr, Ti)O_3$
(a) Vol. **73**, No. 2, p. 573–578 (1982).
See also *13.1*, p. *155*; 22.8.1, p. 497.

A. J. Cohen and L. N. Makar
*Models for Color Centers in Smoky Quartz
(a) Vol. **73**, No. 2, p. 593–596 (1982).
See also *10.2*, p. *115*; 22.6, p. 475.

V. A. Zuikov and V. V. Samartsev
*Reverse Photon Echo as a Method of Investigation of Resonant Medium Parameters
(a) Vol. **73**, No. 2, p. 625–632 (1982).
See also 13.4, p. 175; *20.3*, p. *327*; 22.8.1, p. 497.

B. V. R. Chowdari, A. Rajaratnam, and Y. Ravi Sekhar
Radiation Damage Studies of Dichromate Doped Alums
(a) Vol. **73**, No. 2, p. K231–K235 (1982).
See also *11*, p. *127*; 19, p. 290; 22.8.1, p. 497.

A. Graja, E. Sopa, and L. Dobiášová
*Infrared Study of the Structural Phase Transition in a Ternary 1 D Salt $(TMA^+)(TCNQ)^{2/3-}(I_3^-)^{1/3}$
(a) Vol. **73**, No. 2, p. K237–K241 (1982).
See also *1.2*, p. *22*; 22.9, p. 507.

A. D. Avesov and E. M. Gasanov
*Some Radiation-Optical Properties of Fluorphlogopite with Admixture of Vanadium
(a) Vol. **73**, No. 2, p. K259–K261 (1982).
See also 11, p. 127; 22.8, p. 490.

J. Horák, P. Lošťák, L. Šiška, and M. Stordeur
*The Nature of Silver Impurity Atoms in Antimony Telluride
(b) Vol. **114**, No. 1, p. 39–45 (1982).
See also 13.1, p. 155; *14.3*, p. *210*; 15, p. 248; 22.7, p. 479.

V. L. Bonch-Bruevich and To Thi Bich Nguyet
*Low Temperature Conductivity and Optical Transitions of a Two-Dimensional Disordered Semiconductor under Weak Localization Conditions
(b) Vol. **114**, No. 1, p. 67–71 (1982).
See also 2, p. 53; *14.3*, p. *211*; 22, p. 391.

F. J. Lohmeier and F. Fischer
*Optical Properties of AsO_2^- Impurities in Alkali Halide Crystals
(b) Vol. **114**, No. 1, p. 95–102 (1982).
See also 13.4, p. 176; *20.3*, p. *327*; 22.5.2, p. 462.

C. Patel, W. F. Sherman, and G. R. Wilkinson
Lattice Dynamics of InP under Hydrostatic Pressures
(b) Vol. **114**, No. 1, p. 169–174 (1982).
See also *6*, p. *80*; 12.1, p. 137; 22.2.2, p. 426.

S. Mochizuki
*Infrared Optical Properties of Cerium Dioxide
(b) Vol. **114**, No. 1, p. 189–199 (1982).
See also *6*, p. *80*; 22.6, p. 475.

J. Baumann and K. Sieber
*Simple Approach to Temperature and Pressure Dependence of F-Centre Absorption Energies in Alkali Halides
(b) Vol. **114**, No. 1, p. K1–K4 (1982).
See also *10.2*, p. *116*; 13.4, p. 176; 22.5.2, p. 462.

P. I. Khadzhi, S. A. Moskalenko, A. H. Rotaru, and G. D. Shibarshina
*Reflection and Refraction of Wave Packets Forming Polariton Solitons in Crystals
(b) Vol. **114**, No. 1, p. K25–K29 (1982).
See also *13.5*, p. *179*.

W. Ulrici
*Photoinduced Transient Optical Absorption of Cr^{4+} in Semi-Insulating GaAs:Cr
(b) Vol. **114**, No. 1, p. K87–K90 (1982).
See also *13.4*, p. *176*; 22.2.1, p. 423.

R. Rauch and G. Schwotzer
Disturbed Colour Centres in Oxygen- and Alkali-Doped Alkaline Earth Fluoride Crystals after X-Ray Irradiation at 77 and 295 K
(a) Vol. **74**, No. 1, p. 123–132 (1982).
See also *10.2*, p. *116*; 22.5.3, p. 466.

M. N. Khan and E. E. Khawaja
*The Electrical and Optical Properties of Glasses of the Li_2O–GeO_2 and Na_2O–GeO_2 Systems
(a) Vol. **74**, No. 1, p. 273–277 (1982).
See also 2, p. 54; *14.3*, p. *211*; 22.6.1, p. 478.

C. DE BLASI, G. MICOCCI, A. RIZZO, and A. TEPORE
*Photoconductivity of InSe Single Crystals
(a) Vol. **74**, No. 1, p. 291–296 (1982).
See also *16*, p. *254*; 22.3, p. 434.

T. S. SHI, S. N. SAHU, G. S. OEHRLEIN, A. HIRAKI, and J. W. CORBETT
Models for the Hydrogen-Related Defect–Impurity Complexes and Si–H Infrared Bands in Crystalline Silicon
(a) Vol. **74**, No. 1, p. 329–341 (1982).
See also *4*, p. *65*; *10.2*, p. *116*; 11, p. 128; 22.1.2, p. 414.

G. M. RAMĀNS, J. V. GABRUSENOKS, and A. A. VEISPĀLS
*Structure of Tungstic Acids and Amorphous and Crystalline WO_3 Thin Films
(a) Vol. **74**, No. 1, p. K41–K44 (1982).
See also 2, p. 54; 6, p. 80; *8*, p. *90*; 22.6, p. 476; 22.8, p. 490.

A. TWARDOWSKI and J. GINTER
The Influence of Exchange Spin Splitting on Exciton in $Cd_{1-x}Mn_xTe$ Mixed Crystals
(b) Vol. **114**, No. 2, p. 331–336 (1982).
See also *13.5.1*, p. *185*; 18, p. 264; 22.4.4, p. 449.

AL. L. EFROS, L. M. KANSKAYA, S. I. KOKHANOVSKII, and R. P. SEISYAN
Diamagnetic Excitons and the Indium Antimonide Energy Band Parameters
(b) Vol. **114**, No. 2, p. 373–382 (1982).
See also *13.1*, p. *155*; 13.5.1, p. 185; 22.2.3, p. 429.

P. KLEINERT
Coherent-Potential Approximation for Vibrations in $Ga_{1-x}In_xSb$ and $Ga_{1-x}In_xAs$ Mixed Crystals
(b) Vol. **114**, No. 2, p. 459–466 (1982).
See also *6*, p. *81*; 22.2.4, p. 431.

A. KREISSL, K. BOHNERT, V. G. LYSSENKO, and C. KLINGSHIRN
*Experimental Investigation on the Complex Dielectric Function of CdS in the Exciton and in the Plasma Limit
(b) Vol. **114**, No. 2, p. 537–544 (1982).
See also *13.5*, p. *180*; 22.4.1, p. 441.

D. FRÖHLICH, K. REIMANN, and P. KOIDL
*Investigation of Mixed-Mode Polariton Dispersion in $AgGaS_2$
(b) Vol. **114**, No. 2, p. 553–559 (1982).
See also *13.5.1*, p. *185*; 22.8, p. 491.

A. SELMI, R. LE TOULLEC, and R. FAYMONVILLE
*Influence of the Spin Disorder on the Plasma Edge in n-Type $HgCr_2Se_4$
(b) Vol. **114**, No. 2, p. K97–K99 (1982).
See also *13.5.2*, p. *187*; 22.8, p. 491.

A. ANEDDA, E. FORTIN, F. LEDDA, and A. SERPI
*Photoelectrical Properties of Cr_2S_3
(b) Vol. **114**, No. 2, p. K143–K146 (1982).
See also *16*, p. *254*; 22, p. 392.

20.2 Lasers

Original Papers and Short Notes

E. P. Pokatilov and S. N. Klymin
Non-Linear Optical Properties of the Degenerate Band Charge Carrier Gas in the Magnetic Field
(b) Vol. **110**, No. 2, p. 709 – 716 (1982).
See also 13, p. 142; *20*, p. *294*.

A. A. Kaminskii, N. R. Agamalyan, G. A. Denisenko, S. E. Sarkisov, and P. P. Fedorov
*Spectroscopy and Laser Emission of Disordered GdF_3–CaF_2:Nd^{3+} Trigonal Crystals
(a) Vol. **70**, No. 2, p. 397 – 406 (1982).
See also 20.1, p. 304; *20.3*, p. *322*; 22.8, p. 483.

B. R. Sood
*Suitability of NaF as a Phaser Material
(a) Vol. **71**, No. 1, p. K117 – K119 (1982).
See also *6*, p. *74*; 10.2, p. 109; 22.5.2, p. 457.

A. A. Kaminskii, A. G. Petrosyan, G. A. Denisenko, T. I. Butaeva, V. A. Federov, and S. E. Sarkisov
*Spectroscopic Properties and 3 μm Stimulated Emission of Er^{3+} Ions in the $(Y_{1-x}Er_x)_3Al_5O_{12}$ and $(Lu_{1-x}Er_x)_3Al_5O_{12}$ Garnet Crystal Systems
(a) Vol. **71**, No. 2, p. 291 – 312 (1982).
See also 20.1, p. 307; *20.3*, p. *323*; 22.8, p. 485.

N. D. Vieira, Jr., I. M. Ranieri, and S. P. Morato
*Room Temperature Visible Laser Action of F Aggregated Centers in LiF:Mg, OH Crystals
(a) Vol. **73**, No. 1, p. K115 – K117 (1982).
See also 10.2, p. 114; 22.5.2, p. 461.

H. Endert and W. Melle
Laser-Induced Damage in KDP Crystals. The Influence of Growth Ghosts and Growth Bands
(a) Vol. **74**, No. 1, p. 141 – 148 (1982).
See also *1.4*, p. *37*; *11*, p. *127*; 22.8.1, p. 498.

20.3 Luminescence (X-Ray Spectra See 20)

Review Article

R. Mach and G. O. Müller
*Physical Concepts of High-Field, Thin-Film Electroluminescence Devices
(a) Vol. **69**, No. 1, p. 11–66 (1982).
See also 14.3.1, p. 213; *14.3.3*, p. *219*; 22.4.1, p. 439; 22.4.2, p. 442.

Original Papers and Short Notes

Zh. Egemberdiev, S. Zazubovich, and V. Nagirnyi
*Polarized Luminescence of Lead Ions Associated with Interstitial Radiation Defects in $KCl:PbCl_2$ Crystals
(b) Vol. **109**, No. 1, p. 173–182 (1982).
See also *10.2*, p. *103*; 22.5.2, p. 454.

U. Hahn, R. Haensel, and N. Schwentner
*Kr and Xe Guest Atoms in Ar Matrix-: Emission Spectra, Excitation Spectra, and Lifetimes
(b) Vol. **109**, No. 1, p. 233–244 (1982).
See also *13.4*, p. *162*; 20.1, p. 298; 23, p. 509.

A. A. Alybakov, N. Toichiev, and Sh. Akchalov
*Investigation of EPR and Thermoluminescence Glow-Curves of X-Irradiated $KCl.CrO_4^{2-}$, Ca^{2+} Crystals
(b) Vol. **109**, No. 1, p. 295–301 (1982).
See also *19*, p. *282*; 22.5.2, p. 454.

Y. V. G. S. Murti and N. Sucheta
*Thermoluminescence of Coloured Alkali Halides
(b) Vol. **109**, No. 1, p. 325–334 (1982).
See also 13.4, p. 163; 22.5.2, p. 454.

A. P. Galushka, G. E. Davidyuk, and N. S. Bogdanyuk
Interaction of 1.2 MeV Electron Irradiation Produced Defects with Cu Impurity Atoms in CdS Single Crystals
(a) Vol. **69**, No. 1, p. 71–76 (1982).
See also *11*, p. *118*; *14.3*, p. *202*; 16, p. 249; 20.1, p. 299; 22.4.1, p. 439.

G. E. Archangelskii, F. Karel, J. Mareš, S. Pačesová, and J. Pastrňák
*The Luminescence and EPR Spectra of Manganese Activated AlN
(a) Vol. **69**, No. 1, p. 173–183 (1982).
See also *19*, p. *283*; 22.2, p. 418.

A. Zehe, L. Hesse, and R. Reyes S.
*Analysis of Ghost Peak Formation in the Emission Spectra of GaAs p–n Junctions
(a) Vol. **69**, No. 1, p. 281–285 (1982).
See also *14.3.3*, p. *219*; 22.2.1, p. 420.

S. Bhan
*Thermoluminescence of NaF Harshaw Single Crystals Irradiated to Medium Doses
(a) Vol. **69**, No. 1, p. 367–376 (1982).
See also *10.2*, p. *104*; 22.5.2, p. 454.

K. D. Glinchuk, K. Lukat, and V. I. Vovnenko
Effect of Electron Irradiation on Recombination Characteristics of Deep Radiative Centres in GaAs
(a) Vol. **69**, No. 1, p. K43–K47 (1982).
See also *11*, p. *118*; *13.4*, p. *164*; 22.2.1, p. 420.

T. Taguchi and D. W. Palmer
*Channeling and Ionoluminescence Spectra of Cubic-Structured ZnS Crystals
(a) Vol. **69**, No. 1, p. K55–K59 (1982).
See also *1*, p. *9*; 22.4.1, p. 439.

G. Blasse
*On the Photoluminescence of $Ba_2NaNb_5O_{15}$
(a) Vol. **69**, No. 1, p. K83–K85 (1982).
See also 22.8, p. 482.

Zh. Egemberdiev, S. Zazubovich, V. Seeman, and V. Nagirnyi
*Luminescence Accompanying the Recombination of Electrons with V_F Centres in KCl Crystals
(b) Vol. **109**, No. 2, p. 473–482 (1982).
See also *10.2*, p. *104*; 19, p. 283; 22.5.2, p. 455.

S. Benci, F. Fermi, and R. Reverberi
*Symmetries of Luminescent Centres in NaCl:Pb^{2+} Excited in the A-Absorption Band
(b) Vol. **109**, No. 2, p. 543–550 (1982).
See also *10.2*, p. *104*; 22.5.2, p. 455.

E. Gutsche
Non-Condon Approximations and the Static Approach in the Theory of Non-Radiative Multiphonon Transitions
(b) Vol. **109**, No. 2, p. 583–597 (1982).
See also *6*, p. *71*; *13.4*, p. *164*; 22, p. 383.

K. Peuker, R. Enderlein, A. Schenk, and E. Gutsche
Theory of Non-Radiative Multiphonon Capture Processes. Solution of Old Controversies
(b) Vol. **109**, No. 2, p. 599–606 (1982).
See also *6*, p. *71*; *13.4*, p. *164*; 22, p. 383.

R. Mach, P. Flögel, L. G. Suslina, A. G. Areshkin, J. Maege, and G. Voigt
*The Influence of Compositional Disorder on Electrical and Optical Properties of ZnS_xSe_{1-x} Single Crystals
(b) Vol. **109**, No. 2, p. 607–615 (1982).
See also 13.5.1, p. 181; *14.3*, p. *203*; 20.1, p. 300; 22.4.4, p. 447.

E. Molva, J. P. Chamonal, and J. L. Pautrat
Shallow Acceptors in Cadmium Telluride
(b) Vol. **109**, No. 2, p. 635–644 (1982).
See also *13.4*, p. *164*; 22.4.3, p. 445.

P. KUKK, Õ. PALMRE, and E. MELLIKOV
The Structure of Recombination Centres in Activated ZnSe Phosphors
(a) Vol. **70**, No. 1, p. 35–42 (1982).
See also *10.2*, p. *106*; *13.4*, p. *166*; 22.4.2, p. 442.

L. N. SKUJA and A. R. SILIN
A Model for the Non-Bridging Oxygen Center in Fused Silica. The Dynamic Jahn-Teller Effect.
(a) Vol. **70**, No. 1, p. 43–49 (1982).
See also *10.2*, p. *106*; *13.4*, p. *166*; 20.1, p. 302; 22.6, p. 469.

X. W. FAN and J. WOODS
*Green Electroluminescence and Photoluminescence in CdS
(a) Vol. **70**, No. 1, p. 325–334 (1982).
See also 13.5.1, p. 182; 22.4.1, p. 440.

K. N. REDDY, M. L. RAO, and V. H. BABU
*Thermoluminescence and Optical Absorption Studies of Z_1-Centres in NaCl Crystals Doped with Samarium
(a) Vol. **70**, No. 1, p. 335–341 (1982).
See also *10.2*, p. *106*; 20.1, p. 303; 22.5.2, p. 456.

X. W. FAN and J. WOODS
*Blue and Violet Electroluminescence Emission in $ZnSe_xS_{1-x}$ MIS Diodes
(a) Vol. **70**, No. 1, p. K27–K30 (1982).
See also 13.5.1, p. 182; *14.3.3*, p. *221*; 22.4.4, p. 447.

S. BHAN
*Luminescence and Optical Investigation of $NaF:C^{3+}$ Single Crystals
(b) Vol. **110**, No. 2, p. 399–405 (1982).
See also *10.2*, p. *107*; 20.1, p. 303; 22.5.2, p. 456.

B. Z. MALKIN, A. A. KAMINSKII, N. R. AGAMALYAN, L. A. BUMAGINA, and T. I. BUTAEVA
Spectra of Rare-Earth Ions in the Crystals Fields of Double Tungstates and Molybdates
(b) Vol. **110**, No. 2, p. 417–422 (1982).
See also *13.4*, p. *166*; 20.1, p. 303; 22.8.1, p. 494.

G. BLASSE and G. J. DIRKSEN
*The Luminescence of Broad-Band Emitters in $LiLaP_4O_{12}$
(b) Vol. **110**, No. 2, p. 487–494 (1982).
See also 22.8, p. 483.

B. BONNO, J. L. LAPORTE et Y. ROUSSET
*Effets des pièges sur la fusion triplet–triplet dans les cristaux de naphthalène
(Effects of Traps on the Triplet–Triplet Annihilation in Naphthalene Crystals)
(b) Vol. **110**, No. 2, p. 581–589 (1982).
See also *13.5.1*, p. *183*; 22.9, p. 504.

W. SCHMID and P. J. DEAN
Bound Exciton Time Decay and the Auger Effect in Zinc Telluride
(b) Vol. **110**, No. 2, p. 591–600 (1982).
See also *13.4*, p. *167*; *13.5.1*, p. *183*; 22.4.3, p. 445.

S. BHAN
*Thermoluminescence of NaF Harshaw Single Crystals Irradiated to Very Low and High Doses
(a) Vol. **71**, No. 1, p. 73–78 (1982).
See also 11, p. 121; 22.5.2, p. 457.

E. CHIMCZAK, M. BERTRANDT-ŻYTKOWIAK, R. CZAJKA, W. S. GORDON, and M. KOZIELSKI
*Spectral Properties of SnO_x–ZnS(A)–Cu_xS–ZnS(A)–Al Electroluminescent Thin Film Structures
(a) Vol. **71**, No. 1, p. 79–82 (1982).
See also 22.4.1, p. 440.

G. OELGART, H. HAEFNER, R. REULKE, and F. JOACHIM
*Degradation of Green Light-Emitting GaP:N Diodes
(a) Vol. **71**, No. 1, p. 89–98 (1982).
See also *14.3.4*, p. *229*; 22.2.2, p. 425.

K. D. GLINCHUK, K. LUKAT, and A. V. PROKHOROVICH
*Photoluminescence of Copper-Compensated p-Type GaAs after Irradiation by 2.2 MeV Electrons
(a) Vol. **71**, No. 1, p. 233–238 (1982).
See also *13.4*, p. *168*; 22.2.1, p. 421.

P. RANSON, P. PERETTI, and Y. ROUSSET
Crystal Heating and Non-Radiative Relaxation Processes of Triplet Excitons in 4–4′ Dichlorobenzophene
(a) Vol. **71**, No. 1, p. K69–K71 (1982).
See also *13.5.1*, p. *183*; 22.9, p. 505.

A. A. KAMINSKII, A. G. PETROSYAN, G. A. DENISENKO, T. I. BUTAEVA, V. A. FEDOROV, and S. E. SARKISOV
*Spectroscopic Properties and 3 μm Stimulated Emission of Er^{3+} Ions in the $(Y_{1-x}Er_x)_3Al_5O_{12}$ and $(Lu_{1-x}Er_x)_3Al_5O_{12}$ Garnet Crystal Systems
(a) Vol. **71**, No. 2, p. 291–312 (1982).
See also 20.1, p. 307; *20.2*, p. *317*; 22.8, p. 485.

H. BARÁNKOVÁ
*Photoluminescence of SSD and LEC Grown InP
(a) Vol. **71**, No. 2, p. K153–K156 (1982).
See also 13.4, p. 170; 22.2.2, p. 425.

K. D. GLINCHUK and K. LUKAT
*Non-Trivial Annealing-Induced Increase of the Near-Intrinsic Emission Intensity in γ-Irradiated GaAs
(a) Vol. **71**, No. 2, p. K187–K190 (1982).
See also 11, p. 123; *13.4*, p. *171*; 22.2.1, p. 421.

M. YAMAWAKI and C. HAMAGUCHI
*Effect of Electric Fields on Luminescence in GaAs
(b) Vol. **112**, No. 1, p. 201–210 (1982).
See also 13.5.1, p. 184; 20.1, p. 307; 22.2.1, p. 422.

A. A. KAMINSKII, S. E. SARKISOV, H. D. KÜRSTEN, and D. SCHULTZE
*Crystal Growth and Spectroscopic Properties of Nd^{3+} Ions in Ferroelectric $Pb_5Ge_3O_{11}$ Crystals
(a) Vol. **72**, No. 1, p. 207–213 (1982).
See also 3, p. 58; 14.4.1, p. 242; *20.1*, p. *308*; 22.8.1, p. 495.

B. P. CHANDRA, R. D. VERMA, A. K. JAISWAL, B. MAJUMDAR, T. R. CHANDRAKER, and S. V. DESHPANDE
*Mechanoluminescence of Halides and Other Inorganic Crystals
(a) Vol. **72**, No. 1, p. 361–367 (1982).
See also 22.5, p. 451.

H. OPYRCHAŁ, K. D. NIERZEWSKI, and B. MACALIK
Effect of γ-Irradiation on Eu^{2+} Ions in KCl Crystals
(b) Vol. **112**, No. 2, p. 429–434 (1982).
See also *10.2*, p. *111*; 11, p. 124; 19, p. 288; 20.1, p. 309; 22.5.2, p. 459.

V. I. MELNIK, K. I. NELIPOVICH, A. N. FAIDYSH, M. T. SHPAK, and L. B. YANKOVSKAYA
*Influence of Secondary Excitons on Intensity and Decay of Phosphorescence in Doped Benzophenone Single Crystals
(b) Vol. **112**, No. 2, p. 435–443 (1982).
See also *13.5.1*, p. *184*; 22.9, p. 506.

A. DAULETBEKOVA, A. AKILBEKOV, and A. ELANGO
Thermo- and Photostimulated Recombinations of $F–H_A$ and $α-I_A$ Centres in KBr with Large Na Concentration
(b) Vol. **112**, No. 2, p. 445–452 (1982).
See also *10.2*, p. *111*; 22.5.2, p. 459.

S. BHAN
*Thermoluminescence of $NaF:Ca^{2+}$ Single Crystals
(b) Vol. **112**, No. 2, p. 507–514 (1982).
See also *10.2*, p. *112*; 22.5.2, p. 459.

A. J. WOJTOWICZ and H. J. ŁOŻYKOWSKI
*Unknown Excited States of the Self-Activated Center in ZnSe
(b) Vol. **112**, No. 2, p. K103–K106 (1982).
See also *13.4*, p. *172*; 22.4.2, p. 443.

L. E. KHOLODENKOV and A. G. MAKHANEK
*Two Photon Luminescence Excitation of Eu^{3+} in Fluorite
(b) Vol. **112**, No. 2, p. K149–K151 (1982).
See also 13.4, p. 172; 22.5.3, p. 466.

P. PIPINYS, A. PROSKURA, and A. RIMEIKA
*Temperature Dependence of Electroluminescence in ZnS Ceramics
(a) Vol. **72**, No. 2, p. 511–514 (1982).
See also *14.3.4*, p. *230*; 22.4.1, p. 441.

D. SEURET, O. VIGIL, and F. LECCABUE
*Electro-Optical Properties of $CuGa_{0.7}In_{0.3}Se_2/ZnSe$ Heterojunction
(a) Vol. **72**, No. 2, p. 661–666 (1982).
See also *14.3.3*, p. *223*; 22.4.2, p. 443; 22.8, p. 488.

M. Blaschke, H. J. Grundmann, and K. Wandel
*Current Dependent Electroluminescence Decay and Degradation Behaviour in Green and Yellow LED's
(a) Vol. **72**, No. 2, p. 687 – 694 (1982).
See also 13.4, p. 173; 14.3.3, p. 224; 22.2.4, p. 431.

K. D. Glinchuk, A. V. Prokhorovich, and N. S. Zayats
*Photoluminescence of Copper-Doped n-Type GaAs after Irradiation by 2.2 MeV Electrons
(a) Vol. **72**, No. 2, p. 715 – 719 (1982).
See also *13.4*, p. *173*; 22.2.1, p. 422.

C. Paracchini and G. Schianchi
*Luminescence of WO_3
(a) Vol. **72**, No. 2, p. K129 – K132 (1982).
See also 22.6, p. 473.

P. Swarup, R. K. Jain, S. N. Verma, S. Charan, and D. M. Tandle
LPE Growth of GaInAlAs on GaAs
(a) Vol. **72**, No. 2, p. K189 – K192 (1982).
See also *2*, p. *51*; *13.1*, p. *153*; 20.1, p. 310; 22.2.4, p. 431.

M. A. S. Sweet
The Analysis of Complex Thermoluminescence from Partial Glow Curves
(a) Vol. **72**, No. 2, p. K217 – K222 (1982).
See also *13.4*, p. *173*; 22.4.1, p. 441.

E. Müller, W. Gebhardt, and V. Gerhardt
*Exciton Transfer between the Manganese Ions in the Semiconductor Alloy $Cd_{1-x}Mn_xTe$ with $x = 0.51$
(b) Vol. **113**, No. 1, p. 209 – 218 (1982).
See also *13.5.1*, p. *184*; 22.4.4, p. 448.

B. Henderson and A. H. Garrison
Ion Implantation in Alkaline Earth Oxides
(b) Vol. **113**, No. 1, p. 267 – 276 (1982).
See also *11*, p. *125*; 19, p. 289; 20.1, p. 310; 22.6, p. 473.

L. Bosi and M. Nimis
*Luminescence Properties of an M_A-Type Center in KCl:$SrCl_2$ and KCl:$CaCl_2$
(b) Vol. **113**, No. 1, p. 347 – 352 (1982).
See also *10.2*, p. *113*; 22.5.2, p. 460.

R. K. Gartia, R. Singh, and K. Y. Singh
*Thermoluminescence Studies of Post-Irradiation Deformation in Ca-Doped NaCl
(b) Vol. **113**, No. 1, p. K57 – K60 (1982).
See also *10.2*, p. *113*; 22.5.2, p. 460.

R. Cases, R. Alcalá, and V. M. Orera
F'-Centers in CsF
(b) Vol. **113**, No. 1, p. K61 – K65 (1982).
See also *10.2*, p. *113*; 20.1, p. 311; 22.5.2, p. 460.

T. Łukaszewicz
Studies on the Diffusion of Impurities in II–VI Compounds Based on the Two-Sample Luminescence Method
(a) Vol. **73**, No. 2, p. 611 – 616 (1982).
See also *9*, p. *93*; 22.4.1, p. 441; 22.4.2, p. 444.

V. A. Zuikov and V. V. Samartsev
*Reverse Photon Echo as a Method of Investigation of Resonant Medium Parameters
(a) Vol. **73**, No. 2, p. 625 – 632 (1982).
See also 13.4, p. 175; *20.1*, p. *313*; 22.8.1, p. 497.

C. Battaglia, S. Bonora, M. Cevolani, G. C. Degli Esposti, and S. Petralia
*Thermoluminescence Sensitization in LiF (TLD 100) at Different Post-Annealing Temperatures
(a) Vol. **73**, No. 2, p. K169 – K173 (1982).
See also 10.2, p. 115; 22.5.2, p. 462.

B. P. Chandra
*Uncommon Mechanoluminescence Spectra of Certain Organic and Inorganic Crystals
(a) Vol. **73**, No. 2, p. K205 – K208 (1982).
See also 22.6, p. 475; 22.9, p. 507.

K. D. Glinchuk, A. V. Prokhorovich, and N. S. Zayats
*Comparison of Irradiation- and Annealing-Induced Changes of the Intrinsic Emission Intensity in Electron-Bombarded Undeformed and Plastically Deformed GaAs
(a) Vol. **73**, No. 2, p. K267 – K270 (1982).
See also 11, p. 127; 22.2.1, p. 423.

T. Minami, H. Nanto, and S. Takata
*Light Emission from Hot Electrons in Zinc Oxide MS Diodes
(a) Vol. **73**, No. 2, p. K293 – K297 (1982).
See also 22.4, p. 438.

G. B. Talapatra and T. N. Misra
*Delayed Luminescence of Organic Mixed Crystals. Delayed Excimer Fluorescence of Pyrene in p-Terphenyl
(b) Vol. **114**, No. 1, p. 73 – 83 (1982).
See also *13.5.1*, p. *185*; 22.9, p. 507.

F. J. Lohmeier and F. Fischer
*Optical Properties of AsO_2^- Impurities in Alkali Halide Crystals
(b) Vol. **114**, No. 1, p. 95 – 102 (1982).
See also 13.4, p. 176; *20.1*, p. *314*; 22.5.2, p. 462.

Ch. Lushchik, J. Kolk, A. Lushchik, N. Lushchik, M. Taiirov, and E. Vasilchenko
*Decay of Excitons into Long-Lived F, H and α, I Pairs in KCl
(b) Vol. **114**, No. 1, p. 103 – 111 (1982).
See also *10.2*, p. *115*; 22.5.2, p. 462.

B. Esser
*On the Theory of Time-Resolved Band-Tail Luminescence in Amorphous Semiconductors
(b) Vol. **114**, No. 1, p. 175 – 187 (1982).
See also 2, p. 54; 22, p. 391.

N. Nishimaki and S. Shimanuki
*New Luminescence from M and N Centers Perturbed by Divalent and Monovalent Impurities in γ-Irradiated NaCl
(b) Vol. **114**, No. 1, p. K59–K62 (1982).
See also *10.2*, p. *116*; 11, p. 127; 22.5.2, p. 463.

V. I. Osinskii, S. A. Malyshev, and M. P. Ryzhkov
*Minority Carrier Transport and Radiative Recombination in $Al_xGa_{1-x}As$ Variable-Composition Structures
(a) Vol. **74**, No. 1, p. 43–50 (1982).
See also *14.3.1*, p. *215*; 22.2.4, p. 431.

H. L. Oczkowski
*Thermoluminescence Model with Intra-Pair Metastable State
(a) Vol. **74**, No. 1, p. 65–74 (1982).
See also *13.4*, p. *176*; 22.4.2, p. 444.

Hj. Bernhardt
*A Contribution to High-Temperature Thermoluminescence Investigations in Quartz Crystals
(a) Vol. **74**, No. 1, p. 159–165 (1982).
See also 10.2, p. 116; 22.6, p. 475.

S. P. Kuila and V. V. Ratnam
*Monochromatic Thermoluminescence Glow Curve Method for Evaluating Trapping Parameters
(a) Vol. **74**, No. 1, p. 255–263 (1982).
See also 10.2, p. 116; *13.4*, p. *177*; 22.5.2, p. 463.

T. Łukaszewicz
The Diffusion of Atoms and Luminescence Centers in II–VI Compounds
(a) Vol. **74**, No. 1, p. 307–312 (1982).
See also *9*, p. *93*; 22.4.2, p. 444.

V. V. Hizhnyakov
Vibrational Relaxation in the Excited Electronic State
(b) Vol. **114**, No. 2, p. 721–730 (1982).
See also *6*, p. *81*; 22.5.2, p. 463.

I. Tale, V. Kortov, and V. Popov
Trap Spectroscopy in Alkali Halides at Synchronous Measurements of Exoelectron Emission and Luminescence by Fractional Glow Technique
(a) Vol. **74**, No. 2, p. 395–401 (1982).
See also *10.2*, p. *117*; 17, p. 257; 22.5.2, p. 464.

S. P. Kuila and V. V. Ratnam
*Thermoluminescence Studies of F_A-Centres in X-Irradiated Li-Doped KBr Single Crystals
(a) Vol. **74**, No. 2, p. 615–619 (1982).
See also *10.2*, p. *117*; 20.1, p. 316; 22.5.2, p. 464.

V. V. Nikolaenko, S. V. Svechnikov, V. P. Sushkov, and N. I. Sypko
*Investigation of Current–Voltage and Current–Luminosity Characteristics of $GaAs_{0.6}P_{0.4}$ Light-Emitting Diodes
(a) Vol. **74**, No. 2, p. 621 – 630 (1982).
See also *14.3.3*, p. *226*; 22.2.4, p. 432.

A. S. Popov and I. N. Ivanov
The Aging Process in Red GaP LED's
(a) Vol. **74**, No. 2, p. 651 – 654 (1982).
See also *13.4*, p. *178*; *14.3.3*, p. *226*; 22.2.2, p. 426.

21. Metals. Alloys

Review Articles

Original Papers and Short Notes

Yu. S. Stark and A. S. Sternberg
Application of BBGKY Equations to the Study of Ordering in Alloys
(b) Vol. **109**, No. 1, p. 403–409 (1982).
See also 1.2, p. 14; *1.3*, p. *25*.

H.-J. Kaufmann and P. P. Pal-Val
Peculiarities of High-Amplitude Dislocation Internal Friction in Molybdenum Single Crystals of High Purity in the Temperature Range 5.9 to 300 K
(a) Vol. **69**, No. 1, p. 127–132 (1982).
See also *10.1*, p. *96*; *12.1*, p. *130*.

M. P. Sheaber, S. K. Sen, and C. L. Bauer
Investigation of Interfacial Reactions in Bimetallic Thin-Film Couples of Al/Cu and Pb/Cu by Measurement of Contact Resistance
(a) Vol. **69**, No. 1, p. 139–149 (1982).
See also *1.5*, p. *38*; 9, p. 91; 14.1, p. 193; 21.1, p. 343.

L. B. Dubrovskaya, S. Z. Nazarova, and A. F. Prekul
Superconducting and Normal Properties of Solid Solutions $Nb_{1-x}Ta_xC_y$
(a) Vol. **69**, No. 1, p. 167–172 (1982).
See also *1.3*, p. *25*; 14.2, p. 199; 14.3, p. 203; 18, p. 259.

A. Loiseau and A. Lasalmonie
Contrast of the Dislocations for $\boldsymbol{g} \cdot \boldsymbol{b} = 0, 1, 2$. Application to α-Titanium
(a) Vol. **69**, No. 1, p. 315–324 (1982).
See also *10.1*, p. *96*.

G. Cocco, L. Schiffini, S. Enzo, and A. Benedetti
A SAXS Study of Coarsening Mechanism of GP Zones in the Al–Li–Mg System
(a) Vol. **69**, No. 1, p. 343–348 (1982).
See also *1.3*, p. *25*; 10.1, p. 96; 21.2, p. 365; 21.3, p. 367.

W. Zahn, P. B. Weber, K. Blüthner, and E.-B. Kley
Submicron High Critical Current Density Niobium-Lead Josephson Junctions
(a) Vol. **69**, No. 1, p. K25–K28 (1982).
See also *14.2*, p. *199*.

V. Shamrai, K. Bohmhammel, and G. Wolf
The Heat Capacity of Nb_3SnH_x in Dependence on the Hydrogen Content
(b) Vol. **109**, No. 2, p. 511–517 (1982).
See also 6, p. 71; *8*, p. *85*; 14.2, p. 199.

H. R. Philipp
$AuAl_2$: Optical Properties and Consideration as a Transparent Electrode Material
(a) Vol. **69**, No. 2, p. 547–551 (1982).
See also 14.1, p. 193; *20.1*, p. *301*; 21.6, p. 374.

H. Mühlbach
Dislocation Relaxation Peaks in High-Purity Molybdenum Single Crystals
(a) Vol. **69**, No. 2, p. 615–624 (1982).
See also *10.1*, p. *96*; *12.1*, p. *130*.

V. Prabhawalkar and B. D. Padalia
L_3-Absorption Edge Studies of Rare Earths in Rare Earth–Aluminium Intermetallics
(b) Vol. **110**, No. 1, p. 357–361 (1982).
See also *20*, p. *294*; 21.4, p. 370.

V. V. Ustinov and D. Z. Khusainov
On the Lineshape of the Conduction Electron Spin Resonance in Thick Metallic Samples under the Anomalous Skin-Effect Conditions
(b) Vol. **110**, No. 1, p. 363–368 (1982).
See also *19*, p. *285*.

I. L. McLaughlin
The WCA Method and Pseudopotentials for Simple Liquid Metals
(b) Vol. **110**, No. 1, p. K45–K50 (1982).
See also *2*, p. *47*; *4*, p. *62*.

E. Kolley and W. Kolley
On Random Itinerant Ferromagnets in the Percolation Limit
(b) Vol. **110**, No. 1, p. K63–K67 (1982).
See also *18.1*, p. *265*; *18.2*, p. *269*.

F. M. Mansy, N. K. Gobran, and G. Said
High-Temperature Creep Mechanism and Dissolution Process in Al–9.5 at% Zn Alloy
(a) Vol. **70**, No. 1, p. 57–61 (1982).
See also *10.1*, p. *97*; 14.1, p. 194; 21.1, p. 345.

T. Troev, A. Krustanov, V. P. Shantarovich, L. G. Aravin, and B. P. Molin
Doppler Broadening and Positron Lifetime Measurements in Al–49 at% Zn at Different Temperatures
(a) Vol. **70**, No. 1, p. 87–92 (1982).
See also *10.1*, p. *97*; 21.1, p. 346.

R. Hamar and J. P. Riquet
Estimate of the Boundary Energy of Metallic Phases
(a) Vol. **70**, No. 1, p. 127–138 (1982).
See also *1.4*, p. *32*; *10.1*, p. *97*; 21.1, p. 346.

J. X. Zhang, T. S. Kê, G. Y. Li, G. Fantozzi, P. F. Gobin, and M. Weller
Low-Temperature Cold Work Internal Friction Peaks in Al–0.5 wt% (0.21 at%) Cu
(a) Vol. **70**, No. 1, p. 159–165 (1982).
See also *10.1*, p. *97*; *12.1*, p. *130*; 21.1, p. 346.

H. Mizubayashi and S. Okuda
Low Temperature Internal Friction in Deformed Ta
(a) Vol. **70**, No. 1, p. 257–268 (1982).
See also *10.1*, p. *97*; *12.1*, p. *130*.

S. A. Mahmoud, M. A. Semary, and Z. Farid
The Effect of Plastic Deformation on the Coherency of the θ'-Phase
(a) Vol. **70**, No. 1, p. K77–K79 (1982).
See also *10.1*, p. *97*; *12.1*, p. *130*; 21.1, p. 346.

J. Heinrichs and P. Magain
Density-Functional Theory of the Pressure Distribution at Metal Surfaces
(b) Vol. **110**, No. 2, p. 495–498 (1982).
See also 12.1, p. 131; *13.3*, p. *158*.

R. M. Agrawal and R. P. S. Rathore
Interactions in Non-Simple F.C.C. Metals
(b) Vol. **110**, No. 2, p. 677–686 (1982).
See also *4*, p. *62*; *6*, p. *73*; 12.1, p. 131; 21.1, p. 347.

R. I. Kuznetsova, N. N. Zhukov, O. A. Kaibyshev, and R. V. Valiev
Mechanism of Superplastic Deformation of Coarse-Grained Materials
(a) Vol. **70**, No. 2, p. 371–378 (1982).
See also *1.4*, p. *33*; *10.1*, p. *98*.

V. I. Chervinskii, M. M. Kantor, I. I. Novikov, and R. M. Sofronova
The Nature of Planar Faults in a Dilute Molybdenum–Boron Alloy
(a) Vol. **70**, No. 2, p. 379–386 (1982).
See also *1.3*, p. *27*; 10.1, p. 98.

J. D. Livingston
Magnetomechanical Properties of Amorphous Metals
(a) Vol. **70**, No. 2, p. 591–596 (1982).
See also 2, p. 47; *12.1*, p. *131*; *18.2*, p. *269*.

B. Horvatić and V. Zlatić
Perturbation Expansion for the Asymmetric Anderson Hamiltonian (II): General Asymmetry
(b) Vol. **111**, No. 1, p. 65–75 (1982).
See also 8, p. 86; *13*, p. *142*; 14.1, p. 194; 15, p. 246.

P. Singh and K. N. Khanna
Entropy of Mixing of Liquid Alloys
(b) Vol. **111**, No. 1, p. 111–116 (1982).
See also *1.3*, p. *28*; 2, p. 48; 8, p. 86; 21.1, p. 348.

N. C. Halder and K. C. Phillips
Is the Electronic Structure of Liquid Metals Free-Electron-Like or Bloch-Electron-Like?
(b) Vol. **111**, No. 1, p. 221–226 (1982).
See also *2*, p. *48*; *13*, p. *142*.

K. R. Zhdanov and F. S. Rakhmenkulov
Influence of Electron–Phonon Interaction on Electronic Heat Capacity of Nb
(b) Vol. **111**, No. 1, p. 361–364 (1982).
See also *6*, p. *73*; *8*, p. *86*.

F. Brosens and J. T. Devreese
Effective Interactions in Jellium Including Exchange
(b) Vol. **111**, No. 1, p. 365–374 (1982).
See also *13*, p. *143*.

V. CHRISTOPH and A. I. KUZEMSKII
Electroconductivity of a Metallic System with a Nonspherical Fermi Surface
(b) Vol. **111**, No. 1, p. K1 – K6 (1982).
See also *14.1*, p. *195*.

A. D. POGREBNYAK, V. A. KUZMINIKH, and K. P. AREFEV
Positron Profiles and Positron Annihilation in Thin Layers
(a) Vol. **71**, No. 1, p. 145 – 151 (1982).
See also *1.5*, p. *39*; *1.6*, p. *43*; 22.2.1, p. 421.

I. G. RATISHVILI
The Low-Temperature Stable States of Some Transition Metal Hydrides
(a) Vol. **71**, No. 1, p. 223 – 232 (1982).
See also *1.3*, p. *28*; 4, p. 62.

Y. MINONISHI, S. ISHIOKA, M. KOIWA, and S. MOROZUMI
The Structure of $\{11\bar{2}1\}$ Twin Boundaries in H.C.P. Crystals
(a) Vol. **71**, No. 1, p. 253 – 258 (1982).
See also *1.4*, p. *34*; 10.1, p. 98.

J.-M. DEROCHETTE
Adsorption of H_2 and N_2 on Ir(100) Studied by Field Emission
(a) Vol. **71**, No. 1, p. K99 – K102 (1982).
See also *1.6*, p. *43*; *17.1*, p. *258*.

V. I. BOGDANOV, A. V. MIKHAILOV, and D. L. FUKS
Thermodynamics of Aluminium- and Magnesium-Base Interstitial Solutions (I): Analysis of Interaction Parameters
(b) Vol. **111**, No. 2, p. 483 – 489 (1982).
See also *1.3*, p. *28*; 21.3, p. 367.

P. VAROTSOS, K. ALEXOPOULOS, and K. NOMICOS
Comments on the Pressure Variation of the Gibbs Energy for Bound and Unbound Defects
(b) Vol. **111**, No. 2, p. 581 – 590 (1982).
See also *8*, p. *87*; 9, p. 92; *10*, p. *94*; 14.4, p. 236; 21.1, p. 349; 21.6, p. 376; 22.5.2, p. 458.

M. TOMÁŠEK and Š. PICK
About the Influence of Certain Perturbation Effects on Surface States
(b) Vol. **111**, No. 2, p. K89 – K91 (1982).
See also *13.3*, p. *159*; 21.1, p. 349.

W. C. JOHNSON and J. K. LEE
An Integral Equation Approach to the Elastic Interaction of Two Precipitates
(a) Vol. **71**, No. 2, p. 589 – 602 (1982).
See also *10.1*, p. *98*; *12.1*, p. *131*.

A. SIKORA and B. MAKIEJ
On the Occurrence of a Directional Asymmetry of the Critical Current in Type I Superconductor Containing Ferromagnetic Particles
(a) Vol. **71**, No. 2, p. K197 – K200 (1982).
See also *14.2*, p. *200*; *18.2*, p. *270*.

J. Kojnok, A. Szász, L. Kertész, and A. S. Sulakov
Surface Corrosion of Mg-Containing Al-Based Alloys
(a) Vol. **72**, No. 1, p. 131 – 134 (1982).
See also *1.6*, p. *43*.

H. Löffler and R. W. Gould
On the Influence of Small Magnesium Additions on the Decomposition Behaviour of an Al–Zn (6 at%) Mother Alloy
(a) Vol. **72**, No. 1, p. 147 – 152 (1982).
See also *10.1*, p. *99*; 14.1, p. 195; 21.1, p. 350.

A. Luft and Ch. Ritschel
Deformation and Slip Behaviour of Cold-Worked Molybdenum Single Crystals at Elevated Temperature (I): The Effect of Crystals Orientation
(a) Vol. **72**, No. 1, p. 225 – 237 (1982).
See also *10.1*, p. *99*.

K. S. Vahvaselkä
Total and Partial Interference Functions, Radial Atomic Distributions, and Electrical Resistivities for Liquid Indium and Liquid Gallium–Indium Alloys
(a) Vol. **72**, No. 1, p. 261 – 271 (1982).
See also *2*, p. *50*; 14.1, p. 195.

Th. Wirth, V. Atzrodt, and H. Lange
Composition Changes of NiSi and PtSi Due to Ar^+ Ion Bombardment Determined from AES Measurements
(a) Vol. **72**, No. 1, p. K89 – K93 (1982).
See also *1.6*, p. *43*; *11*, p. *124*.

R. Podloucky, J. Deutz, R. Zeller, and P. H. Dederichs
Electronic Structure of 3d and 4d Impurities in Mo and Nb
(b) Vol. **112**, No. 2, p. 515 – 524 (1982).
See also *13.4*, p. *172*.

K. Masuda
Calculation of Binding Energies of Small Vacancy Clusters in B.C.C. Transition Metals
(b) Vol. **112**, No. 2, p. 609 – 614 (1982).
See also *10.1*, p. *99*; *13.4*, p. *172*; 21.1.1, p. 361.

O. H. Prasad and M. Suryanarayana
Fourth-Order Elastic Constant, C_{1111}, of Cubic Crystals
(b) Vol. **112**, No. 2, p. 627 – 631 (1982).
See also *12.1*, p. *132*; *12.2*, p. *136*; 21.1, p. 351; 21.6, p. 377; 22.1, p. 393; 22.5.2, p. 459.

K. Elk
Temperature Dependence of the Electrical Conductivity of the Periodic Anderson Model
(b) Vol. **112**, No. 2, p. 659 – 665 (1982).
See also 13, p. 144; *14.1*, p. *196*.

G. A. Malygin
Impurity Atmospheres and Effective Activation Parameters of Dislocation Motion with Special Reference to Al–Mg and Zr–O Alloys
(a) Vol. **72**, No. 2, p. 493 – 501 (1982).
See also *10.1*, p. *99*; 21.1, p. 352; 21.3, p. 368.

V. I. Zubov
On the Theory of Vacancies in Simple Metals
(b) Vol. **113**, No. 1, p. K73–K76 (1982).
See also *10.1*, p. *100*.

P. G. Self, M. P. Shaw, and W. M. Stobbs
Anomalous Contrast in Weak Beam Images of Stacking Faults. Observation and Calculation
(a) Vol. **73**, No. 1, p. 37–48 (1982).
See also *1.4*, p. *36*; 21.1, p. 353; 21.1.1, p. 362.

O. Simmich
Theoretical Considerations on the Diffusion-Controlled Particle Growth in an Isotropic Medium
(a) Vol. **73**, No. 1, p. 107–115 (1982).
See also *1.2*, p. *21*; *10.1*, p. *100*.

B. Gueffroy-Oettel and H. Löffler
The Dependence of the Reversion of Guinier-Preston Zones in an Al–4.5 at% Zn–2.0 at% Mg Alloy on the Pre-Ageing Conditions
(a) Vol. **73**, No. 1, p. 153–156 (1982).
See also *10.1*, p. *100*; 21.1, p. 353; 21.3, p. 368.

R. Tognato
On Internal Premelting in Metals
(a) Vol. **73**, No. 1, p. 237–241 (1982).
See also *10.1*, p. *100*.

G. V. S. Sastry, C. Suryanarayana, and G. van Tendeloo
A Structural Study of Vapour-Deposited Al–Pd Alloys
(a) Vol. **73**, No. 1, p. 267–278 (1982).
See also *1.3*, p. *30*; **1.4, p. 36**; 21.1, p. 354.

A. M. Ghodgaonkar and K. Ramani
Application of the Mayadas-Shatzkes Model for Electrical Conduction in Thin Films with Unlike Surfaces
(a) Vol. **73**, No. 1, p. K21–K25 (1982).
See also *14.1*, p. *197*.

R. A. Varin
On the Validity of Interpretation of Some Grain Boundary Processes Based on the Assumption about the Existence of Unrelaxed Extrinsic Grain Boundary Dislocations
(a) Vol. **73**, No. 1, p. K51–K55 (1982).
See also *10.1*, p. *100*; 21.1, p. 354; 21.1.1, p. 362.

A. V. Verbilo, N. M. Kotov, A. P. Rusakov, and N. A. Kozlovskai
Superconductive Properties of the Chevrel-Phase System $Cu_xMo_6S_8$ with $x = 3.2$ to 3.9
(a) Vol. **73**, No. 1, p. K73–K74 (1982).
See also *14.2*, p. *200*.

G. Haneczok and J. W. Moroń
The Analysis of Internal Friction Curves for a Temperature Gradient Existing along the Sample
(a) Vol. **73**, No. 1, p. K123–K127 (1982).
See also *10.1*, p. *101*; *12.1*, p. *132*.

J. Mertsching
Kinks in a Quasi-One-Dimensional System with a Nearly 1/3-Filled Band
(b) Vol. **113**, No. 2, p. 609–621 (1982).
See also *13.4*, p. *175*.

M. Jirsa
Exchange-Conductivity Broadening of SSWR Line in Metallic Thin Films
(b) Vol. **113**, No. 2, p. 679–684 (1982).
See also *13*, p. *146*; *19*, p. *290*.

R. D. Bijalwan and P. N. Ram
Low-Temperature Lattice Specific Heat of CuAl
(b) Vol. **113**, No. 2, p. 691–696 (1982).
See also *6*, p. *79*; *8*, p. *89*; 21.1, p. 354.

B. I. Craig and P. V. Smith
The Electronic Structure of a Hydrogen Impurity in Aluminium (I): The Jellium Model
(b) Vol. **113**, No. 2, p. 747–756 (1982).
See also *13.4*, p. *175*.

A. M. Roshchupkin and V. Ya. Platkov
The Static and Dynamic Hysteresis Responsible for the Dislocational Amplitude Dependent Internal Friction (I): Theory
(a) Vol. **73**, No. 2, p. 303–313 (1982).
See also *10.1*, p. *101*; 12.1, p. 132.

V. G. Chudinov, N. V. Moseev, B. N. Goshchitskii, and V. I. Protasov
Atomic Configurations of Point Defects in Intermetallic Nb_3Sn Compound
(a) Vol. **73**, No. 2, p. 325–331 (1982).
See also *10.1*, p. *101*; 21.7, p. 380.

M. M. El Oker, E. A. Abou-Saif, S. A. El Sahhar, and A. A. Mohamed
Optical Properties and Structure of Thin Indium Films
(a) Vol. **73**, No. 2, p. 389–394 (1982).
See also *1.5*, p. *41*; *20.1*, p. *313*.

D. E. G. Williams
Spatial Resolution and Dense Random Packing
(a) Vol. **73**, No. 2, p. 539–544 (1982).
See also *1.3*, p. *31*; *2*, p. *53*.

R. V. Nandedkar and P. Delavignette
On the Formation of a New Superstructure in the Zirconium-Aluminium System
(a) Vol. **73**, No. 2, p. K157–K160 (1982).
See also *1.3*, p. *31*.

A. E. GELYASIN and E. A. VASILEV
Influence of Temperature on Structure and Properties of $Mn_{1+x}Sb_{1-x}Sn_x$ Alloys
(a) Vol. **73**, No. 2, p. K221 – K224 (1982).
See also 5, p. 68; *18.2*, p. *272*; *18.3*, p. *277*.

V. P. PAN, I. V. ZOLOTUKHIN, S. E. LITVIN, B. G. NIKITIN, and A. S. SHPIGEL
Electrical Properties of Lanthanum–Aluminium Glass
(a) Vol. **73**, No. 2, p. K287 – K291 (1982).
See also *2*, p. *53*; 8, p. 89; *14.1*, p. *197*; 14.2, p. 201.

V. SESHU BAI and K. V. S. RAMA RAO
Ferromagnetic Resonance Investigations in $MnSb_{1-x}In_x$ Films
(a) Vol. **73**, No. 2, p. K303 – K305 (1982).
See also *18.2.1*, p. *275*; *19*, p. *290*.

W. SCHILLER and J. RICHTER
Temperature Dependence of the Thermoelectric Power of Disordered Alloys
(b) Vol. **114**, No. 1, p. 151 – 160 (1982).
See also 2, p. 54; *14.1*, p. *197*; *15*, p. *248*.

J. L. BRETONNET, J. G. GASSER, A. BATH, and R. KLEIM
Structure Factors and Compressibilities of Liquid Metals
(b) Vol. **114**, No. 1, p. 243 – 253 (1982).
See also *2*, p. *54*; 14.1, p. 197.

B. N. TEWARI, B. S. BHARGAVA, and K. N. KHANNA
Surface Entropy of Liquid Metals
(b) Vol. **114**, No. 1, p. K31 – K34 (1982).
See also *2*, p. *54*; *8*, p. *89*; 21.1, p. 356; 21.2, p. 366; 21.3, p. 368; 21.7, p. 381.

I. N. KUZMENKO, O. A. PALAGINA, and V. V. PUSTOVALOV
Study of Plastic Deformation of Aluminium near the Superconducting Transition
(a) Vol. **74**, No. 1, p. 51 – 57 (1982).
See also *10.1*, p. *101*; 14.2, p. 201.

B. GUEFFROY-OETTEL, R. RAMLAU, and H. LÖFFLER
On the Relation between Precipitation Density and Size and Hardness during the Reversion Treatment of an Al–4.5 at% Zn–2.0 at% Mg Alloy Pre-Aged at 100 °C
(a) Vol. **74**, No. 1, p. 149 – 157 (1982).
See also *10.1*, p. *101*; 21.1, p. 356; 21.3, p. 368.

V. P. KOVERDA, V. N. SKOKOV, and V. P. SKRIPOV
Liquid–Solid Phase Transition in Discontinuous Metal Films
(a) Vol. **74**, No. 1, p. 343 – 351 (1982).
See also *1.2*, p. *23*; *1.5*, p. *41*.

B. I. CRAIG
The Electronic Structure of a Hydrogen Impurity in Aluminium (II): The Bandstructure Model
(b) Vol. **114**, No. 2, p. 337 – 348 (1982).
See also *13.1*, p, *155*. *13.4*, p. *177*.

21.1 Transition Metals (Group Sc to Zn) and Their Alloys

Review Articles

V. V. NEMOSHKALENKO, V. N. ANTONOV, VL. N. ANTONOV, W. JOHN, H. WONN, and P. ZIESCHE
Electronic Structure and Soft X-Ray Emission Spectra of 5d Transition Metals
(b) Vol. **111**, No. 1, p. 11–52 (1982).
Erratum
(b) Vol. **114**, No. 2, p. 741 (1982).
See also *13.1*, p. *148*; *20*, p. *293*; 21, p. 330; 21.6, p. 374.

F. REYNAUD
Order–Disorder Transitions in Substitutional Solid Solutions
(a) Vol. **72**, No. 1, p. 11–59 (1982).
See also *1.2*, p. *14*; *1.3*, p. *25*; 21, p. 330; 21.1.1, p. 359; 21.6, p. 374.

Original Papers and Short Notes

A. ŚLEBARSKI and J. AULEYTNER
Soft X-Ray Emission Spectra of Aluminium and Dysprosium from $DyCO_2$–$DyAl_2$ Intermetallic Compounds
(b) Vol. **109**, No. 1, p. 125–129 (1982).
See also *20*, p. *293*; 21, p. 330; 21.4, p. 369.

S. C. GOEL and V. S. NANDA
A Refined Equation of State for Monovalent Metals
(b) Vol. **109**, No. 1, p. 183–188 (1982).
See also *8*, p. *84*; 21.6, p. 374.

H. SØRUM, O. M. WENG, and J. BREMER
The CrK Series X-Ray Emission Spectra
(b) Vol. **109**, No. 1, p. 335–342 (1982).
See also 13.1, p. 148; *20*, p. *293*.

M. REIFFERS, K. FLACHBART, J. ŠÁNOŠ, A. B. BEZNOSOV, and G. ESKA
Transport Properties of $PrNi_5$ at Low Temperatures
(b) Vol. **109**, No. 1, p. 369–373 (1982).
See also *8*, p. *84*; *14.1*, p. *193*; 21.4, p. 369.

A. URBANIAK-KUCHARCZYK
Spin Correlation Function in Nickel
(b) Vol. **109**, No. 1, p. K59–K61 (1982).
See also *18.2*, p. *267*.

M. P. SHEARBER, S. K. SEN, and C. L. BAUER
Investigation of Interfacial Reactions in Bimetallic Thin-Film Couples of Al/Cu and Pb/Cu by Measurement of Contact Resistance
(a) Vol. **69**, No. 1, p. 139–149 (1982).
See also *1.5*, p. *38*; 9, p. 91; 14.1, p. 193; 21, p. 331.

H. J. BLYTHE, F. WALZ, and H. KRONMÜLLER
Investigation of the Magnetic Aftereffect in Dilute Fe–Ni Alloys after Low-Temperature Neutron Irradiation
(a) Vol. **69**, No. 1, p. 237 – 248 (1982).
See also *10.1*, p. *96*; 11, p. 118; *18.2*, p. *267*; 21.1.1, p. 359.

M. HIDA, H. MAEDA, N. KAMIJO, and H. TERAUCHI
EXAFS Study on the Local Structure of Copper Crystals Deformed by Heavy Cold Working
(a) Vol. **69**, No. 1, p. 297 – 303 (1982).
See also *1.3*, p. *25*.

A. APOSTOLOV, H. HRISTOV, T. MIDLAG, M. MIKHOV, and V. SKUMRIEV
The Bulk Amorphous Alloy Tb_4CO_3 – An Example for a System with a Transition to a "Spin-Glass Like" State
(a) Vol. **69**, No. 1, p. K7 – K10 (1982).
See also *2*, p. *46*; *18.1*, p. *265*; 18.2, p. 267; 21.4, p. 369.

B. SCHMIDT, V. N. ANTONOV, and E. MROSAN
Cyclotron Masses and Many-Body Effects in Transition Metals
(b) Vol. **109**, No. 2, p. 627 – 633 (1982).
See also *13.2*, p. *157*.

R. MEYER and H. KRONMÜLLER
Micromagnetic Theory of Phase Transitions in Inhomogeneous Ferromagnets (V): Magnetic Phase Transition of Amorphous $Fe_{10}Ni_{70}B_{16}P_4$
(b) Vol. **109**, No. 2, p. 693 – 703 (1982).
Erratum
(b) Vol. **114**, No. 2, p. 741 (1982).
See also *18.2*, p. *268*.

I. ANSARA, A. PASTUREL, and K. H. J. BUSCHOW
Enthalpy Effects in Amorphous Alloys and Intermetallic Compounds in the System Zr–Cu
(a) Vol. **69**, No. 2, p. 447 – 453 (1982).
See also 2, p. 46; *8*, p. *85*.

G. VAN TENDELOO and S. AMELINCKX
Electron Diffraction and Electron Microscopic Study of the Alloy Au_2Mn and Its Formation Mechanism
(a) Vol. **69**, No. 2, p. 589 – 602 (1982).
See also *1.3*, p. *26*; 1.4, p. 32; 21.6, p. 375.

L. BROSSARD, H. BERNAS, L. THOMÉ, A. TRAVERSE, and P. NEDELLEC
Influence of Disorder on Phonon Resistivity of Ion-Implanted Nickel Hydride
(a) Vol. **69**, No. 2, p. 727 – 732 (1982).
See also *11*, p. *119*; *14.1*, p. *193*.

J. M. ALAMEDA and F. LÓPEZ
Transverse Susceptibility and Structure Constant in Cobalt Thin Films
(a) Vol. **69**, No. 2, p. 757 – 765 (1982).
See also *18.2.1*, p. *275*.

A. HAUET, P. DUVAL, and P. MALETRAS
Short-Range Order and Coherent Precipitation in the α-Phase of Copper–Aluminium Alloy
(a) Vol. **69**, No. 2, p. K135–K138 (1982).
See also *1.3*, p. *26*; *1.4*, p. *32*; 21, p. 332.

A. H. ABOU EL ELA and S. MAHMOUD
Optical Absorption of Thin Zinc Films
(a) Vol. **69**, No. 2, p. K193–K197 (1982).
See also 13.1, p. 149; *20.1*, p. *301*.

H. FIGIEL
The Y Transfered Hyperfine Fields in $Y_2(Co_{1-x}Mn_x)_{17}$ Compounds
(a) Vol. **69**, No. 2, p. K219–K221 (1982).
See also 18.2, p. 268; *19*, p. *284*; 21, p. 332.

A. I. NASKIDASHVILI, I. A. NASKIDASHVILI, L. S. TOPCHYAN, V. N. KOZHANOV, and E. P. ROMANOV
Heat Capacity Anomalies in the Intermetallic Compound V_2Zr
(a) Vol. **69**, No. 2, p. K239–K242 (1982).
See also *1.2*, p. *15*; *8*, p. *85*.

P. PAUL, S. P. PURI, D. RAJ, and R. K. PURI
Force Constant Change in Sn–Impurity Systems: ^{107}Pd:^{119}Sn, ^{197}Au:^{119}Sn, and $^{63.5}$Cu:^{119}Sn
(b) Vol. **110**, No. 1, p. 81–87 (1982).
See also *5*, p. *66*; 6, p. 72; 10.1, p. 97; 21.6, p. 375.

H. SOLBRIG
The Excitation Map of Surface Barrier Resonances (I): The Excitation Map for Ni(001) in the (10) Azimuth
(b) Vol. **110**, No. 1, p. 191–201 (1982).
See also *13.3*, p. *158*.

P. PUREUR, J. V. KUNZLER, W. H. SCHREINER, and D. E. BRANDÃO
Electrical Resistivity of the Cobalt Rich Co–Fe Alloys
(a) Vol. **70**, No. 1, p. 11–15 (1982).
See also *14.1*, p. *193*; 21.1.1, p. 360.

V. E. ARKHIPOV, V. I. VORONIN, A. E. KARKIN, and A. V. MIRMELSHTEIN
Radiation Disordering in V_3Si
(a) Vol. **70**, No. 1, p. 17–21 (1982).
See also *1.3*, p. *26*; *11*, p. *120*; 14.2, p. 199.

F. M. MANSY, N. K. GOBRAN, and G. SAID
High-Temperature Creep Mechanism and Dissolution Process in Al–9.5 at% Zn Alloy
(a) Vol. **70**, No. 1, p. 57–61 (1982).
See also *10.1*, p. *97*; 14.1, p. 194; 21, p. 333.

G. S. MURTHY and D. H. SASTRY
Impression Creep of Zinc and the Rate-Controlling Dislocation Mechanism of Plastic Flow at High Temperatures
(a) Vol. **70**, No. 1, p. 63–71 (1982).
See also *10.1*, p. *97*; 12.1, p. 130.

A. Beckmann, E. Kolley, and W. Kolley
On Superconductivity in Narrow-Band Alloys
(b) Vol. **110**, No. 2, p. 557–564 (1982).
See also **6**, p. 73; *14.2*, p. *199*.

R. M. Agrawal and R. P. S. Rathore
Interactions in Non-Simple F.C.C. Metals
(b) Vol. **110**, No. 2, p. 677–686 (1982).
See also *4*, p. *62*; *6*, p. *73*; 12.1, p. 131; 21, p. 334.

J. van Landuyt, G. van Tendeloo, and S. Amelinckx
New Types of Periodic Antiphase Boundary Structures Derived from the DO_{22} Structure in the Gold–Manganese System
(a) Vol. **70**, No. 2, p. 407–414 (1982).
See also *1.3*, p. *27*; 1.4, p. 33; 21.6, p. 375.

Y. J. Wang and H. Kronmüller
The Influence of the Surface Conditions on the Magnetic Properties in Amorphous Alloys $Fe_{40}Ni_{40}B_{20}$ and $Co_{58}Ni_{10}Fe_4Si_{11}B_{16}$
(a) Vol. **70**, No. 2, p. 415–421 (1982).
See also 1.6, p. 42; *18.2*, p. *269*; 21.1.1, p. 360.

X. Z. Dong and H. Kronmüller
Magnetic Domain Structure of Some Nearly Non-Magnetostrictive Amorphous Alloys under Elastic Stresses
(a) Vol. **70**, No. 2, p. 451–462 (1982).
See also *18.2*, p. *269*; 21.1.1, p. 360.

H. Eichenmüller
New Investigations of the Chemical Nature of the Centres of the Exoelectron Emission on Metal Surfaces
(a) Vol. **70**, No. 2, p. 483–487 (1982).
See also 1.6, p. 42; *17*, p. *255*.

A. I. Lotkov, V. N. Grishkov, and V. V. Fadin
The Influence of Palladium on the Martensitic Transformation of the Intermetallic Compound TiNi
(a) Vol. **70**, No. 2, p. 513–517 (1982).
See also *1.3*, p. *27*; *14.1*, p. *194*.

D. Lepski and P. Burck
Precipitate Structure Analysis by Diffraction Contrast Image Matching for a Metastable Semicoherent Carbide Phase in Molybdenum (II): Image Matching
(a) Vol. **70**, No. 2, p. 571–581 (1982).
See also *1.2*, p. *16*; *1.4*, p. *33*; 10.1, p. 98; 12.1, p. 131.

O. N. Awasthi and B. K. Basu
On Magnetoelastic Dispersion in a Nickel Single Crystal
(a) Vol. **70**, No. 2, p. 713–717 (1982).
See also *7*, p. *82*; 18.2, p. 269.

G. L. WHITTLE, S. J. CAMPBELL, and A. M. STEWART
A Mössbauer Study of the Magnetic Hyperfine Field Distribution in the Amorphous Alloy $Fe_{32}Ni_{36}Cr_{14}P_{12}B_6$
(a) Vol. **71**, No. 1, p. 245–251 (1982).
See also 2, p. 48; *5*, p. *67*; 18.2, p. 270; 21.1.1, p. 361.

D. SCHRYVERS, J. VAN LANDUYT, G. VAN TENDELOO, and S. AMELINCKX
High Resolution Electron Microscopy of Coincidence Patterns in the Ordered Pd_3Mn Alloy
(a) Vol. **71**, No. 1, p. K9–K11 (1982).
See also *1.1*, p. *12*; *1.4*, p. *34*; 21.6, p. 376.

K. SCHRÖDER and D. ONENGUT
Effect of Stress Waves on the Magnetization of Magnesiowüstite with Magnesioferrite Precipitates
(a) Vol. **71**, No. 1, p. K93–K97 (1982).
See also *18.2*, p. *270*; 21.7, p. 379.

A. ANDREEFF, TH. FRAUENHEIM, E. A. GOREMYCHKIN, H. GRIESSMANN, B. LIPPOLD, W. MATZ, O. D. CHISTYAKOV, and E. M. SAVITSKII
Crystal Field in the Laves Phase Compound $PrNi_2$
(b) Vol. **111**, No. 2, p. 507–512 (1982).
See also *1*, p. *10*; 8, p. 87; *13.4*, p. *169*; 15, p. 246; 18.1, p. 266; 21.4, p. 370.

P. VAROTSOS, K. ALEXOPOULOS, and K. NOMICOS
Comments on the Pressure Variation of the Gibbs Energy for Bound and Unbound Defects
(b) Vol. **111**, No. 2, p. 581–590 (1982).
See also *8*, p. *87*; 9, p. 92; *10*, p. *94*; 14.4, p. 236; 21, p. 335; 21.6, p. 376; 22.5.2, p. 458.

M. TOMÁŠEK and Š. PICK
About the Influence of Certain Perturbation Effects on Surface States
(b) Vol. **111**, No. 2, p. K89–K91 (1982).
See also *13.3*, p. *159*; 21, p. 335.

A. R. PATEL, N. C. PANDYA, N. C. CHOURASIA, and G. K. SHIVAKUMAR
Annealing Characteristics and Lattice Distortion Energy Spectra of Chromium–Copper Alloy Films
(a) Vol. **71**, No. 2, p. 323–327 (1982).
See also 10.1, p. 98; *14.1*, p. *195*.

H. NAKASHIMA and K. HASHIMOTO
Barrier Height for Cobalt on Silicon
(a) Vol. **71**, No. 2, p. K201–K203 (1982).
See also *14.3.3*, p. *223*; *14.3.4*, p. *230*; 22.1.2, p. 408.

A. V. ANDREEV, E. N. TARASOV, A. V. DERYAGIN, and S. M. ZADVORKIN
Influence of Symmetry of the Crystalline Structure on the Magnetic Anisotropy of Tb_2Co_7
(a) Vol. **71**, No. 2, p. K245–K247 (1982).
See also *1.2*, p. *19*; 1.3, p. 29; *18.2*, p. *271*; 21.4, p. 371.

M. RAYCHAUDHURI and S. CHATTERJEE
Electronic Energy Bands of Some AgMg, CuAl, and AlLi Alloys
(b) Vol. **112**, No. 1, p. 105–114 (1982).
See also *13.1*, p. *152*; 21, p. 336; 21.2, p. 365; 21.3, p. 367; 21.6, p. 376.

J. GOEBBELS, H. R. KHAN, K. LÜDERS, and CH. J. RAUB
NMR on Metastable Pt–P and Ni–Pt–P
(b) Vol. **112**, No. 1, p. 115–118 (1982).
See also 2, p. 49; *19*, p. *288*.

V. S. STEPANYUK, O. A. EVDOKIMENKO, A. A. KATSNELSON, I. S. SHPOTIN, and L. I. YASTREBOV
Model of Self-Consistency Simulation in Constructing a Non-Self-Consistent MT Potential for Metals
(b) Vol. **112**, No. 1, p. 211–218 (1982).
See also *13.1*, p. *152*.

A. SAHLING, P. FRACH, and E. HEGENBARTH
The Heat Capacity of $LaNi_2$, $LaNi_5$, $PrNi_2$, and $PrNi_5$ in the Temperature Range between 0.3 and 7 K
(b) Vol. **112**, No. 1, p. 243–250 (1982).
See also *8*, p. *87*; 21.4, p. 371.

W. DÄUMER, A. KETTSCHAU, H. R. KHAN, K. LÜDERS, CH. J. RAUB, H. RIESEMEIER, and G. ROTH
Superconductivity of Solid Solutions $V_{90}X_{10}$ (X = Cr, Mn, Nb, Mo, Pd, Ta, W, Re, Pt, Au)
(b) Vol. **112**, No. 1, p. K67–K71 (1982).
See also *14.2*, p. *200*; 21, p. 336; 21.6, p. 376.

R. V. NANDEDKAR, K. VARATHARAJAN, S. PANCHAPAKESAN, and A. K. TYAGI
Microhardness, Microstructure, and Blistering of Helium Irradiated Nickel, Alloy Nickel 201, and Nimonic 90
(a) Vol. **72**, No. 1, p. 89–94 (1982).
See also *11*, p. *123*; *12.1*, p. *131*.

H. LÖFFLER and R. W. GOULD
On the Influence of Small Magnesium Additions on the Decomposition Behaviour of an Al–Zn (6 at%) Mother Alloy
(a) Vol. **72**, No. 1, p. 147–152 (1982).
See also *10.1*, p. *99*; 14.1, p. 195; 21, p. 337.

H. VAN SWIGENHOVEN, I. M. SLATS, and K. H. J. BUSCHOW
Phase Transformations in Amorphous Zr–Ni Alloys
(a) Vol. **72**, No. 1, p. 153–160 (1982).
See also 1.3, p. 29; *2*, p. *50*.

H. J. KOLKMAN
Deviations from Schmid's Law in Thin Copper Single Crystals
(a) Vol. **72**, No. 1, p. 189–198 (1982).
See also *10.1*, p. *99*.

H. MEHRER, G. FLIK, J. HORVÁTH, and H. KRONMÜLLER
Crystallization and Aging Effects in Some Amorphous Ferromagnets
(a) Vol. **72**, No. 1, p. 215–223 (1982).
See also *2*, p. *50*; *18.2*, p. *271*; 21.1.1, p. 361.

A. H. VAN OMMEN
Vacancies in NiGa
(a) Vol. **72**, No. 1, p. 273–280 (1982).
See also 9, p. 92; *10.1*, p. *99*.

I. M. Reda, J. Hafner, P. Pongratz, A. Wagendristel, H. Bangert, and P. K. Bhat
Amorphous Cu–Ag Films with High Stability
(a) Vol. **72**, No. 1, p. 313 – 324 (1982).
See also *1.3*, p. *29*; *2*, p. *50*; 14.1, p. 196; 21.6, p. 377.

Q. B. Yang
Coincidence Coefficients of Two Space Lattices and Their Lattice Planes
(a) Vol. **72**, No. 1, p. 343 – 351 (1982).
See also *1.3*, p. *29*; 21.1.1, p. 361.

A. del Moral, J. F. D. Montenegro, J. S. Abell, and M. R. Ibarra
On the Magnetization and Magnetostriction of ErZn
(a) Vol. **72**, No. 1, p. 353 – 360 (1982).
See also 12.1, p. 132; *18.1*, p. *266*; 18.2, p. 271; 21.4, p. 371.

T. A. Ryumshina and V. I. Zaitsev
Influence of Compressibility Anisotropy on the Dislocation Structure of Zinc, Bismuth, and Antimon Bicrystals
(a) Vol. **72**, No. 1, p. K11 – K13 (1982).
See also *10.1*, p. *99*; 21.7, p. 380.

Y. Aoki and K. Ikeda
Electrical Resistivity Study of the F.C.C. Co–V Alloys Quenched from 1200 °C
(a) Vol. **72**, No. 1, p. K41 – K43 (1982).
See also *14.1*, p. *196*.

R. H. Rautioaho
An Interatomic Pair Potential for Copper. Calculation of Stacking Fault Energy
(b) Vol. **112**, No. 2, p. 391 – 398 (1982).
See also 6, p. 76; *10.1*, p. *99*.

O. H. Prasad and M. Suryanarayana
Fourth-Order Elastic Constant, C_{1111}, of Cubic Crystals
(b) Vol. **112**, No. 2, p. 627 – 631 (1982).
See also *12.1*, p. *132*; *12.2*, p. *136*; 21, p. 337; 21.6, p. 377; 22.1, p. 393; 22.5.2, p. 459.

R. Khanna
Force Model Dependence and Asymptotic Behaviour of the Static Lattice Green Function for H.C.P. Crystals
(b) Vol. **112**, No. 2, p. 667 – 674 (1982).
See also *6*, p. *77*; 10.1, p. 99; 21.3, p. 368.

S. Matthies and G. W. Vinel
On the Reproduction of the Orientation Distribution Function of Texturized Samples from Reduced Pole Figures Using the Conception of a Conditional Ghost Correction
(b) Vol. **112**, No. 2, p. K111 – K114 (1982).
See also *1.3*, p. *30*; *1.4*, p. *35*.

S. Matthies and G. W. Vinel
An Example Demonstrating a New Reproduction Method of the ODF of Texturized Samples from Reduced Pole Figures
(b) Vol. **112**, No. 2, p. K115 – K120 (1982).
See also *1.3*, p. *30*; *1.4*, p. *35*.

T. K. Boletskaya, B. E. Egorushkin, and V. P. Fadin
Electron Structure and Optical Conductivity of NiTi
(b) Vol. **113**, No. 1, p. 307 – 311 (1982).
See also *13.1*, p. *154*; 14.1, p. 196; 20, p. 296.

M. R. Khan, I. K. Schuller, and Ch. M. Falco
Sputter Epitaxy of Ag and Ni Films
(a) Vol. **73**, No. 1, p. 23 – 29 (1982).
See also *1.5*, p. *40*; *3*, p. *58*; 21.6, p. 377.

J. Strzeszewski and K. Świderczak
Electron Microscopic Studies of Crystallization and Magnetic Domain Structure of Mn–Ga Thin Films
(a) Vol. **73**, No. 1, p. 31 – 36 (1982).
See also *1.2*, p. *20*; 3, p. 59; *18.2.1*, p. *275*.

P. G. Self, M. P. Shaw, and W. M. Stobbs
Anomalous Contrast in Weak Beam Images of Stacking Faults. Observation and Calculation
(a) Vol. **73**, No. 1, p. 37 – 48 (1982).
See also *1.4*, p. *36*; 21, p. 339; 21.1.1, p. 362.

H. Kirihata
Exoelectron Emission from Metals (Cu, Zn, Au) Excited by Low-Energy Electrons below 200 eV
(a) Vol. **73**, No. 1, p. 71 – 79 (1982).
See also 13.3, p. 160; *17*, p. *256*; 21.6, p. 377.

R. Damköhler and Th. Heumann
Experimental Investigation on the Diffusion Behaviour of Copper-Rich Copper–Nickel Alloys
(a) Vol. **73**, No. 1, p. 117 – 127 (1982).
See also *9*, p. *92*.

B. Gueffroy-Oettel and H. Löffler
The Dependence of the Reversion of Guinier-Preston Zones in an Al–4.5 at% Zn–2.0 at% Mg Alloy on the Pre-Ageing Conditions
(a) Vol. **73**, No. 1, p. 153 – 156 (1982).
See also *10.1*, p. *100*; 21, p. 339; 21.3, p. 368.

Bai-Xin Liu, M.-A. Nicolet, and S. S. Lau
Ti–Au Metallic Glasses Formed by Ion Mixing
(a) Vol. **73**, No. 1, p. 183 – 188 (1982).
See also *1.5*, p. *41*; 2, p. 52; *11*, p. *126*; 21.6, p. 377.

C. R. Brooks, P. J. Meschter, and T. G. Kollie
The Magnetic Heat Capacity of the Configurationally Disordered Ni–25 at% Fe Alloy
(a) Vol. **73**, No. 1, p. 189 – 198 (1982).
See also *8*, p. *88*; 12.1, p. 132; 14.1, p. 197; 18.2, p. 272; 21.1.1, p. 362.

D. DĂDÂRLAT, T. PETRIŞOR, and I. POP
The Magnetic Behaviour of Diluted CrGe Alloys
(b) Vol. **113**, No. 2, p. K111–K114 (1982).
See also *18.4*, p. *281*.

T. SOMA, Y. IWAYAMA, and H. MATSUO KAGAYA
Equation of State of Cu-Based Binary Alloys
(b) Vol. **113**, No. 2, p. K131–K133 (1982).
See also *1.2*, p. *21*; *12.1*, p. *132*.

G. S. KANDAUROVA, L. G. ONOPRIENKOO, and N. I. SOKOLOVSKAYA
The Effect of Twin Microstructure on the Magnetic Properties of Alloys with CuAuI Superlattice Ordering
(a) Vol. **73**, No. 2, p. 351–360 (1982).
See also *1.4*, p. *36*; *18.2*, p. *272*; 21.1.1, p. 362.

V. KUCKHERMANN, G. THUMMES, and H. H. MENDE
Influence of the Real Fermi Surface on the Longitudinal Magnetoresistance Size Effect of Single Crystal Copper Whiskers
(a) Vol. **73**, No. 2, p. 439–445 (1982).
See also *14.1*, p. *197*; 18, p. 263.

M. EIZENBERG
Silicide Formation Resulting from the Interfacial Reaction of Silicon and Thin Films of Ir–V Alloys and Bilayers
(a) Vol. **73**, No. 2, p. 483–495 (1982).
See also *1.3*, p. *30*; *14.3.4*, p. *232*; 22.1.2, p. 413.

R. A. JAGO and P. L. ROSSITER
A Re-Examination of Long-Range Ordering at Fe–25 at% Ni
(a) Vol. **73**, No. 2, p. 497–502 (1982).
See also *1.3*, p. *31*; 1.4, p. 36; 21.1.1, p. 362; 22.8.2, p. 501.

K. K. MISRA and A. N. NIGAM
Thermoelectric Behaviour of Copper Wires Electroplated with Nickel
(a) Vol. **73**, No. 2, p. 559–563 (1982).
See also *15*, p. *248*.

H. RATAJCZAK and I. GOŚCIAŃSKA
Thermal Hall Hysteresis Loops in Thin Amorphous HoCo Films
(a) Vol. **73**, No. 2, p. 633–639 (1982).
See also 2, p. 53; *14.1*, p. *197*; *18.3*, p. *277*; 21.4, p. 371.

B. RÖNNEFARTH, R. MATTHEIS, and H. DINTNER
Deposition and Structure of Different Composite Ni–Cr Thin Films
(a) Vol. **73**, No. 2, p. K153–K156 (1982).
See also *1.3*, p. *31*; *1.5*, p. *41*; 2, p. 53.

G. SAID, H. H. HASSAN, G. A. HASSAN, and A. A. YOUSSEF
Effect of Type and Amount of Pre-Torsional Deformation on the Creep Characteristics of Zinc Wire Samples
(a) Vol. **73**, No. 2, p. K175–K177 (1982).
See also *10.1*, p. *101*.

V. I. PECHERSKAYA and D. N. BOLSHUTKIN
Electrical Resistance and Thermal Expansion of Fe–18Cr–(10 to 25) Ni Alloys in the Paramagnetic Temperature Range
(a) Vol. **74**, No. 1, p. 285–290 (1982).
See also *8*, p. *89*; *14.1*, p. *198*; 21.1.1, p. 363.

G. A. PETRAKOVSKII, A. G. KLIMENKO, F. S. RAKHMENKULOV, and A. V. BARANOV
Specific Heat and Low Field Magnetic Properties of $Co(S_{1-x}Se_x)_2$
(a) Vol. **74**, No. 1, p. K31–K35 (1982).
See also *8*, p. *89*; *18*, p. *263*.

L. TAKÁCS, A. VÉRTES, and H. LEIDHEISER, Jr.
Effect of Stresses on the Mössbauer Line-Intensites in Nickel Electrodeposits of [100] Texture
(a) Vol. **74**, No. 1, p. K45–K48 (1982).
See also *5*, p. *69*; *12.1*, p. *133*; 18.2, p. 274.

M. JURCZYK, B. SCZYMAŃSKI, A. WRZECIONO, and A. J. JANICKI
Some Aspects of Magnetism in Amorphous $Co_{90}Zr_{10}$ Alloy
(a) Vol. **74**, No. 1, p. K69–K72 (1982).
See also *2*, p. *54*; *18.2*, p. *274*.

M. ČERŇANSKÝ
Broadening and Shift of Diffraction Lines Due to Partial Coherence of X-Rays
(b) Vol. **114**, No. 2, p. 365–372 (1982).
See also *1.4*, p. *37*; *10.1*, p. *101*.

A. LODDER, P. M. BOERRIGTER, and P. J. BRASPENNING
Cluster Model Residual Resistivities in 35 Copper Alloys
(b) Vol. **114**, No. 2, p. 405–411 (1982).
See also *14.1*, p. *198*.

L. WEISS and H. ESCHRIG
Phonon Anomalies in Copper
(b) Vol. **114**, No. 2, p. 419–427 (1982).
See also *6*, p. *80*; 13.2, p. 157.

L. SKÁLA, L. KÜNNE, H.-G. FRITSCHE, and H. MÜLLER
Cluster Model for Interstitial Hydrogen Bonding in Nickel
(b) Vol. **114**, No. 2, p. 439–448 (1982).
See also *4*, p. *65*; 10.1, p. 102; 13.4, p. 177.

P. RENNERT, U. H. GLÄSER, W. HERGERT, and A. N. TIMOSHEVSKII
Electronic Structure of Intermetallic Compounds by Interpolation Schemes
(b) Vol. **114**, No. 2, p. 449–458 (1982).
See also *13.1*, p. *156*.

H. BROSS
Conduction Spin Density and Knight Shift of Copper
(b) Vol. **114**, No. 2, p. 529–535 (1982).
See also *13.2*, p. *157*; 19, p. 291.

21.1.1 Fe and Fe Alloys

Review Article

F. Reynaud
Order–Disorder Transitions in Substitutional Solid Solutions
(a) Vol. **72**, No. 1, p. 11–59 (1982).
See also *1.2*, p. *14*; *1.3*, p. *25*; 21, p. 330; 21.1, p. 343; 21.6, p. 374.

Original Papers and Short Notes

H. J. Blythe, F. Walz, and H. Kronmüller
Investigation of the Magnetic Aftereffect in Dilute Fe–Ni Alloys after Low-Temperature Neutron Irradiation
(a) Vol. **69**, No. 1, p. 237–248 (1982).
See also *10.1*, p. *96*; 11, p. 118; *18.2*, p. *267*; 21.1, p. 344.

S. B. Fisher and P. K. Madden
The Calculation of the Effect of Point Defect Trapping on Void Growth
(a) Vol. **69**, No. 1, p. 257–268 (1982).
See also *10.1*, p. *96*; 11, p. 118.

H. J. Blythe, F. Dworschak, I. M. Richardson, and F. Walz
Magnetic Relaxation Spectra of Some Dilute Binary Iron-Based Alloys after Low-Temperature Electron Irradiation
(a) Vol. **69**, No. 1, p. K97–K101 (1982).
See also *10.1*, p. *96*; 11, p. 118; *18.2*, p. *267*.

H. J. Blythe, F. Walz, I. M. Richardson, and F. Dworschak
Low-Temperature Magnetic Relaxation in a Dilute Iron–Silicon Alloy after Electron Irradiation at 44 K
(a) Vol. **69**, No. 2, p. 563–567 (1982).
See also *11*, p. *119*; *18.2*, p. *268*.

P. K. Madden and S. B. Fisher
Void Swelling Studies of Stainless Steel Containing Silicon and Titanium
(a) Vol. **69**, No. 2, p. 569–588 (1982).
See also *1.4*, p. *32*; 11, p. 119.

J. Pichon, J. Salardenne, and A. S. Barrière
Study of Electronic Localized States Responsible of the Polarization of FeF_3 Thin Films
(a) Vol. **69**, No. 2, p. 699–706 (1982).
See also 13.4, p. 165; *14.3.1*, p. *213*; *14.3.3*, p. *220*.

T. Kanaizuka
Interpretation of Mössbauer Spectra of β-FeB and Its Low Temperature Modification α-FeB
(a) Vol. **69**, No. 2, p. 739–744 (1982).
See also 1.2, p. 15; *5*, p. *66*; *18.2*, p. *268*.

V. V. NEMOSHKALENKO, A. G. LESNIK, YU. V. KUDRYAVTSEV, I. V. LEZHNENKO, L. I. NIKOLAEV, and L. S. VOSKREKASENKO
Electronic Structure of Fe–45 Al Films
(b) Vol. **110**, No. 1, p. 239–243 (1982).
See also *1.5*, p. *38*; *13*, p. *141*; 20, p. 294.

P. PUREUR, J. V. KUNZLER, W. H. SCHREINER, and D. E. BRANDÃO
Electrical Resistivity of the Cobalt-Rich Co–Fe Alloys
(a) Vol. **70**, No. 1, p. 11–15 (1982).
See also *14.1*, p. *193*; 21.1, p. 345.

S. PARVIAINEN and M. LEHTINEN
Magnetic Contribution to the Thermal Expansion in Some $SiFe_xCo_{1-x}$ Alloys
(a) Vol. **70**, No. 1, p. 79–85 (1982).
See also *8*, p. *86*; *18.2*, p. *269*; 21.1, p. 346.

L. M. CASPERS, A. VAN VEEN, M. R. YPMA, and G. J. VAN DER KOLK
Helium Precipitation in α-Fe
(a) Vol. **70**, No. 1, p. 109–125 (1982).
See also *10.1*, p. *97*.

A. SH. CHAVCHANIDZE, I. S. SHPOTIN, and V. S. STEPANYUK
The Perturbation of Electron Density around Mn in α-Fe Based Dilute Solid Solution
(b) Vol. **110**, No. 2, p. K155–K159 (1982).
See also *13.4*, p. *167*.

Y. J. WANG and H. KRONMÜLLER
The Influence of the Surface Conditions on the Magnetic Properties in Amorphous Alloys $Fe_{40}Ni_{40}B_{20}$ and $Co_{58}Ni_{10}Fe_5Si_{11}B_{16}$
(a) Vol. **70**, No. 2, p. 415–421 (1982).
See also 1.6, p. 42; *18.2*, p. *269*; 21.1, p. 347.

X. Z. DONG and H. KRONMÜLLER
Magnetic Domain Structure of Some Nearly Non-Magnetostrictive Amorphous Alloys under Elastic Stresses
(a) Vol. **70**, No. 2, p. 451–462 (1982).
See also *18.2*, p. *269*; 21.1, p. 347.

R. KUMAR, VISHWAMITTAR, K. CHANDRA, and S. K. DATE
Mössbauer and Magnetic Studies of $Fe_{0.968-x}Mn_xSi_{0.032}$ Ternary Systems
(a) Vol. **70**, No. 2, p. 545–553 (1982).
See also *1.3*, p. *27*; 5, p. 67; 12.1, p. 131; 18, p. 260.

E. A. VASILEV and V. A. VIRCHENKO
Solid Solutions in the $(Fe_{1.22}Sb)_{1-x}(Fe_{1.68}Sn_x)$ System
(a) Vol. **70**, No. 2, p. K141–K143 (1982).
See also *1.3*, p. *28*; *18.2*, p. *270*.

C. K. ONG
Dipole Force Tensors of Vacancy-Type Defects in α-Iron
(b) Vol. **111**, No. 1, p. 331–336 (1982).
See also *10.1*, p. *98*.

J. A. MISIAK
Low Angle Electron Deflection Studies of Domain Wall Structure in Thin NiFe Films
(a) Vol. **71**, No. 1, p. 43–48 (1982).
See also *18.2.1*, p. *275*; 21.1, p. 348.

G. L. WHITTLE, S. J. CAMPBELL, and A. M. STEWART
A Mössbauer Study of the Magnetic Hyperfine Field Distribution in the Amorphous Alloy $Fe_{32}Ni_{36}Cr_{14}P_{12}B_6$
(a) Vol. **71**, No. 1, p. 245–251 (1982).
See also 2, p. 48; *5*, p. *67*; 18.2, p. 270; 21.1, p. 349.

B. CHALUPA and M. VRÁNA
Neutron Diffraction Topography of Fe(Si) Single Crystals
(a) Vol. **71**, No. 1, p. K13–K15 (1982).
See also *1.4*, p. *34*.

PH. BAUER and J. M. GENIN
Dipolar Contribution and Quadrupole Effect in $GdFe_2$
(a) Vol. **71**, No. 2, p. K149–K152 (1982).
See also *5*, p. *67*; 18.3, p. 276.

H. MEHRER, G. FLIK, J. HORVÁTH, and H. KRONMÜLLER
Crystallization and Aging Effects in Some Amorphous Ferromagnets
(a) Vol. **72**, No. 1, p. 215–223 (1982).
See also *2*, p. *50*; *18.2*, p. *271*; 21.1, p. 350.

Q. B. YANG
Coincidence Coefficients of Two Space Lattices and Their Lattice Planes
(a) Vol. **72**, No. 1, p. 343–351 (1982).
See also *1.3*, p. *29*; 21.1, p. 351.

K. MASUDA
Calculation of Binding Energies of Small Vacancy Clusters in B.C.C. Transition Metals
(b) Vol. **112**, No. 2, p. 609–614 (1982).
See also *10.1*, p. *99*; *13.4*, p. *172*; 21, p. 337.

B. RAUSCHENBACH and K. HOHMUTH
Formation of Amorphous Metal–Metalloid Compounds by Ion Implantation
(a) Vol. **72**, No. 2, p. 667–678 (1982).
See also *2*, p. *51*; *11*, p. *125*; 21, p. 338; 21.1, p. 352.

S. PARVIAINEN
Magnetic Contribution to the Thermal Expansion of $TiFe_{0.5}Co_{0.5}$
(a) Vol. **72**, No. 2, p. K141–K144 (1982).
See also *8*, p. *88*; *18.2*, p. *271*; 21.1, p. 352.

K. D. MOORE, P. L. JONES, and F. H. COCKS
A Positron Annihilation Study of the Precipitation Hardening Effects in a Maraging Steel
(a) Vol. **72**, No. 2, p. K223–K227 (1982).
See also *10.1*, p. *100*.

J. P. JERNOT and J. L. CHERMANT
Model to Predict the Tensile Strength of Sintered Materials
(a) Vol. **73**, No. 2, p. 579–586 (1982).
See also 7, p. 83; *12.1*, p. *133*; 14.1, p. 197.

Z. Z. AKSELROD, B. A. KOMMISSAROVA, L. N. KRYUKOVA, A. A. OPALENKO, G. K. RYASNY, and A. A. SOROKIN
Pressure Dependence of the Hyperfine Magnetic Field at ^{181}Ta in the Intermetallic Compound $(Zr_{0.5}Hf_{0.5})Fe_2$
(a) Vol. **73**, No. 2, p. K299–K302 (1982).
See also *18.2*, p. *272*.

H. J. BLYTHE
Magnetic Disaccommodation in Ordered Ni_3Fe Following Either Cold-Work or Low-Temperature Electron Irradiation
(a) Vol. **73**, No. 2, p. K311–K315 (1982).
See also *11*, p. *127*; *18.2*, p. *273*; 21.1, p. 356.

Y. SOMEYA
Phase Diagram of Random Mixtures with Competing Spin Anisotropies: $Fe_{1-x}Co_xCl_2$ and $K_2Mn_{1-x}Fe_xF_4$
(a) Vol. **74**, No. 1, p. 167–172 (1982).
See also *18.4*, p. *281*; 21.1, p. 356.

H. J. BLYTHE, F. WALZ, and F. DWORSCHAK
Magnetic Relaxation in Dilute Iron–Nickel Alloys Following Low-Temperature Electron Irradiation
(a) Vol. **74**, No. 1, p. 181–184 (1982).
See also *10.1*, p. *101*; 11, p. 128; *18.2*, p. *273*; 21.1, p. 356.

J. DEGAUQUE and B. ASTIE
Evolution of the Configuration of Magnetic Domains Interacting with Structural Defects in High Purity Iron
(a) Vol. **74**, No. 1, p. 201–210 (1982).
See also *10.1*, p. *101*; *18.2*, p. *273*.

R. A. JAGO, P. E. CLARK, and P. L. ROSSITER
The Santa Catharina Meteorite and the Equilibrium State of Fe–Ni Alloys
(a) Vol. **74**, No. 1, p. 247–254 (1982).
See also *1.3*, p. *31*; *5*, p. *68*; 18.2, p. 273; 21.1, p. 356.

V. I. PECHERSKAYA and D. N. BOLSHUTKIN
Electrical Resistance and Thermal Expansion of Fe–18Cr–(10 to 25) Ni Alloys in the Paramagnetic Temperature Range
(a) Vol. **74**, No. 1, p. 285–290 (1982).
See also *8*, p. *89*; *14.1*, p. *198*; 21.1, p. 357.

K. CHOUDHURY, B. K. DAS, C. K. MAJUMDAR, and M. ADHIKARI
Study of Corrosion of Steel in Acid Medium by Mössbauer Spectroscopy
(a) Vol. **74**, No. 1, p. K27–K30 (1982).
See also *5*, p. *69*.

21.2 Alkali Metals

Original Papers and Short Notes

Yu. Ya. Boguslavskii
Equation for the Melting Curve of Solids under High Pressure
(b) Vol. **109**, No. 1, p. 145 – 151 (1982).
See also *1.2*, p. *14*; *8*, p. *84*; 12.1, p. 130; 23, p. 509.

G. Cocco, L. Schiffini, S. Enzo, and A. Benedetti
A SAXS Study of Coarsening Mechanism of GP Zones in the Al–Mi–Mg System
(a) Vol. **69**, No. 1, p. 343 – 348 (1982).
See also *1.3*, p. *25*; 10.1, p. 96; 21, p. 331; 21.3, p. 367.

S. C. Mishra and P. L. Srivastava
Thermodynamic Properties of Liquid Alkali Metals and Their Equiatomic Alloys
(b) Vol. **109**, No. 2, p. 457 – 461 (1982).
See also 2, p. 46; *8*, p. *85*.

L. Skála
The CNDO/BW Study of Lithium Clusters (II): B.C.C. Clusters
(b) Vol. **109**, No. 2, p. 733 – 741 (1982).
See also *4*, p. *61*; *13*, p. *140*.

R. V. Gopala Rao and B. M. Satpathy
Partial Structure of Liquid Sodium–Cesium Alloys
(b) Vol. **110**, No. 1, p. 273 – 279 (1982).
See also *2*, p. *46*.

L. Skála
The CNDO/BW Study of Lithium Clusters (III): Different Crystallographic Structures
(b) Vol. **110**, No. 1, p. 299 – 306 (1982).
See also *4*, p. *61*; *13*, p. *141*.

T. Soma, Y. Kimura, and H.-Matsuo Kagaya
Elastic-Stiffness Constants under Pressure of Cs–K, Cs–Rb, and Rb–K Solid Solutions
(b) Vol. **111**, No. 2, p. 679 – 683 (1982).
See also *1.3*, p. *28*; *12.1*, p. *131*.

L. E. Oliveira
On the Long-Wavelength Plasmon in the Alkali Metals. A Nearly-Free-Electron Pseudopotential Calculation
(b) Vol. **111**, No. 2, p. 685 – 690 (1982).
See also *13.5.2*, p. *187*; 14.4, p. 236.

M. Raychaudhuri and S. Chatterjee
Electronic Energy Bands of Some AgMg, CuAl, and AlLi Alloys
(b) Vol. **112**, No. 1, p. 105 – 114 (1982).
See also *13.1*, p. *152*; 21, p. 336; 21.1, p. 349; 21.3, p. 367; 21.6, p. 376.

T. Soma, Y. Kimura, and H.-M. Kagaya
Compression Effect on the Lattice Dynamics of Alkali Metals
(b) Vol. **112**, No. 1, p. 151–154 (1982).
See also *6*, p. *76*; 12.1, p. 131.

W. Gerber, W. Hoppe, O. Brümmer, and G. Dlubek
Estimation of Fermi Surface Parameters from the One-Dimensional Fourier Transform of the Positron Annihilation Angular Correlation Curves
(b) Vol. **112**, No. 1, p. 231–235 (1982).
See also *13.2*, p. *157*; 21, p. 336; 21.3, p. 367.

T. Soma, Y. Kimura, and H. M. Kagaya
Equation of State of the Alloy System Composed of Alkali Metal Elements
(b) Vol. **112**, No. 2, p. K107–K109 (1982).
See also *1.3*, p. *29*; 12.1, p. 132.

Mariana Weissmann and Norah V. Cohan
Charge Transfers in Structurally Disordered Alloys
(b) Vol. **113**, No. 1, p. 395–399 (1982).
See also *2*, p. *52*; *13*, p. *145*; 21.6, p. 377.

F. Brouers and J. Franz
Monte-Carlo Calculation of Density of States and Localization in Alloys Including Short-Range Order
(b) Vol. **113**, No. 2, p. 431–437 (1982).
See also *2*, p. *52*; *13*, p. *145*; 21.6, p. 377.

B. N. Tewari, B. S. Bhargava, and K. N. Khanna
Surface Entropy of Liquid Metals
(b) Vol. **114**, No. 1, p. K31–K34 (1982).
See also *2*, p. *54*; *8*, p. *89*; 21, p. 341; 21.1, p. 356; 21.3, p. 368; 21.7, p. 381.

21.3 Alkali-Earth Metals

Original Papers and Short Notes

G. VAN TENDELOO and S. AMELINCKX
High Resolution Electron Microscopic and Electron Diffraction Study of the Au–Mg System (II): The Y-Phase and Some Observations on the $Au_{77}Mg_{23}$ Phase
(a) Vol. **69**, No. 1, p. 103–120 (1982).
See also *1.3*, p. *25*; *1.4*, p. *32*; 21.6, p. 374.

G. COCCO, L. SCHIFFINI, S. ENZO, and A. BENEDETTI
A SAXS Study of Coarsening Mechanism of GP Zones in the Al–Li–Mg System
(a) Vol. **69**, No. 1, p. 343–348 (1982).
See also *1.3*, p. *25*; 10.1, p. 96; 21, p. 331; 21.2, p. 365.

M. SEN and S. CHATTERJEE
Electronic Energy Bands of Nearly Free Electron Metallic Glasses. An Application to $Mg_{0.7}Zn_{0.3}$ Amorphous Alloy
(b) Vol. **111**, No. 1, p. 353–360 (1982).
See also 2, p. 48; *13.1*, p. *150*; 21.1, p. 348.

V. I. BOGDANOV, A. V. MIKHAILOV, and D. L. FUKS
Thermodynamics of Aluminium- and Magnesium-Base Interstitial Solutions (I): Analysis of Interaction Parameters
(b) Vol. **111**, No. 2, p. 483–489 (1982).
See also *1.3*, p. *28*; 21, p. 335.

C. MINIER, J. LAUZIER, C. ESNOUF, and G. FANTOZZI
A Study of the Peaking Effect in a Hexagonal Structure Metal: Magnesium
(a) Vol. **71**, No. 2, p. 381–386 (1982).
See also *10.1*, p. *98*; 11, p. 122; *12.1*, p. *131*.

M. RAYCHAUDHURI and S. CHATTERJEE
Electronic Energy Bands of Some AgMg, CuAl, and AlLi Alloys
(b) Vol. **112**, No. 1, p. 105–114 (1982).
See also *13.1*, p. *152*; 21, p. 336; 21.1, p. 349; 21.2, p. 365; 21.6, p. 376.

V. I. BOGDANOV, A. V. MIKHAILOV, and D. L. FUKS
Thermodynamics of Aluminium- and Magnesium-Base Interstitial Solutions (II): Energy of Mixing of Binary and Ternary Alloys
(b) Vol. **112**, No. 1, p. 171–178 (1982).
See also *1.3*, p. *29*; 21, p. 336.

W. GERBER, W. HOPPE, O. BRÜMMER, and G. DLUBEK
Estimation of Fermi Surface Parameters from the One-Dimensional Fourier Transform of the Positron Annihilation Angular Correlation Curves
(b) Vol. **112**, No. 1, p. 231–235 (1982).
See also *13.2*, p. *157*; 21, p. 336; 21.2, p. 366.

B. M. LOGINOV and A. A. PREDVODITELEV
Computer Simulation of Dislocation Motion through a Flexible and Reactionable Dislocation Forest of Different Density in NaCl and Mg Crystals
(a) Vol. **72**, No. 1, p. 69–77 (1982).
See also *10.1*, p. *98*; *10.2*, p. *110*; 22.5.2, p. 458.

R. Khanna
Force Model Dependence and Asymptotic Behaviour of the Static Lattice Green Function for H.C.P. Crystals
(b) Vol. **112**, No. 2, p. 667 – 674 (1982).
See also *6*, p. *77*; 10.1, p. 99; 21.1, p. 351.

M. Pstrokoński and W. Chomka
Measurement of the Internal Friction and the Shear Modulus in Cold-Worked Ordered and Disordered Ag_3Mg Alloys
(a) Vol. **72**, No. 2, p. 475 – 482 (1982).
See also *10.1*, p. *99*; *12.1*, p. *132*; 21.6, p. 377.

G. A. Malygin
Impurity Atmospheres and Effective Activation Parameters of Dislocation Motion with Special Reference to Al–Mg and Zr–O Alloys
(a) Vol. **72**, No. 2, p. 493 – 501 (1982).
See also *10.1*, p. *99*; 21, p. 337; 21.1, p. 352.

B. Gueffroy-Oettel and H. Löffler
The Dependence of the Reversion of Guinier-Preston Zones in an Al–4.5 at% Zn–2.0 at% Mg Alloy on the Pre-Ageing Conditions
(a) Vol. **73**, No. 1, p. 153 – 156 (1982).
See also *10.1*, p. *100*; 21, p. 339; 21.1, p. 353.

A. J. Fendrik, A. M. Monti, and E. J. Savino
Self-Interstitial in Hexagonal Close Packed Metals
(b) Vol. **113**, No. 2, p. 709 – 714 (1982).
See also *10.1*, p. *101*.

B. N. Tewari, B. S. Bhargava, and K. N. Khanna
Surface Entropy of Liquid Metals
(b) Vol. **114**, No. 1, p. K31 – K34 (1982).
See also *2*, p. *54*; *8*, p. *89*; 21, p. 341; 21.1, p. 356; 21.2, p. 366; 21.7, p. 381.

B. Gueffroy-Oettel, R. Ramlau, and H. Löffler
On the Relation between Precipitation Density and Size and Hardness during the Reversion Treatment of an Al–4.5 at% Zn–2.0 at% Mg Alloy Pre-Aged at 100 °C
(a) Vol. **74**, No. 1, p. 149 – 157 (1982).
See also *10.1*, p. *101*; 21, p. 341; 21.1, p. 356.

D. J. González, J. A. Alonso, and M. P. Iñiguez
Cellular Density Functional Theory of the Heat of Formation of Disordered Simple Alloys. Influence of the Cell Partitioning
(b) Vol. **114**, No. 2, p. 495 – 501 (1982).
See also *2*, p. *55*; *13*, p. *147*.

R. Gerlach and H. Löffler
Clustering Processes Proceeding in Two Al–2.0 at% Zn–1.3 at% Mg Alloys after Direct Quench to the Ageing Temperature
(a) Vol. **74**, No. 2, p. 517 – 528 (1982).
See also *1.3*, p. *31*; *10.1*, p. *102*; 21, p. 342; 21.1, p. 358.

21.4 Rare-Earth Metals

Original Papers and Short Notes

L. V. PANIA, M. A. SAVCHENKO, and A. V. STEPHANOVICH
The Theory of the Superconductivity Phase in Rare Earth Metal Compounds
(b) Vol. **109**, No. 1, p. 37–48 (1982).
See also *14.2*, p. *199*.

A. ŚLEBARSKI and J. AULEYTNER
Soft X-Ray Emission Spectra of Aluminium and Dysprosium from $DyCo_2$–$DyAl_2$ Intermetallic Compounds
(b) Vol. **109**, No. 1, p. 125–129 (1982).
See also *20*, p. *293*; 21, p. 330; 21.1, p. 343.

M. REIFFERS, K. FLACHBART, Š. JÁNOŠ, A. B. BEZNOSOV, and G. ESKA
Transport Properties of $PrNi_5$ at Low Temperatures
(b) Vol. **109**, No. 1, p. 369–373 (1982).
See also *8*, p. *84*; *14.1*, p. *193*; 21.1, p. 343.

M. DRULIS and B. STALIŃSKI
Low-Temperature Specific Heats of Non-Stoichiometric Neodymium Hydrides
(a) Vol. **69**, No. 1, p. 385–392 (1982).
See also 1.2, p. 15; *8*, p. *85*.

A. APOSTOLOV, H. HRISTOV, T. MIDLAG, M. MIKHOV, and V. SKUMRIEV
The Bulk Amorphous Alloy Tb_4Co_3 – An Example for a System with a Transition to a "Spin-Glass Like" State
(a) Vol. **69**, No. 1, p. K7–K10 (1982).
See also *2*, p. *46*; *18.1*, p. *265*; 18.2, p. 267; 21.1, p. 344.

A. J. MCEVOY and W. GISSLER
ESCA Spectra and Electronic Properties of Some Ruthenium Compounds
(a) Vol. **69**, No. 1, p. K91–K96 (1982).
See also 13, p. 140; *17*, p. *255*; 22.6, p. 469.

P. TRZASKOMA, J. ZIELIŃSKI, and M. SZOPA
On the Role of Phonons for the Pressure Induced Change of Valence
(b) Vol. **109**, No. 2, p. K129–K132 (1982).
See also *1.2*, p. *15*; *6*, p. *71*.

A. A. DAKHEL
X-Ray Kβ and $L_{2,3}$ Emission Bands of Silicon in La–Si Compounds
(b) Vol. **109**, No. 2, p. K133–K134 (1982).
See also *20*, p. *294*; 22.1.2, p. 402.

H. KONGUETSOF and J. K. YAKINTHOS
Magnetic Properties of RE_5Ga_3 Compounds (RE = Pr, Nd, Gd, Tb, Dy, Ho, Er, Tm, and Y)
(a) Vol. **69**, No. 2, p. K157–K161 (1982).
See also *18*, p. *260*.

A. V. Andreev, E. N. Tarasov, A. V. Deryagin, and S. M. Zadvorkin
Influence of Symmetry of the Crystalline Structure on the Magnetic Anisotropy of Tb_2Co_7
(a) Vol. **71**, No. 2, p. K245–K247 (1982).
See also *1.2*, p. *19*; 1.3, p. 29; *18.2*, p. *271*; 21.1, p. 349.

A. Sahling, P. Frach, and E. Hegenbarth
The Heat Capacity of $LaNi_2$, $LaNi_5$, $PrNi_2$, and $PrNi_5$ in the Temperature Range between 0.3 and 7 K
(b) Vol. **112**, No. 1, p. 243–250 (1982).
See also *8*, p. *87*; 21.1, p. 350.

J. P. Jernot, M. Coster, and J. L. Chermant
Model to Describe the Elastic Modulus of Sintered Materials
(a) Vol. **72**, No. 1, p. 325–332 (1982).
See also *12.1*, p. *132*; *14.1*, p. *196*.

A. del Moral, J. F. D. Montenegro, J. S. Abell, and M. R. Ibarra
On the Magnetization and Magnetostriction of ErZn
(a) Vol. **72**, No. 1, p. 353–360 (1982).
See also 12.1, p. 132; *18.1*, p. *266*; 18.2, p. 271; 21.1, p. 351.

Z. Drzazga and T. Mydlarz
Magnetic Properties of the 3d Metal-Rich $HoCo_5$ Compounds
(a) Vol. **72**, No. 2, p. 771–776 (1982).
See also *18.3*, p. *277*; 21.1, p. 352.

E. Burzo, D. Seitabla, and M. Chipară
Magnetic Properties and Resonance Study of $Gd(Co_xAl_{1-x})_3$ Compounds
(b) Vol. **113**, No. 1, p. 87–92 (1982).
See also *18.1*, p. *266*; *18.3*, p. *277*; 19, p. 289; 21, p. 338; 21.1, p. 352.

A. V. Andreev, A. V. Deryagin, V. N. Moskalev, and N. V. Mushnikov
On Crystalline Structure of Hydrides of $ErFe_2$ and $HoFe_2$
(a) Vol. **73**, No. 1, p. K69–K71 (1982).
See also *1.2*, p. *21*; *1.3*, p. *30*; 18.2, p. 272; 21.1.1, p. 362.

A. Castets, D. Gignoux, J. C. Gomez-Sal, F. Rodriguez-Gonzalez, and E. Roudaut
Magnetic Properties and Structure of Tb_2Pt
(a) Vol. **73**, No. 2, p. 475–481 (1982).
See also *1*, p. *10*; *18.2*, p. *272*; 18.4, p. 281; 21.6, p. 378.

H. Ratajczak and I. Gościańska
Thermal Hall Hysteresis Loops in Thin Amorphous HoCo Films
(a) Vol. **73**, No. 2, p. 633–639 (1982).
See also 2, p. 53; *14.1*, p. *197*; *18.3*, p. *277*; 21.1, p. 355.

S. Adam, Gh. Adam, and A. Corciovei
Effective Interatomic Forces from One-Particle Crystal Field Parameters
(b) Vol. **114**, No. 1, p. 85–93 (1982).
See also *4*, p. *64*; 13, p. 146.

21.5 Actinides

Original Papers and Short Notes

H. Matsui and M. Tamaki
Electrical Resistivity of β-U_2N_3
(a) Vol. **71**, No. 1, p. K121–K125 (1982).
See also *14.1*, p. *195*; 18.2, p. 270.

A. Blaise, J. M. Fournier, and D. Damien
Magnetic Properties of Plutonium Diantimonides
(a) Vol. **71**, No. 2, p. K147–K148 (1982).
See also 18.1, p. 266; *18.4*, p. *280*; 21.7, p. 380.

21.6 Noble Metals

Review Articles

Original Papers and Short Notes

G. van Tendeloo and S. Amelinckx
Electron Diffraction and Electron Microscopic Study of the Alloy Au_2Mn and Its Formation Mechanism
(a) Vol. **69**, No. 2, p. 589 – 602 (1982).
See also *1.3*, p. *26*; 1.4, p. 32; **21.1, p. 344.**

E. Harnik, S. Kovnovich, and T. Chernobelskaya
Secondary Acoustoelectric Current in Ultrathin Metal Films Deposited on a Piezoelectrie Substrate
(a) Vol. **69**, No. 2, p. K179 – K181 (1982).
See also 1.5, p. 38; *7*, p. *82*.

P. Paul, S. P. Puri, D. Raj, and R. K. Puri
Force Constant Change in Sn–Impurity Systems: $^{107}Pd:^{119}Sn$, $^{197}Au:^{119}Sn$, and $^{63.5}Cu:^{119}Sn$
(b) Vol. **110**, No. 1, p. 81 – 87 (1982).
See also *5*, p. *66*; 6, p. 72; 10.1, p. **97**; **21.1, p. 345.**

G. van Tendeloo, J. van Landuyt, and S. Amelinckx
High Resolution Electron Microscopic and Electron Diffraction Study of Gold–Manganese Alloys with a Composition in the Vicinity of Au_3Mn
(a) Vol. **70**, No. 1, p. 145 – 158 (1982).
See also *1.3*, p. *26*; 1.4, p. 33; **21.1, p. 346.**

W. Kleinn
Determination of the Residual Resistivity from Current–Voltage Characteristics of Short Support-Free Samples of Epitaxial Gold Films
(a) Vol. **70**, No. 1, p. 219 – 232 (1982).
See also *1.5*, p. *38*; *14.1*, p. *194*.

J. Teuho and K. Mäkinen
On the Out-of-Step Period in $Au_3Zn(H)$
(a) Vol. **70**, No. 1, p. K1 – K3 (1982).
See also *1.3*, p. *27*; *1.4*, p. *33*; **21.1, p. 346.**

V. I. Bogdanov, A. V. Ruban, and D. L. Fuks
Covalent Contribution to the Bond Energy of Phases with Ll_2 Structure
(b) Vol. **110**, No. 2, p. 431 – 435 (1982).
See also 1.2, p. 16; *1.3*, p. *27*; *4*, p. *62*; **21.1, p. 346.**

R. Blessing and H. Pagnia
Electron Emission from Gold Island Films
(b) Vol. **110**, No. 2, p. 537 – 542 (1982).
See also *17*, p. *255*.

G. A. Iferov and Yu. N. Zhukova
The Impact-Parameter Dependence of Energy Losses in Proton Scattering by Gold Atoms
(b) Vol. **110**, No. 2, p. 653 – 658 (1982).
See also *11*, p. *120*.

J. van Landuyt, G. van Tendeloo, and S. Amelinckx
New Types of Periodic Antiphase Boundary Structures Derived from the DO_{22} Structure in the Gold–Manganese System
(a) Vol. **70**, No. 2, p. 407 – 414 (1982).
See also *1.3*, p. *27*; 1.4, p. 33; 21.1, p. 347.

R. DIMMICH and F. WARKUSZ
Optical Reflectance and Electrical Resistance of Polycrystalline Thin Metallic Films
(a) Vol. **72**, No. 1, p. 117–122 (1982).
See also *14.3.1*, p. *215*; *20.1*, p. *308*.

I. M. REDA, J. HAFNER, P. PONGRATZ, A. WAGENDRISTEL, H. BANGERT, and P. K. BHAT
Amorphous Cu–Ag Films with High Stability
(a) Vol. **72**, No. 1, p. 313–324 (1982).
See also *1.3*, p. *29*; *2*, p. *50*; 14.1, p. 196; 21.1, p. 351.

O. H. PRASAD and M. SURYANARAYANA
Fourth-Order Elastic Constant, C_{1111}, of Cubic Crystals
(b) Vol. **112**, No. 2, p. 627–631 (1982).
See also *12.1*, p. *132*; *12.2*, p. *136*; 21, p. 337; 21.1, p. 351; 22.1, p. 393; 22.5.2, p. 459.

M. PSTROKOŃSKI and W. CHOMKA
Measurements of the Internal Friction and the Shear Modulus in Cold-Worked Ordered and Disordered Ag_3Mg Alloys
(a) Vol. **72**, No. 2, p. 475–482 (1982).
See also *10.1*, p. *99*; *12.1*, p. *132*; 21.3, p. 368.

M. S. BAPNA
An Estimate of the Precipitation Hardening by Coherent Particles
(a) Vol. **72**, No. 2, p. 799–807 (1982).
See also *1.4*, p. *35*; *10.1*, p. *100*; 21, p. 338; 21.1, p. 352.

MARIANA WEISSMANN and NORAH V. COHAN
Charge Transfers in Structurally Disordered Alloys
(b) Vol. **113**, No. 1, p. 395–399 (1982).
See also *2*, p. *52*; *13*, p. *145*; 21.2, p. 366.

M. R. KHAN, I. K. SCHULLER, and CH. M. FALCO
Sputter Epitaxy of Ag and Ni Films
(a) Vol. **73**, No. 1, p. 23–29 (1982).
See also *1.5*, p. *40*; *3*, p. *58*; 21.1, p. 353.

H. KIRIHATA
Exoelectron Emission from Metals (Cu, Zn, Au) Excited by Low-Energy Electrons below 200 eV
(a) Vol. **73**, No. 1, p. 71–79 (1982).
See also 13.3, p. 160; *17*, p. *256*; 21.1, p. 353.

BAI-XIN LIU, M-A. NICOLET, and S. S. LAU
Ti–Au Metallic Glasses Formed by Ion Mixing
(a) Vol. **73**, No. 1, p. 183–188 (1982).
See also *1.5*, p. *41*; 2, p. 52; *11*, p. *126*; 21.1, p. 353.

F. BROUERS and J. FRANZ
Monte-Carlo Calculation of Density of States and Localization in Alloys Including Short-Range Order
(b) Vol. **113**, No. 2, p. 431–437 (1982).
See also *2*, p. *52*; *13*, p. *145*; 21.2, p. 366.

21.7 Semi-Metals
(Semiconducting Alloys See 22.7)

Original Papers and Short Notes

W. Kraak, R. Herrmann, and H. Haupt
Investigation of the Charge Carrier Scattering in Bi under High Hydrostatic Pressure
(b) Vol. **109**, No. 2, p. 785–792 (1982).
See also 6, p. 71; *14*, p. *190*.

U. Pietsch
The Validity of Vegard's Rule for the Solid Solution System $Bi_{1-x}Sb_x$
(b) Vol. **110**, No. 1, p. K5–K9 (1982).
See also *1.3*, p. *26*; 4, p. 62.

M. S. Rahman Khan
In Situ Study of the Electrical Resistivity and Temperature Coefficient of Resistance of Bismuth Films
(b) Vol. **110**, No. 1, p. K41–K44 (1982).
See also *14.1*, p. *193*.

H. Müller
Simulation of Oriented Crystallization in Evaporated Antimony Thin Films
(a) Vol. **70**, No. 1, p. 249–255 (1982).
See also *1.5*, p. *39*; *3*, p. *57*.

Z. Kletowski
Thermopower of Monocrystalline $RESn_3$ Compounds (RE = La, Ce, Pr, Nd, Sm, Eu, and Gd)
(a) Vol. **70**, No. 1, p. K51–K53 (1982).
See also *15*, p. *245*; 21.4, p. 370.

W. Braune, B. Fellmuth, N. Kubicki, and R. Herrmann
Composition Dependence of the Electron Fermi Surfaces Tilt in $Bi_{1-x}Sb_x$ Alloys ($0 \leqq x \lesssim 0.22$)
(b) Vol. **110**, No. 2, p. 549–556 (1982).
See also *13.2*, p. *157*.

K. Schröder and D. Onengut
Effect of Stress Waves on the Magnetization of Magnesiowüstite with Magnesioferrite Precipitates
(a) Vol. **71**, No. 1, p. K93–K97 (1982).
See also *18.2*, p. *270*; 21.1, p. 349.

V. Damodara Das and S. Vaidehi
Variation of Energy Gap and Resistivity Minimum Position with Thickness in Bismuth Thin Films
(a) Vol. **71**, No. 2, p. 351–356 (1982).
See also 14.1, p. 195; *14.3.1*, p. *214*.

B. N. Tewari, B. S. Bhargava, and K. N. Khanna
Surface Entropy of Liquid Metals
(b) Vol. **114**, No. 1, p. K31 – K34 (1982).
See also *2*, p. *54*; *8*, p. *89*; 21, p. 341; 21.1, p. 356; 21.2, p. 366; 21.3, p. 368.

B. I. Belevtsev, Yu. F. Komnik, V. I. Odnokozov, and A. V. Fomin
Discovery of Low-Temperature Metastable State in Amorphous Bismuth–Hydrogen Systems
(b) Vol. **114**, No. 2, p. 467 – 474 (1982).
See also *1.5*, p. *41*; 2, p. 55; *14.2*, p. *201*.

22. Semiconductors and Ionic Crystals

Review Articles

M. Prasad
Quantum-Statistical Theories of the Cyclotron Resonance Lineshape in Parabolic Semiconductors
(b) Vol. **109**, No. 1, p. 11–35 (1982).
See also 13.4, p. 162; *19*, p. *282*.

F. Bechstedt
Electronic Relaxation Effects in the Core Level Spectra of Solids
(b) Vol. **112**, No. 1, p. 9–49 (1982).
See also 4, p. 61; *13.1*, p. *148*; 17, p. 255; 20, p. 293; 21, p. 330; 23, p. 509.

H. Böttger and V. V. Bryksin
Hopping Conductivity in Ordered and Disordered Systems (III)
(b) Vol. **113**, No. 1, p. 9–49 (1982).
See also 2, p. 45; 6, p. 70; 13, p. 139; 13.5.3, p. 188; *14.3*, p. *202*.

P. P. Seregin, A. R. Regel, A. A. Andreev, and F. S. Nasredinov
A Study of Local Structure of Chalcogenide Vitreous Semiconductors and of States of Impurity Atoms in Them by Means of Mössbauer and Electron Spectroscopies
(a) Vol. **74**, No. 2, p. 373–394 (1982).
See also 1.2, p. 14; *2*, p. *45*; 5, p. 66; *13.4*, p. *162*.

Original Papers and Short Notes

L. Nedbal
Higher-Order Test of the Kenkre-Knox Relation for Excitonic Optical and Transport Properties of Molecular Crystals
(b) Vol. **109**, No. 1, p. 109–118 (1982).
See also *13.5.1*, p. *181*; 14.3, p. 202; 20.1, p. 298.

V. I. Litvinov, K. A. Valuev, and K. D. Tovtsyuk
Redistribution of Impurities in the Field of a Strong Electromagnetic Waves
(b) Vol. **109**, No. 1, p. 189–194 (1982).
See also *13.4*, p. *162*; 14.3, p. 202; 20, p. 293.

V. Čápek and R. W. Munn
From Band to Hopping Diffusion of Excitons and Charge Carriers in Molecular Crystals (II): Results for Local and Weak Nonlocal Electron–Phonon Coupling
(b) Vol. **109**, No. 1, p. 245–253 (1982).
See also 6, p. 70; *13.5.1*, p. *181*.

V. A. Kozlov and I. M. Nefedov
Modified Self-Scattering Procedure for Monte-Carlo Simulation of Hot Electrons
(b) Vol. **109**, No. 1, p. 393–401 (1982).
See also *13*, p. *139*; *14.3.4*, p. *227*.

H. van Cong, S. Brunet, and S. Charar
Fermi-Dirac Integrals and Fermi Energy for Degenerate Narrow-Gap Semiconductors
(b) Vol. **109**, No. 1, p. K1–K5 (1982).
See also *13*, p. *139*; *14.3*, p. *202*.

A. V. Vaisleib, V. P. Oleinikov, and Yu. B. Rozenfeld
Local and Pseudolocal Jahn-Teller States of Crystals with T Level Impurity
(b) Vol. **109**, No. 1, p. K35–K38 (1982).
See also *13.4*, p. *163*.

G. M. Shmelev, G. I. Tsurkan, and E. M. Ephstein
Photostimulated Radioelectrical Transverse Effect in Semiconductors
(b) Vol. **109**, No. 1, p. K53–K58 (1982).
See also *16*, p. *249*.

Hj. Matzke and C. Politis
The Fracture Properties of NbC, Nb(C, N), and NbN by Hertzian Indentation
(a) Vol. **69**, No. 1, p. 269–279 (1982).
See also *10.2*, p. *104*; 12.2, p. 134.

C. Manolikas
Electron Microscopic Study of the Ordered and Disordered Phases in Ga_2Se_3
(a) Vol. **69**, No. 1, p. 393–405 (1982).
See also *1.3*, p. *26*; *1.4*, p. *32*.

C. Gunkel, D. Schalch and A. Scharmann
Electron Beam Induced Leakage Currents in Silicon Nitride Thin Films
(a) Vol. **69**, No. 1, p. K131–K134 (1982).
See also *11*, p. *118*; *14.3.1*, p. *213*.

Nguyen Ba An, Nguyen Van Hieu, Nguyen Toan Thang, and Nguyen Ai Viet
On the Role of the Coulomb Interaction of Charge Carriers in the Electronic Raman Scattering on Donor Levels in Direct Band Gap Semiconductors
(b) Vol. **109**, No. 2, p. 463–472 (1982).
See also *13.4*, p. *164*; 13.5.1, p. 181; *20.1*, p. *300*.

E. Gutsche
Non-Condon Approximations and the Static Approach in the Theory of Non-Radiative Multiphonon Transitions
(b) Vol. **109**, No. 2, p. 583–597 (1982).
See also *6*, p. *71*; *13.4*, p. *164*; 20.3, p. 319.

K. Peuker, R. Enderlein, A. Schenk, and E. Gutsche
Theory of Non-Radiative Multiphonon Capture Processes. Solution of Old Controversies
(b) Vol. **109**, No. 2, p. 599–606 (1982).
See also *6*, p. *71*; *13.4*, p. *164*; 20.3, p. 319.

F. Sukegawa, M. Kobayashi, and I. Yokota
Collective Motion in Silver Chalcogenides
(b) Vol. **109**, No. 2, p. 717–725 (1982).
See also *6*, p. *71*; *13.5.2*, p. *187*; 14, p. 190.

C. Trallero Giner
Long-Wave Secondary Radiation in Polar Semiconductors as a Function of Temperature
(b) Vol. **109**, No. 2, p. 817–824 (1982).
See also *6*, p. *71*; *20*, p. *294*.

V. D. Krevchik and A. Ya. Yafasov
Two-Photon Interband Absorption in Quantized Semiconductor Films with Participation of Deep Impurity Centers
(b) Vol. **109**, No. 2, p. K97–K101 (1982).
See also 13.1, p. 149; *13.4*, p. *164*; *20.1*, p. *300* .

C. M. Vodenicharov, A. A. Stanchev, and S. G. Christov
A New Method for Determination of the Metal–Insulator (Semiconductor) Work Function by Photoelectric Measurements
(a) Vol. **69**, No. 2, p. 541–546 (1982).
See also *14.3.1*, p. *213*; *14.3.3*, p. *219*; 16, p. 250.

L. Pasemann
EBIC Contrast of Surface-Parallel Dislocations in a Planar Semiconductor Diode with Large Surface Recombination
(a) Vol. **69**, No. 2, p. K199–K203 (1982).
See also *10.2*, p. *105*; 14.3.3, p. 220.

J. Sinkkonen
Photoadmittance of an Inhomogeneous Semiconductor
(a) Vol. **69**, No. 2, p. K233–K237 (1982).
See also *14.3.4*, p. *228*; 16, p. 250.

T. D. Clark, K. K. Bajaj, W. M. Theis, and D. E. Phelps
Theory of Donor–Bound Exciton Complexes in Semiconductors
(b) Vol. **110**, No. 1, p. 341–347 (1982).
See also *13.4*, p. *165*; *13.5.1*, p. *182*.

M. I. Auslender, N. G. Bebenin, A. B. Zolotovitskii, and V. P. Kalashnikov
On the Density of States of a Broad-Band Ferromagnetic Semiconductor at $T \geqq T_c$
(b) Vol. **110**, No. 1, p. 369–377 (1982).
See also *13*, p. *141*; 18.2, p. 269.

S. Ghosh and V. K. Agarwal
Excitation of Acoustohelicon Waves Due to Modulation of a Laser Beam in Longitudinally Magnetised Semiconductors
(b) Vol. **110**, No. 1, p. K37–K40 (1982).
See also *7*, p. *82*; 13, p. 141; 18, p. 260; *20*, p. *294*.

A. M. Berezhkovskii and A. A. Ovchinnikov
Influence of Impurities on the Wannier-Stark Ladder in a Semiconductor in a Strong Electric Field
(b) Vol. **110**, No. 2, p. 455–459 (1982).
See also *13*, p. *142*.

M. van der Meer, R. Schuchardt and R. Keiper
Strong-Field Hopping in Disordered Semiconductors. A Problem of Directed Percolation
(b) Vol. **110**, No. 2, p. 571–580 (1982).
See also 2, p. 47; *14.3*, p. *205*.

A. A. Ignatov and V. I. Shashkin
Current–Voltage Characteristics of the Semiconductor Superlattice Interacting with a Strong High-Frequency Field
(b) Vol. **110**, No. 2, p. K117–K119 (1982).
See also *14.3.4*, p. *228*.

V. B. Ptashnik, T. Yu. Dunaeva, and Yu. M. Baikov
The Electrical Conductivity of Solid Lithium Hybride Doped with Sulphur
(b) Vol. **110**, No. 2, p. K121–K124 (1982).
See also *10.2*, p. *107*; *14.4*, p. *235*.

E. S. Alekseev, R. G. Arkhipov, and S. V. Popova
Band Structure of Hexagonal Tungsten Carbide
(b) Vol. **110**, No. 2, p. K151–K154 (1982).
See also *13.1*, p. *150*.

J. Shanker and J. P. Singh
Third- and Fourth-Order Elastic Constants for Silver Halides, Alkaline Earth Oxides, and Chalcogenides of Pb and Sn
(a) Vol. **70**, No. 2, p. 677–681 (1982).
See also *12.2*, p. *135*; 22.5.1, p. 453; 22.6, p. 470.

F. Schauer, M. Jedlička, and J. Polcer
Amorphous–Crystalline Heterojunction with Built-in Interface Charge (I): Steady-State Transport
(a) Vol. **70**, No. 2, p. 755–762 (1982).
See also 2, p. 48; 13.3, p. 159; *14.3.3*, p. *221*; 22.4.2, p. 443.

P. P. Seregin and F. S. Nasredinov
Study of Gold and Platinum Impurity Atom State in Vitreous Arsenic Selenide
(a) Vol. **70**, No. 2, p. 769–775 (1982).
See also *2*, p. *48*; 5, p. 67; *10.2*, p. *108*; 14.3, p. 206; 17, p. 256.

V. Kumar, A. R. Chetal, and K. S. Srivastava
Calculation of the Density of Compounds from the Chemical Shift of the X-Ray Absorption Edge
(a) Vol. **70**, No. 2, p. K107–K111 (1982).
See also *13*, p. *142*; *20*, p. *295*; 21.1, p. 348.

J. Gazsó
A Refinement upon the Method of Evaluating the Density of the Localized States from the Field Effect
(a) Vol. **70**, No. 2, p. K181–K183 (1982).
See also 2, p. 48; *13.4*, p. *168*; 14.3.1, p. 214; *14.3.4*, p. *228*.

H. Finkenrath, G. Franz, and N. Uhle
Optical Absorption in $PbSe_{1-x}Te_x$ Due to Transitions between Band States and Localized Energy Levels
(b) Vol. **111**, No. 1, p. 125–130 (1982).
See also 13.1, p. 150; *20.1*, p. *304*.

I. M. Dykman and L. A. Prokhniskii
Influence of Electron Heating on Reflection and Attenuation Coefficients of Electromagnetic Waves in Semiconductors with Superlattice
(b) Vol. **111**, No. 2, p. 449–454 (1982).
See also *13*, p. *143*; *14.3*, p. *206*; 20, p. 295.

R. S. Singh, V. K. Jain, and J. Shanker
Relative Stability and Structural Phase Transitions in Alkaline Earth Chalcogenide Crystals
(b) Vol. **111**, No. 2, p. 519–523 (1982).
See also *1.2*, p. *17*; 4, p. 63.

E. Sigmund and K. Lassmann
The Influence of the Dynamic Jahn-Teller Effect on the Phonon-Scattering Mechanism at Acceptors in Semiconductors with Cubic Symmetry
(b) Vol. **111**, No. 2, p. 631–637 (1982).
See also *6*, p. *74*; *13.4*, p. *170*.

T. G. Petrova
Electron–Magnon Impurity States in the Energetic Spectrum of the Ferromagnetic Semiconductor
(b) Vol. **111**, No. 2, p. K141–K145 (1982).
See also *13.4*, p. *170*; 13.5.4, p. 189; *18.2*, p. *270*.

F. Schauer, M. Jedlička, and J. Macků
Amorphous–Crystalline Heterojunction with Built-in Interface Charge (II): Non-Steady-State Transport
(a) Vol. **71**, No. 2, p. 335–343 (1982).
See also 2, p. 49; 13.3, p. 159; *14.3.3*, p. *222*; 22.4.2, p. 443.

W. J. Kordalski and B. M. Wilamowski
Method for Determination of Charge Density Distribution in Silicon Nitride of MNOS Structures
(a) Vol. **71**, No. 2, p. 387–398 (1982).
See also 14.3.2, p. 217; *14.3.3*, p. *222*.

J. M. Dusseau and S. Abdel-Maksoud
High Electric Field Effect in Boron–Silicon Compounds
(a) Vol. **71**, No. 2, p. K191–K196 (1982).
See also *14.3.4*, p. *230*.

V. Holý
The Coherence Description of the Dynamical X-Ray Diffraction from Randomly Disordered Crystals (II): Some Numerical Results
(b) Vol. **112**, No. 1, p. 161–169 (1982).
See also *1.1*, p. *12*; 10.2, p. 110.

F. Kh. Abdullaev, A. A. Abdumalikov, and A. S. Rakhmatov
The Dynamics of Solitons in One-Dimensional Molecular Crystals
(b) Vol. **112**, No. 1, p. K5–K10 (1982).
See also *6*, p. *76*; *13.5*, p. *179*.

M. Mäenpää, I. Suni, D. Sigurd, M. Finetti, and M-A. Nicolet
Stable Metallization Systems for Solar Cells
(a) Vol. **72**, No. 2, p. 763–769 (1982).
See also *14.3.3*, p. *224*; 14.3.4, p. 230.

J. M. López and J. A. Alonso
Semiempirical Study of Metastable Alloys Obtained by Ion Implantation in Metals and Semiconductors
(a) Vol. **72**, No. 2, p. 777–781 (1982).
See also *1.3*, p. *30*; *11*, p. *125*; 21, p. 338.

K. C. Sharma and D. R. Sharma
On the Gunn Instability near Threshold
(a) Vol. **72**, No. 2, p. K137–K140 (1982).
See also *13*, p. *144*; *14.3.4*, p. *230*.

T. Suzuki, Y. Hirose, and H. Hirose
A Model of the Photodoping Mechanism in the Ag/As_2S_3 System
(a) Vol. **72**, No. 2, p. K165–K168 (1982).
See also *10.2*, p. *113*; 14.3.2, p. 217; *16*, p. *253*.

G. Paasch and H. Übensee
A Modified Local Density Approximation. Electron Density in Inversion Layers
(b) Vol. **113**, No. 1, p. 165–178 (1982).
See also *13.3*, p. *160*.

S. Takács and I. Janetka
Influence of Resonance Tunneling on Resistivity in Semiconductors with Ionized Impurities and on Tunneling Current between Superconductors
(b) Vol. **113**, No. 1, p. 313–326 (1982).
See also 14.2, p. 200; *14.3.4*, p. *231*.

S. F. Nee, T. W. Nee, S. F. Fan, and D. W. Lynch
Wavelength Modulated Reflectance Spectrum of ZrS_3
(b) Vol. **113**, No. 1, p. K5–K8 (1982).
See also 13.1, p. 154; *20.1*, p. *310*.

R. S. Agrawal, Y. K. Sharma, V. K. Gautam, and C. S. Agrawal
Exact J–U Characteristic of an Insulator with Traps Lying above the Fermi Level in Non-Constant Mobility Regime
(a) Vol. **73**, No. 1, p. K27–K30 (1982).
See also *14.3.3*, p. *224*; *14.3.4*, p. *231*.

H.-D. Geiler, K. Hehl, and D. Stock
A Model of Energy Deposition into Semiconductors during Laser Annealing
(a) Vol. **73**, No. 1, p. K57–K62 (1982).
See also *11*, p. *126*.

E. G. Tishkovsky
Random Walk and Interaction of Vacancies with Imperfections in Diamond Lattice as a Discrete Markovian Process
(b) Vol. **113**, No. 2, p. 459–466 (1982).
See also *10.2*, p. *114*.

V. L. Bonch-Bruevich and To Thi Bich Nguyet
Low Temperature Conductivity and Optical Transitions of a Two-Dimensional Disordered Semiconductor under Weak Localization Conditions
(b) Vol. **114**, No. 1, p. 67 – 71 (1982).
See also 2, p. 53; *14.3*, p. *211*; *20.1*, p. *314*.

E. Heiner
The Dielectric Response of a Charge Density Wave (CDW) Distorted Electron–Phonon System in the Precritical Region (II): Macroscopic Dielectric Function and Conductivity
(b) Vol. **114**, No. 1, p. 121 – 124 (1982).
See also *13*, p. *146*; 14.4, p. 238.

B. Esser
On the Theory of Time-Resolved Band-Tail Luminescence in Amorphous Semiconductors
(b) Vol. **114**, No. 1, p. 175 – 187 (1982).
See also 2, p. 54; *20.3*, p. *327*.

E. Z. Imamov and V. D. Krevchik
On the Diamagnetism of Deep Impurities in Quantized Semiconductor Films
(b) Vol. **114**, No. 1, p. 201 – 207 (1982).
See also 13.4, p. 176; *18*, p. *263*.

V. K. Jain and J. Shanker
Analysis of Interionic Potentials in Alkali Chalcogenide Crystals
(b) Vol. **114**, No. 1, p. 271 – 275 (1982).
See also *4*, p. *65*; 12.2, p. 137.

J. Auth, U. Henniger, and R. Keiper
On the Formulation of Transport Equations in Semiconductors with Smooth Macroscopic Inhomogeneity
(b) Vol. **114**, No. 1, p. K75 – K78 (1982).
See also *14.3*, p. *211*.

J. C. Bernède, A. Khelil, M. Kettaf, and A. Conan
Transition from S- to N-Type Differential Negative Resistance in Al–Al_2O_3–$Ag_{2-x}Se_{1+x}$ Thin Film Structures
(a) Vol. **74**, No. 1, p. 217 – 224 (1982).
See also *14.3.1*, p. *216*; *14.3.4*, p. *232*.

V. M. Bazovkin, G. L. Kurushev, and V. G. Polovinkin
On the Theory of the MIS High-Frequency Capacitance. The Effect of Degeneration and Non-Parabolicity
(a) Vol. **74**, No. 1, p. 297 – 306 (1982).
See also *14.3.3*, p. *225*.

A. M. Andriesh, A. I. Buzdugan, L. I. Zelenina, and S. D. Shutov
Dark Decay of Surface Potential in Vitreous As_2S_3
(a) Vol. **74**, No. 1, p. K79 – K82 (1982).
See also *2*, p. *54*; *14.3.1*, p. *216*.

V. M. Fomin amd E. P. Pokatilov
Conductivity of Semiconductors Influenced by Strong Electromagnetic Radiation
(b) Vol. **114**, No. 2, p. 615 – 625 (1982).
See also 6, p. 81; 13, p. 147; *14.3.4*, p. *232*.

22.1 Elements

Review Article

R. J. J. Zijlstra
Distribution Model of Flicker Noise in Semiconductors
(a) Vol. **73**, No. 1, p. 11–21 (1982).
See also *14.3*, p. *202*; 14.3.3, p. 219; 14.3.4, p. 227; 22.2, p. 418.

Original Papers and Short Notes

N. R. S. Tait and D. W. L. Tolfree
Optical Absorption Measurements on Carbon Foils
(a) Vol. **69**, No. 1, p. 201–206 (1982).
See also *20.1*, p. *299*.

A. Ghosh, L. K. Banerjee, and A. N. Basu
Cohesion, Phase Transition, Elastic and Dielectric Properties of Group IV Semiconductors
(b) Vol. **109**, No. 2, p. 799–806 (1982).
See also *1.2*, p. *15*; *4*, p. *61*; 12.2, p. 134; 14.4, p. 234; 22.1.1, p. 395; 22.1.2, p. 401.

S. Myhra
Acceptor Point Defects in n-Type α-Sn
(b) Vol. **110**, No. 1, p. 97–103 (1982).
See also *13.4*, p. *165*; *14.3*, p. *204*; 15, p. 245.

G. B. Bachelet, C. Tanner, and M. Schluter
Static Atomic Polarizabilities from Norm-Conserving Pseudopotentials
(b) Vol. **110**, No. 1, p. 313–322 (1982).
See also *4*, p. *62*; *13*, p. *141*; 22.1.2, p. 403.

K. Bewilogua, D. Dietrich, G. Holzhüter, and Ch. Weissmantel
Structure of Amorphous Carbon Films
(a) Vol. **71**, No. 1, p. K57–K59 (1982).
See also *1.5*, p. *39*; *2*, p. *49*.

Diep-The-Hung and J. L. Motchane
Magnetic Properties of Oxygen Monolayers Adsorbed on Graphite
(b) Vol. **111**, No. 2, p. 639–648 (1982).
See also *18.4*, p. *280*.

C. Raisin, Y. Bertrand, and J. Robin
Photoemission Studies of Crystalline Arsenic
(b) Vol. **112**, No. 1, p. 261–267 (1982).
See also *13.1*, p. *152*; 17, p. 256.

O. H. Prasad and M. Suryanarayana
Fourth-Order Elastic Constant, C_{1111}, of Cubic Crystals
(b) Vol. **112**, No. 2, p. 627–631 (1982).
See also *12.1*, p. *132*; *12.2*, p. *136*; 21, p. 337; 21.1, p. 351; 21.6, p. 377; 22.5.2, p. 459.

22.1.1 Germanium

Original Papers and Short Notes

J. METZDORF and F. R. KESSLER
Quantitative Analysis of the Nonresonant Interband Faraday Rotation in Ge and Si
(b) Vol. **109**, No. 1, p. 283–294 (1982).
See also *18*, p. *259*; *20*, p. *293*; 22.1.2, p. 400.

T. SOMA and H. MATSUO
Lindeman's Melting Law and the Effect of Pressure on the Melting Point of Si and Ge
(b) Vol. **109**, No. 1, p. 387–391 (1982).
See also *6*, p. *70*; *8*, p. *84*; 12.2, p. 134; 22.1.2, p. 400.

S. MILSHTEIN and A. SENDERICHIN
Model of Single Charged Dislocation in Tetravalent Semiconductors
(b) Vol. **109**, No. 1, p. 429–436 (1982).
See also *10.2*, p. *103*; *13.4*, p. *163*; 22.1.2, p. 400.

D. A. ARONOV, R. MAMATKULOV, and V. V. RUBINOV
The Effects of Field and Temperature Heating in Intrinsic Semiconductors with Non-Injecting Contacts
(a) Vol. **69**, No. 1, p. 159–165 (1982).
See also *14.3.4*, p. *227*; 16, p. 249.

YA. S. BUDZHAK and O. I. MAVRIN
The Solid Solubility of Ga and In in $Ge_{1-x}Si_x$
(a) Vol. **69**, No. 1, p. K61–K64 (1982).
See also *1.3*, p. *26*; 22.1.2, p. 401.

A. GHOSH, L. K. BANERJEE, and A. N. BASU
Cohesion, Phase Transition, Elastic and Dielectric Properties of Group IV Semiconductors
(b) Vol. **109**, No. 2, p. 799–806 (1982).
See also *1.2*, p. *15*; *4*, p. *61*; 12.2, p. 134; 14.4, p. 234; 22.1, p. **393**; 22.1.2, p. 401.

A. KUMAR and G. C. SRIVASTAVA
Thermal Conductivity Analysis of Germanium at Low Temperatures
(b) Vol. **109**, No. 2, p. K123–K127 (1982).
See also *8*, p. *85*.

S. J. PEARTON
Energy Levels of Some Rare-Earth Related Impurities in Germanium
(b) Vol. **109**, No. 2, p. K135–K138 (1982).
See also *13.4*, p. *164*.

A. V. DVURECHJENSKII, B. P. KASHNIKOV, L. S. SMIRNOV, R. GRÖTZSCHEL, and R. KLABES
Synthesis of Solid Solutions at Electron Pulse Processing of Ge–Si Structures
(a) Vol. **69**, No. 2, p. K183–K187 (1982).
See also 1.5, p. 38; *3*, p. *57*; *11*, p. *119*; 22.1.2, p. 402.

K. S. DUBEY
Role of Boundary Scattering in Phonon Drag Thermoelectric Power and Lattice Thermal Conductivity of a Doped Semiconductor at Low Temperatures: Application to Phosphorus Doped Ge
(b) Vol. **111**, No. 1, p. K49–K54 (1982).
See also *6*, p. *74*; 8, p. 86; 15, p. 246.

V. K. MALYUTENKO, YU. M. MALOSOVSKII, A. P. SASHCHUK, and A. CHOVET
Increase of the Sensitivity of Devices Based on Carrier Concentration Effects. The Optimum Thickness Revisited
(a) Vol. **71**, No. 1, p. 217–222 (1982).
See also *14.3.3*, p. *222*; 22.2.3, p. 427.

T. SOMA and H. MATSUO
Debye-Waller Factor of Si and Ge
(b) Vol. **111**, No. 2, p. K93–K97 (1982).
See also *6*, p. *75*; 22.1.2, p. 407.

V. G. KOHN, M. V. KOVALCHUK, R. M. IMAMOV, B. G. ZAKHAROV, and E. F. LOBANOVICH
The Investigation of Disturbed Surface Layer Structure by the External Photoeffect under X-Ray Diffraction Conditions
(a) Vol. **71**, No. 2, p. 603–610 (1982).
See also *1.6*, p. *43*; *17*, p. *256*; 22.1.2, p. 408; 22.2.1, p. 421.

S. R. SINGH and S. SINGH
Phonon Scattering in Plastically Deformed Ge
(b) Vol. **112**, No. 1, p. 51–54 (1982).
See also *6*, p. *75*; 10.2, p. 110.

G. MADORE, J. D. N. CHEEKE, D. HUET, and J.-P. MANEVAL
Photo-Excited Acoustic Phonons in InSb and Ge
(b) Vol. **112**, No. 1, p. K1–K4 (1982).
See also *6*, p. *76*; 22.2.3, p. 428.

K. GERMANOVA, K. MARINOVA, and A. TRAYANOV
On the Electronic Subbands in n-Inversion Layers on (111) Ge
(b) Vol. **112**, No. 2, p. 575–580 (1982).
See also *13.1*, p. *153*; *14.3.2*, p. *217*.

A. D. POGREBNYAK, V. A. KUZMINIKH, and K. P. AREFIEV
Positron Annihilation and Profiles of Displaced Atoms in Electron Irradiated Ge
(b) Vol. **112**, No. 2, p. K79–K83 (1982).
See also *11*, p. *124*.

V. V. ARISTOV, V. G. KOHN, V. I. POLOVINKINA, and A. A. SNIGIREV
Change of Interference Pattern in the X-Ray Section Topography of Single Crystals with Increasing Dimensions of an Incoherent Source
(a) Vol. **72**, No. 2, p. 483–491 (1982).
See also *1.1*, p. *12*; *10.2*, p. *112*; 22.1.2, p. 410.

V. A. MOKRITSKII, E. M. KURITSIN, B. A. DANILCHENKO, and V. I. SHAKHOVTSOV
Epitaxial Germanium Layers Grown by Liquid Phase in γ-Radiation Field
(a) Vol. **72**, No. 2, p. K121–K124 (1982).
See also 1.5, p. 40; *3*, p. *58*; *11*, p. *125*.

A. R. Badzian
On the Geometrical Condition for X-Ray Scattering by Thermal Waves in Crystals
(a) Vol. **74**, No. 2, p. 637–649 (1982).
See also *1.1*, p. *13*; 6, p. 81; 22, p. 392; 22.1.2, p. 416; 22.5.2, p. 464.

22.1.2 Silicon

Original Papers and Short Notes

K. SEN and R. S. SRIVASTAVA
SIS Solar Cell Theory and Calculations
(a) Vol. **69**, No. 1, p. 413–418 (1982).
See also *14.3.3*, p. *219*; *16*, p. *249*.

YA. S. BUDZHAK and O. I. MAVRIN
The Solid Solubility of Ga and In in $Ge_{1-x}Si_x$
(a) Vol. **69**, No. 1, p. K61–K64 (1982).
See also *1.3*, p. *26*; 22.1.1, p. 395.

S. K. KRAWCZYK
Simple Equation Describing *I–U* Characteristics of MIS Solar Cells
(a) Vol. **69**, No. 1, p. K77–K81 (1982).
See also *14.3.3*, p. *219*; *16*, p. *250*.

A. V. DVURECHENSKII and I. A. RYAZANTSEV
Substitutional Doping of Vaccum-Evaporated Amorphous Silicon by Ion Implantation
(a) Vol. **69**, No. 1, p. K117–K120 (1982).
See also 2, p. 46; *11*, p. *118*; *14.3.1*, p. *213*.

M. SEILENTHAL and K.-S. REBANE
Photovoltage Spectroscopy of Fast Surface States in Silicon MOS Structures
(a) Vol. **69**, No. 1, p. K121–K123 (1982).
See also *13.3*, p. *158*; 16, p. 250.

V. CHÁB, J. ZEMEK, S. KOC, and P. MIKUŠÍK
Photoemission Study of Oxygen Adsorbed on Sputtered a-Si(H) and a-Si
(b) Vol. **109**, No. 2, p. 495–501 (1982).
See also 2, p. 46; *13.3*, p. *158*; *17*, p. *255*.

M. HÖHNE
Electron Paramagnetic Resonance of Gold in Silicon (II): Cluster Centres
(b) Vol. **109**, No. 2, p. 525–534 (1982).
See also 13.4, p. 164; *19*, p. *283*.

G. BOSMAN and R. J. J. ZIJLSTRA
Thermal Noise, Diffusion, and the Mobility of Hot Holes in Silicon
(b) Vol. **109**, No. 2, p. 683–691 (1982).
See also *14.3.4*, p. *227*.

A. GHOSH, L. K. BANERJEE, and A. N. BASU
Cohesion, Phase Transition, Elastic and Dielectric Properties of Group IV Semiconductors
(b) Vol. **109**, No. 2, p. 799–806 (1982).
See also *1.2*, p. *15*; *4*, p. *61*; 12.2, p. 134; 14.4, p. 234; 22.1, p. 393; 22.1.1, p. 395.

E. BARTA
On the Optically Determined Relaxation Time of Heavily Doped p-Type Silicon
(b) Vol. **109**, No. 2, p. K85–K88 (1982).
See also *14.3*, p. *203*; *20.1*, p. *300*.

H. Kostal and L. F. Kurtenok
Temperature Dependence of the Multivalued Sasaki Effect
(b) Vol. **109**, No. 2, p. K109 – K113 (1982).
See also *13*, p. *141*; *14.3.4*, p. *227*.

A. A. Dakhel
K-Ray $K\beta$ and $L_{2,3}$ Emission Bands of Silicon in La–Si Compounds
(b) Vol. **109**, No. 2, p. K133 – K134 (1982).
See also *20*, p. *294*; 21.4, p. 369.

V. I. Kushnir and E. V. Suvorov
Investigation of the Dynamic Diffraction Focusing of X-Rays by a Homogeneously Bent Crystal in the Symmetric Laue Case
(a) Vol. **69**, No. 2, p. 483 – 490 (1982).
See also *1.1*, p. *11*.

E. Glaser, G. Götz, N. Sobolev, and W. Wesch
Investigations of Radiation Damage Production in Ion Implanted Silicon
(a) Vol. **69**, No. 2, p. 603 – 614 (1982).
See also *11*, p. *119*; 14.3, p. 204; 19, p. 284; 20.1, p. 301.

S. Loualiche, C. Lucas, P. Baruch, J. P. Gailliard, and J. C. Pfister
Theoretical Model for Radiation Enhanced Diffusion and Redistribution of Impurities. Comparison with Experiments
(a) Vol. **69**, No. 2, p. 663 – 676 (1982).
See also *9*, p. *91*; 11, p. 119.

L. Castellani, P. Gondi, C. Patuelli, and R. Berti
On the EBIC Contrast of Dislocations in Si
(a) Vol. **69**, No. 2, p. 677 – 685 (1982).
See also *1.4*, p. *32*; *10.2*, p. *105*.

K. Naukkarinen and T. Tuomi
Explanation of Contrast Effects in Synchrotron X-Ray Topographs of Bent Crystals by Means of Multiple Diffraction
(a) Vol. **69**, No. 2, p. 687 – 697 (1982).
See also *1.1*, p. *11*; 12.2, p. 134.

H. Föll
Lattice Imaging of Silicide–Silicon Interfaces and the Interpretation of Interfacial Defects
(a) Vol. **69**, No. 2, p. 779 – 788 (1982).
See also *1.6*, p. *42*; *10.2*, p. *105*.

A. V. Dvurechenskii, B. P. Kashnikov, L. S. Smirnov, R. Grötzschel, and R. Klabes
Synthesis of Solid Solution at Electron Pulse Processing of Ge–Si Structures
(a) Vol. **69**, No. 2, p. K183 – K187 (1982).
See also 1.5, p. 38; *3*, p. *57*; *11*, p. *119*; 22.1.1, p. 395.

P. A. Cherdantsev, I. P. Chernov, and A. P. Mamontov
Chain Annihilation of Defects in Semiconductor Crystals
(a) Vol. **69**, No. 2, p. K215 – K218 (1982).
See also *10.2*, p. *105*; *11*, p. *119*; 22.2.1, p. 420.

K. Weber, M. Grünewald, W. Fuhs, and P. Thomas
Field Effect in a-Si:H Films. Influence of Annealing and Light Exposure
(b) Vol. **110**, No. 1, p. 133–142 (1982).
See also *2*, p. *46*; 13.4, p. 165; *14.3.1*, p. *214*.

G. B. Bachelet, C. Tanner, and M. Schluter
Static Atomic Polarizabilities from Norm-Conserving Pseudopotentials
(b) Vol. **110**, No. 1, p. 313–322 (1982).
See also *4*, p. *62*; *13*, p. *141*; 22.1, p. 393.

E. G. Sieverts and S. H. Muller
Heat-Treatment Centers in Silicon and Their g-Values
(b) Vol. **110**, No. 1, p. K89–K92 (1982).
See also *13.4*, p. *166*; *19*, p. *285*.

J. Kočka, J. Stuchlík, O. Štika, M. Láznička, O. Renner, and E. Krouský
Physical Properties of a-Si:H Prepared at about 1 Torr
(a) Vol. **70**, No. 1, p. 93–99 (1982).
See also *2*, p. *47*; *14.3.1*, p. *214*; 20.1, p. 302.

M. Renninger
Imagine of Lattice Defects by Double Diffractometric Transmission Topography
(a) Vol. **70**, No. 1, p. 183–188 (1982).
See also *10.2*, p. *106*.

L. Pasemann, H. Blumtritt, and R. Gleichmann
Interpretation of the EBIC Contrast of Dislocations in Silicon
(a) Vol. **70**, No. 1, p. 197–209 (1982).
See also *1.4*, p. *33*; *10.2*, p. *106*; 14.3.3, p. 220.

J. Kilmer, E. R. Chenette, C. M. van Vliet, and P. H. Handel
Absence of Temperature Fluctuations in $1/f$ Noise Correlation Experiments in Silicon
(a) Vol. **70**, No. 1, p. 287–294 (1982).
See also *14.3.3*, p. *220*.

Tran Chot
Some New Experimental Results of a Study of Metal–n-Type Semiconductor Schottky Barrier Contacts
(a) Vol. **70**, No. 1, p. 311–316 (1982).
See also *14.3.4*, p. *228*; 22.2.1, p. 421; 22.2.2, p. 424; 22.4.1, p. 440; 22.4.2, p. 443; 22.4.3, p. 445.

V. Kalyanaraman and V. Kumar
Properties of the Gold Related Acceptor Level in Silicon
(a) Vol. **70**, No. 1, p. 317–323 (1982).
See also *13.4*, p. *166*; 14.3.3, p. 220.

K. B. Kadyrakunov, E. V. Nidaev, and L. S. Smirnov
Behaviour of a Built-in Charge in a Dielectric at Pulsed Annealing of Si–SiO_2 Structures
(a) Vol. **70**, No. 1, p. K15–K17 (1982).
See also *13.3*, p. *158*; *14.4*, p. *235*; 22.6, p. 469.

W. R. FAHRNER, K. HEIDEMANN, and P. SCHÖTTLE
Results of Ion Implantation into Silicon in the 100 MeV Range (I): Oxygen and Boron Implantation
(a) Vol. **70**, No. 2, p. 463–472 (1982).
See also *11*, p. *121*; 14.3.3, p. 221.

O. BOSTANJOGLO
Time-Resolved TEM of Pulsed Crystallization of Amorphous Si and Ge Films
(a) Vol. **70**, No. 2, p. 473–481 (1982).
See also *1.5*, p. *39*; *3*, p. *57*; 22.1.1, p. 396.

H. KERKOW, B. LUKASCH, and R. PIPPIG
Disorder Production in Ion Implanted Silicon
(a) Vol. **70**, No. 2, p. 489–495 (1982).
See also *11*, p. *121*.

P. ZAUMSEIL and U. WINTER
Triple Crystal Diffractometer Investigations of Silicon Crystals with Different Collimator-Analyzer Arrangements
(a) Vol. **70**, No. 2, p. 497–505 (1982).
See also *1.1*, p. *11*; 10.2, p. 108.

J. MARCZEWSKI, R. NITECKI, and M. LIBERADZKA
The Influence of the Negative Bias-Temperature Test on the Properties of the Si–SiO_2 Interface
(a) Vol. **70**, No. 2, p. 555–561 (1982).
See also *13.3*, p. *159*; *14.3.2*, p. *217*; 20.1, p. 304.

N. M. RAVINDRA, K. SUNIL KUMAR, and V. K. SRIVASTAVA
Temperature and Voltage Dependence of the Varrier Height in SnO_2/Si Solar Cells
(a) Vol. **70**, No. 2, p. 623–630 (1982).
See also *13.3*, p. *159*; 14.3.3, p. 221; 16, p. 251.

Z. ŠOUREK and R. BUBÁKOVÁ
The Deformation by Stress Relaxation at the Edge Dislocation
(a) Vol. **70**, No. 2, p. 641–648 (1982).
See also *10.2*, p. *108*.

M. SERVIDORI, A. ZANI, and G. GARULLI
Residual Lattice Disorder in Self-Implanted Silicon after Pulsed Laser Irradiation
(a) Vol. **70**, No. 2, p. 691–701 (1982).
See also *10.2*, p. *108*; 11, p. 121.

V. S. LYSENKO, R. N. LITOVSKII, A. N. NAZAROV, and V. P. KULICHKOV
Temperature Dependence of Capacitance–Voltage Characteristics in Implanted MOS Structures
(a) Vol. **70**, No. 2, p. 719–728 (1982).
See also 11, p. 121; *14.3.3*, p. *221*.

Y. NISHINO, H. SAKA, and T. IMURA
Dislocation Configurations Characteristic of Deformed Czochralski-Grown Silicon Crystals
(a) Vol. **70**, No. 2, p. 729–737 (1982).
See also *10.2*, p. *108*.

M. Weidner and F. Ritter
Spreading Resistance Profile Correction Scheme for Thin Layers on a Conducting Boundary
(a) Vol. **71**, No. 1, p. 153–158 (1982).
See also *14.3.2*, p. *217*.

N. D. Zakharov, M. Pasemann, and V. N. Rozhanski
Observations of Point Defects in Silicon by Means of Dark-Field Lattice Plane Imaging
(a) Vol. **71**, No. 1, p. 275–281 (1982).
See also *10.2*, p. *109*.

G. A. Stanciu, I. M. Popescu, and C. M. Stoichiţa
Evolution of the Photocurrent in a Region with Inversion Layer of the pnp Transistor
(a) Vol. **71**, No. 1, p. K17–K19 (1982).
See also *14.3.3*. p. *222*; *16*, p. *251*.

N. I. Akulovich, V. V. Petrov, and V. I. Utenko
On the Electrical Activity of Transmutated Phosphorus in Neutron Transmutation Doped Silicon
(a) Vol. **71**, No. 1, p. K63–K67 (1982).
See also *11*, p. *122*; 13.4, p. 169; *14.3*, p. *206*.

J. P. Schaffer, J. H. Lind, P. L. Jones, and F. H. Cocks
Photoconductivity of Ion-Plated Amorphous Hydrogenated Silicon
(a) Vol. **71**, No. 1, p. K111–K115 (1982).
See also *2*, p. *49*; *16*, p. *252*.

R. Klabes, M. Voelskow, H. Woittennek, E. V. Nidaev, and L. S. Smirnov
Dopant Redistribution after Flash Lamp Annealing
(a) Vol. **71**, No. 1, p. K127–K130 (1982).
See also *11*, p. *122*.

H. Zeile and K. Lassmann
Ultrasonic Investigation of the Acceptor Ground State of Si(B) (II): The Distribution of Splittings
(b) Vol. **111**, No. 2, p. 555–564 (1982).
See also 7, p. 83; *13.4*, p. *169*.

I. C. da Cunha Lima, M. Fabbri, and A. Ferreira da Silva
Diagonal and Off-Diagonal Disorder in Doped Semiconductors
(b) Vol. **111**, No. 2, p. K69–K74 (1982).
See also *8*, p. *87*; *14.3*, p. *207*.

T. Soma and H. Matsuo
Debye-Waller Factor of Si and Ge
(b) Vol. **111**, No. 2, p. K93–K97 (1982).
See also *6*, p. *75*; 22.1.1, p. 397.

V. M. Kaganer, V. L. Indenbom, M. Vrána, and B. Chalupa
Laue-to-Bragg Transition in Extremely Asymmetric Dynamic Neutron Diffraction
(a) Vol. **71**, No. 2, p. 371–380 (1982).
See also *1.1*, p. *12*.

E. I. Rau, A. Y. Sasov, G. W. Spivak, J. Dziesiaty, and K. Wencel
Determination of Local Parameters of p–n Junctions Using the EBIV Mode of SEM
(a) Vol. **71**, No. 2, p. 429–440 (1982).
See also *14.3.3*, p. *222*.

E. F. Krimmel, A. G. K. Lutsch, and E. Doering
Contribution to Electron Beam Annealing of High-Dose Ion Implanted Polysilicon
(a) Vol. **71**, No. 2, p. 451–456 (1982).
See also *11*, p. *122*; *14.3.1*, p. *215*.

V. G. Kohn, M. V. Kovalchuk, R. M. Imamov, B. G. Zakharov, and E. F. Lobanovich
The Investigation of Disturbed Surface Layer Structure by the External Photoeffect under X-Ray Diffraction Conditions
(a) Vol. **71**, No. 2, p. 603–610 (1982).
See also *1.6*, p. *43*; *17*, p. *256*; 22.1.1, p. 397; 22.2.1, p. 421.

G. Kowalski and J. Gronkowski
On the Intermediary Image in X-Ray Section Topography
(a) Vol. **71**, No. 2, p. 611–617 (1982).
See also *1.4*, p. *34*; *10.2*, p. *109*.

V. S. Lysenko, T. N. Sytenko, V. I. Zimenko, and O. V. Snitko
Investigation of Traps in the Transition Region of $Si–SiO_2$ Structures at Cryogenic Temperatures
(a) Vol. **71**, No. 2, p. 619–626 (1982).
See also 11, p. 123; *13.4*, p. *170*; *14.3.3*, p. *222*.

K. Sen
The Effect of Interface States on the Performance of MIS Solar Cells
(a) Vol. **71**, No. 2, p. 641–648 (1982).
See also 13.3, p. 159; *14.3.3*, p. *223*; *16*, p. *252*.

A. V. Dvurechenskii, I. A. Ryazantsev, and V. A. Dravin
Doping of Amorphous Silicon by Manganese Ion Implantation
(a) Vol. **71**, No. 2, p. K133–K136 (1982).
See also 2, p. 49; *11*, p. *123*; 13.4, p. 170; *14.3.1*, p. *215*.

P. Becker, H. Siegert, and L. Bliek
Investigation of Thermoelectric Effects in Silicon by X-Ray Interferometry
(a) Vol. **71**, No. 2, p. K139–K141 (1982).
See also *15*, p. *247*.

H. Nakashima and K. Hashimoto
Barrier Height for Cobalt on Silicon
(a) Vol. **71**, No. 2, p. K201–K203 (1982).
See also *14.3.3*, p. *223*; *14.3.4*, p. *230*; 21.1, p. 349.

H. Lemke
Anonaly of Ion Pairs of Shallow Acceptors with Manganese in Silicon
(a) Vol. **71**, No. 2, p. K215–K217 (1982).
See also *13.4*, p. *171*.

R. Fellenberg, P. Streubel, and A. Meisel
Chemical Shifts of Auger Lines in Solids on the Example of the $KL_{23}L_{23}$ Transition in Silicon and Its Compounds
(b) Vol. **112**, No. 1, p. 55–60 (1982).
See also *4*, p. *63*; 17, p. 256.

V. D. Akhmetov and V. V. Bolotov
Kinetics of Accumulation of Radiation Defects and Annihilation of Vacancies and Interstitials in Carbon- and Boron-Containing Silicon
(a) Vol. **72**, No. 1, p. 61–68 (1982).
See also *11*, p. *123*.

W. R. Fahrner, D. Bräunig, and E. Borchert
MOS and Thyristor Damage Parameters after 2.5 and 15 MeV High Temperature Electron Irradiation
(a) Vol. **72**, No. 1, p. 79–88 (1982).
See also *11*, p. *123*; *14.3.3*, p. *223*.

H. Lemke
Properties of Some Impurity Complexes of Zinc in Silicon
(a) Vol. **72**, No. 1, p. 177–187 (1982).
See also *13.4*, p. *171*; 14.3, p. 208.

A. V. Dvurechenskii, R. Grötzschel, N. M. Igonina, B. P. Kashnikov, and N. I. Komolova
Impurity Profiles at Multi-Pulse Electron-Beam Annealing of Ion-Implanted Silicon
(a) Vol. **72**, No. 1, p. 301–304 (1982).
See also *11*, p. *123*.

L. Pelikán, V. Rybka, P. Krejčí, V. Hnatowicz, and J. Kvítek
Study of Boron Implantation in Ag–Si Layer Structures
(a) Vol. **72**, No. 1, p. 369–373 (1982).
See also *11*, p. *123*.

V. V. Korobtsov, V. G. Zavodinskii, and A. V. Zotov
Epitaxial Regrowth of Amorphous Si Deposited on Si (111)
(a) Vol. **72**, No. 1, p. 391–398 (1982).
See also *1.5*, p. *40*; 2, p. 50; *3*, p. *58*.

L. S. Wieluński, B. M. Paine, B. X. Liu, C.-D. Lien, and M.-A. Nicolet
Ion Mixing of Ni and Pt Layers on Si
(a) Vol. **72**, No. 1, p. 399–404 (1982).
See also 9, p. 92; *11*, p. *123*.

D. E. Ioannou
Comments on the $E_V + 0.45$ eV Quenched-in Level in Silicon
(a) Vol. **72**, No. 1, p. K33–K36 (1982).
See also *10.2*, p. *111*; *13.4*, p. *171*.

P. K. Khabibullaev, L. P. Khiznichenko, N. S. Tulyaganova, and L. G. Zotova
Transformation of Self-Interstitials in Silicon under Electronic and Thermal Stimulation
(a) Vol. **72**, No. 1, p. K69–K72 (1982).
See also *10.2*, p. *111*; 11, p. 124.

S. J. Pearton
Hydrogen Passivation of γ-Induced Point Defects in Silicon
(a) Vol. **72**, No. 1, p. K73–K75 (1982).
See also *10.2*, p. *111*; *14.3*, p. *208*.

S. Milshtein
C–U Measurements on Heavily Dislocated Si
(a) Vol. **72**, No. 1, p. K99–K103 (1982).
See also *10.2*, p. *111*; *13.4*, p. *171*; 14.3, p. 208.

T. Rang
Simple Relationship between the Breakdown Voltage, Concentration, and Junction Depth for Diffused pn Junction
(a) Vol. **72**, No. 1, p. K117–K119 (1982).
See also *14.3.3*, p. *223*.

V. I. Kuznetsov and P. F. Lugakov
Study of Rearrangements of Intrinsic Defects at Annealing of Proton-Irradiated Silicon
(b) Vol. **112**, No. 2, p. 457–462 (1982).
See also *10.2*, p. *111*; 11, p. 124; *13.4*, p. *172*.

J. Kreissl and W. Gehlhoff
EPR Investigation of Manganese–Boron Pairs in Silicon
(b) Vol. **112**, No. 2, p. 695–704 (1982).
See also *13.4*, p. *172*; *19*, p. *289*.

N. M. Olekhnovich, A. L. Karpei, V. M. Skidan, and L. D. Puzenkova
Dynamical Oscillations of X-Ray Integrated Intensity in Laue Diffraction and Their Temperature Dependences for Si Dislocation Crystals
(a) Vol. **72**, No. 2, p. 463–467 (1982).
See also *10.2*, p. *112*.

V. V. Aristov, V. G. Kohn, V. I. Polovinkina, and A. A. Snigirev
Change of Interference Pattern in the X-Ray Section Topography of Single Crystals with Increasing Dimensions of an Incoherent Source
(a) Vol. **72**, No. 2, p. 483–491 (1982).
See also *1.1*, p. *12*; *10.2*, p. *112*; 22.1.1, p. 397.

M. Ogino, Y. Oana, and M. Watanabe
The Diffusion Coefficient of Germanium in Silicon
(a) Vol. **72**, No. 2, p. 535–541 (1982).
See also *9*, p. *92*.

H. S. Kim, R. Perez, and S. S. Sheinin
Experimental Evidence Showing that the Use of Symmetrical Laue Diffraction Conditions in the Dynamical Theory of Electron Diffraction Can Give Significant Errors in Diffraction Contrast Calculations
(a) Vol. **72**, No. 2, p. 561–568 (1982).
See also *1.1*, p. *13*.

N. V. Vershinina and N. N. Gerasimenko
Some Peculiarities of Radiation-Induced Changes in the Properties of Si–SiO_2 Structures Irradiated at Elevated Temperatures
(a) Vol. **72**, No. 2, p. 569–574 (1982).
See also *11*, p. *124*; 13.4, p. 173; *14.3.2*, p. *217*.

K. Tempelhoff and N. van Sung
Formation of Vacancy Agglomerates in Quenched Dislocation-Free Silicon
(a) Vol. **72**, No. 2, p. 617–622 (1982).
See also *10.2*, p. *112*.

V. V. Kveder, Yu. A. Osipyan, W. Schröter, and G. Zoth
On the Energy Spectrum of Dislocations in Silicon
(a) Vol. **72**, No. 2, p. 701–713 (1982).
See also 10.2, p. 112; *13.4*, p. *173*.

A. G. Revesz and S. H. Wemple
The Optical Properties of Noncrystalline Silicon and $Si_{1-x}H_x$ Films
(a) Vol. **72**, No. 2, p. 721–729 (1982).
See also *1.5*, p. *40*; 2, p. 51; *4*, p. *63*; 20, p. 296.

V. J. Rao, W. A. Anderson, and G. Rajeswaran
The Role of Hydrogen in Improving Photovoltaic Response of Solar Cells on Polycrystalline Silicon
(a) Vol. **72**, No. 2, p. 745–749 (1982).
See also *14.3.3*, p. *224*; *16*, p. *253*; 20.1, p. 309.

S. K. Krawczyk and A. Jakubowski
Investigation of Lateral Non-Uniformities in MOS Structures by the Surface Photovoltage Technique
(a) Vol. **72**, No. 2, p. K149–K153 (1982).
See also 13.3, p. 160; *14.3.3*, p. *224*; *16*, p. *253*.

E. M. Lawson and S. J. Pearton
Hydrogen Passivation of Laser-Induced Acceptor Defects in p-Type Silicon
(a) Vol. **72**, No. 2, p. K155–K158 (1982).
See also *10.2*, p. *113*; *14.3*, p. *209*.

V. E. Borisenko and V. A. Labunov
Annealing of Antimony Implanted Silicon with Halogen Lamp Irradiation
(a) Vol. **72**, No. 2, p. K173–K176 (1982).
See also *11*, p. *125*; *14.3*, p. *209*.

E. V. Nidaev and A. I. Popov
Influence of the Surface on the Stability of Radiation Defects Introduced by Electron Irradiation
(a) Vol. **72**, No. 2, p. K193–K195 (1982).
See also *11*, p. *125*; 13.4, p. 173; 14.3.3, p. 224.

V. D. Akhmetov, V. V. Bolotov, L. S. Smirnov, O. I. Zhumaeva, P. Gaworzewski, and K. Schmalz
Accumulation of Radiation Defects in Oxygen-Containing Silicon Thermally Treated at 600 °C
(a) Vol. **72**, No. 2, p. K211–K216 (1982).
See also *11*, p. *125*; 13.4, p. 173; 20.1, p. 310.

M. S. GILENKO, P. M. KARAGEORG-ALKALAEV, and A. YU. LEIDERMAN
On the Noise of Semiconductor Diode Structures with Strong Carrier Accumulation
(a) Vol. **73**, No. 1, p. K147 – K151 (1982).
See also *14.3.3*, p. *225*; *14.3.4*, p. *231*.

J. A. VERGÉS, D. GLÖTZEL, M. CARDONA, and O. K. ANDERSEN
Absolute Hydrostatic Deformation Potentials of Tetrahedral Semiconductors
(b) Vol. **113**, No. 2, p. 519 – 534 (1982).
See also *13.1*, p. *155*; 22.1.1, p. 398; 22.2.1, p. 422.

H. VAN CONG, S. CHARAR, S. BRUNET, M. AVEROUS, and J. L. BIRMAN
Heavy Doping Effects on the Energy Band Structure of n-Type Crystals
(b) Vol. **113**, No. 2, p. 757 – 778 (1982).
See also *13.1*, p. *155*; 13.4, p. 175.

YU. V. BULGAKOV, K. LENKEIT, and G. I. GORBUNOVA
Resonance Dechanneling of Hydrogen and Helium Ions with Hundreds of keV Energies in the ⟨110⟩–(110) Axial-to-Planar Transitions in Silicon
(b) Vol. **113**, No. 2, p. K103 – K105 (1982).
See also *11*, p. *126*.

H. HEDLER, W. ANDRÄ, and G. GÖTZ
Activation of High As and Sb Concentrations in Silicon by Laser Irradiation
(a) Vol. **73**, No. 2, p. 333 – 338 (1982).
See also *1.2*, p. *22*; 11, p. 126; 14.3, p. 210.

C. DONOLATO and P. VENTURI
Influence of the Generation Distribution on the Calculated EBIC Contrast of Line Defects
(a) Vol. **73**, No. 2, p. 377 – 387 (1982).
See also *10.2*, p. *114*; 11, p. 127.

P. ZAUMSEIL and U. WINTER
Triple Crystal Diffractometer Investigations of Imperfections in Silicon Crystals with Laue-Case Diffraction
(a) Vol. **73**, No. 2, p. 455 – 466 (1982).
See also *10.2*, p. *115*.

M. EIZENBERG
Silicide Formation Resulting from the Interfacial Reaction of Silicon and Thin Films of Ir–V Alloys and Bilayers
(a) Vol. **73**, No. 2, p. 483 – 495 (1982).
See also *1.3*, p. *30*; *14.3.4*, p. *232*; 21.1, p. 355.

TRAN CHOT
On the Location of the Interface Fermi Level in Metal–Semiconductor Schottky-Barrier Contacts
(a) Vol. **73**, No. 2, p. 551 – 558 (1982).
See also *13.3*, p. *160*; *14.3.2*, p. *218*; 14.3.4, p. 232; 22.1.1, p. 398; 22.2, p. 419; 22.4, p. 438; 22.8, p. 490.

V. YU. BALANDIN, A. V. DVURECHENSKII, and L. N. ALEKSANDROV
The Polycenter Crystallization of Amorphous Silicon Layers at Pulse Annealing
(a) Vol. **73**, No. 2, p. 587 – 592 (1982).
See also 2, p. 53; *3*, p. *59*.

I. G. Salisbury
Weak-Beam Contrast from Stacking Faults in Silicon
(a) Vol. **74**, No. 1, p. 353–359 (1982).
See also *1.4*, p. *37*.

F.-G. Kirscht, D. Tänzer, and C. Hansch
Thermomechanical Behaviour of Stress-Reduced Silicon Wafers
(a) Vol. **74**, No. 1, p. K1–K3 (1982).
See also *10.2*, p. *116*.

R. Klabes, J. Mattäi, M. Voelskow, and S. Mutze
Pulsed Incoherent Light Annealing of Arsenic and Phosphorus Implanted Polycrystalline Silicon
(a) Vol. **74**, No. 1, p. K5–K7 (1982).
See also 9, p. 93; *11*, p. *128*; *14.3.1*, p. *216*.

N. Sieber, R. Klabes, M. Voelskow, F. Fenske, and K. H. Stegemann
Annealing Characteristics of SiO_2–Si Structures after Incoherent Light Pulse Processing
(a) Vol. **74**, No. 1, p. K9–K12 (1982).
See also *11*, p. *128*; 13.3, p. 161; *14.3.2*, p. *218*.

A. A. Dakhel
X-Ray Study of Nd–Si and Tb–Si Compounds
(a) Vol. **74**, No. 1, p. K73–K77 (1982).
See also *20*, p. *297*; 21.4, p. 372.

M. Asche, Z. S. Gribnikov, V. M. Ivastchenko, H. Kostial, and V. V. Mitin
Conductivity and Transverse Fields in n-Si for Currents in the {110} and {100} Planes
(b) Vol. **114**, No. 2, p. 429–438 (1982).
See also *13*, p. *147*; *14.3.4*, p. *232*.

D. Mergel and R. Labusch
Variable Reconstruction of Dislocation Cores in Si
(b) Vol. **114**, No. 2, p. 545–551 (1982).
See also *10.2*, p. *117*; *13.4*, p. *177*; 14.3, p. 211; 16, p. 254.

G. B. Demidovich, S. N. Karyagin, and V. F. Kiselev
Spin Centres on Clean Germanium and Silicon Surfaces and Their Interaction with Adsorbed Molecules
(b) Vol. **114**, No. 2, p. 705–713 (1982).
See also *13.3*, p. *161*; 19, p. 291; 22.1.1, p. 398.

V. V. Sidorik and A. N. Zhevno
On the Identification of Paramagnetic Defects: The Nature of the Si-B3 Center
(b) Vol. **114**, No. 2, p. K119–K124 (1982).
See also *13.4*, p. *177*.

W. R. Fahrner, H. G. Bach, D. Bräunig, and M. Knoll
Implantation Profiles of Buried Layers (MOS pn) from HF or LF $C(U)$ Curves
(a) Vol. **74**, No. 2, p. 421–429 (1982).
See also *11*, p. *128*; *14.3.3*, p. *225*.

22.1.3 Group VI Elements

Original Papers and Short Notes

F. Henneberger, N. S. Averkiev, and R. J. Rasulov
On a New Photogalvanic Effect Due to Free-Carrier Absorption
(b) Vol. **109**, No. 1, p. 343–351 (1982).
See also 6, p. 70; 13.1, p. 148; *16*, p. *249*.

I. E. Bolotov and V. Yu. Kolosov
Electron Microscope Investigation of Crystals Based on Bend-Contour Arrangement (I): Relationship between Bend-Contour Arrangement and Bend Geometry
(a) Vol. **69**, No. 1, p. 85–96 (1982).
See also *1.5*, p. *38*; 10.2, p. 103.

P. Koštial
Surface Wave Investigations of Se Amorphous Thin Films Deposited on $LiNbO_3$
(a) Vol. **70**, No. 1, p. K55–K58 (1982).
See also *2*, p. *47*; *7*, p. *82*.

C. Viger and G. Fleury
Spectral Response of Amorphous Se ar 233 K
(a) Vol. **71**, No. 1, p. K103–K105 (1982).
See also *2*, p. *49*; *16*, p. *251*.

I. E. Bolotov, V. Yu. Kolosov, and A. V. Kozhyn
Electron Microscope Investigation of Crystals Based on Bend-Contour Arrangement (II): Bending Phenomena of the Crystal Growth in an Amorphous Film
(a) Vol. **72**, No. 2, p. 645–654 (1982).
See also *1.5*, p. *40*; 10.2, p. 112.

O. Bostanjoglo and G. Hoffmann
Time-Resolved TEM of Transient Effects in Pulse Annealing of Ge and Ge–Te Films
(a) Vol. **73**, No. 1, p. 95–105 (1982).
See also *1.2*, p. *21*; *1.5*, p. *40*; 3, p. 59; 22.1.1, p. 398.

V. Balynas and A. Krotkus
On the Origin of S-Shaped Current–Voltage Characteristics in p-Type Tellurium
(a) Vol. **73**, No. 1, p. K101–K103 (1982).
See also *14.3.4*, p. *231*.

V. Ya. Prokhorenko, B. I. Sokolovskii, V. A. Alekseev, A. S. Basin, S. V. Stankus, and V. M. Sklyarchuk
The Semiconductor–Metal Transition in Liquid Tellurium
(b) Vol. **113**, No. 2, p. 453–458 (1982).
See also 2, p. 52; *14.3*, p. *210*; *15*, p. *247*.

J. Szczyrbowski
The Exponential Tail of Light Absorption in Disordered Solids
(b) Vol. **113**, No. 2, p. 715–720 (1982).
See also 2, p. 52; *20.1*, p. *312*; 22, p. 390; 22.1.1, p. 398; 22.2.1, p. 423; 22.3, p. 434; 22.6, p. 474.

22.2 III-V Compounds

Review Articles

S. Chandra and R. K. Pandey
Semiconductor Photoelectrochemical Solar Cells
(a) Vol. **72**, No. 2, p. 415–454 (1982).
See also *14.3.3*, p. *219*; *16*, p. *249*; 22.4, p. 437.

R. J. J. Zijlstra
Distribution Model of Flicker Noise in Semiconductors
(a) Vol. **73**, No. 1, p. 11–21 (1982).
See also *14.3*, p. *202*; 14.3.3, p. 219; 14.3.4, p. 227; 22.1, p. 393.

Original Papers and Short Notes

G. E. Archangelskii, F. Karel, J. Mareš, S. Pačesová, and J. Pastrňák
The Luminescence and EPR Spectra of Manganese Activated AlN
(a) Vol. **69**, No. 1, p. 173–183 (1982).
See also *19*, p. *283*; *20.3*, p. *318*.

G. Eichhorn and U. Rensch
MOCDV of AlN on Silicon
(a) Vol. **69**, No. 1, p. K3–K6 (1982).
See also *1.5*, p. *38*; *3*, p. *57*.

D. Vignaud, J. L. Farvacque, and D. Ferre
Anisotropy of the Free Carrier Piezoelectric Scattering by Dislocations in III–V Compounds
(b) Vol. **110**, No. 2, p. 601–609 (1982).
See also *10.2*, p. *107*; *14.3*, p. *205*.

U. Pietsch
The Influence of Anharmonic Core Vibrations in the X-Ray Bond Charge Analysis of A^NB^{8-N} Compounds
(b) Vol. **111**, No. 1, p. K7–K12 (1982).
See also *4*, p. *62*; *13*, p. *143*; 22.4, p. 437; 22.5, p. 450.

O. Hildebrand
Anomalous Impurity Diffusion in III–V Compounds: The Consequence of Self-Induced Field Effects
(a) Vol. **72**, No. 2, p. 575–584 (1982).
See also *9*, p. *92*; 14.3.3, p. 223.

A. P. Dmitriev, M. P. Mikhailova, and I. N. Yassievich
High Energy Distribution Function in an Electric Field and Electron Impact Ionization in $A^{III}B^V$ Semiconductors
(b) Vol. **113**, No. 1, p. 125–135 (1982).
See also *13.1*, p. *153*.

TRAN CHOT
On the Location of the Interface Fermi Level in Metal–Semiconductor Schottky-Barrier Contacts
(a) Vol. **73**, No. 2, p. 551 – 558 (1982).
See also *13.3*, p. *160*; *14.3.2*, p. *218*; 14.3.4, p. 232; 22.1.1, p. 398; 22.1.2, p. 413; 22.4, p. 438; 22.8, p. 490.

R. K. SINGH and R. D. SINGH
Study of Lattice Statics and Allied Properties of Zincblende-Type Compound Semiconductors
(b) Vol. **114**, No. 1, p. 235 – 242 (1982).
See also *4*, p. *64*; 6, p. 80; 12.2, p. 137; 22.4, p. 438.

22.2.1 Arsenides

Original Papers and Short Notes

TRAN CHOT
Some New Experimental Results of a Study of Metal–n-Type Semiconductor Schottky Barrier Contacts
(a) Vol. **70**, No. 1, p. 311–316 (1982).
See also *14.3.4*, p. *228*; 22.1.2, p. 403; 22.2.2, p. 424; 22.4.1, p. 440; 22.4.2, p. 443; 22.4.3, p. 445.

A. D. POGREBNYAK, V. A. KUZMINIKH, and K. P. AREFEV
Positron Profiles and Positron Annihilation in Thin Layers
(a) Vol. **71**, No. 1, p. 145–151 (1982).
See also *1.5*, p. *39*; *1.6*, p. *43*; 21, p. 335.

J. WOOD, M. J. HOWES, and D. V. MORGAN
A Study of Geometric Magnetoresistance in Transferred Electron Diodes Using a Numerical Solution of the Boltzmann Equation
(a) Vol. **71**, No. 1, p. 207–216 (1982).
See also *14.3.3*, p. *222*.

K. D. GLINCHUK, K. LUKAT, and A. V. PROKHOROVICH
Photoluminescence of Copper-Compensated p-Type GaAs after Irradiation by 2.2 MeV Electrons
(a) Vol. **71**, No. 1, p. 233–238 (1982).
See also *13.4*, p. *168*; *20.3*, p. *323*.

I. THURZO and T. LALINSKÝ
On the Relationship between Charge- and Current-Based DLTS
(a) Vol. **71**, No. 1, p. K83–K87 (1982).
See also *13.4*, p. *169*; *14.3.3*, p. *222*.

V. A. KISELEV
Pseudo-Exciton Effect in the Schottky-Barrier Electroreflectance
(b) Vol. **111**, No. 2, p. 461–468 (1982).
See also 13.5.1, p. 183; *14.3.4*, p. *229*; *20.1*, p. *305*.

A. I. BOBRYSHEVA, S. A. MOSKALENKO, and V. T. ZYUKOV
The Angular Dependences of Two-Photon Transitions to Surface States
(b) Vol. **111**, No. 2, p. K75–K80 (1982).
See also *13.3*, p. *159*; *13.5.1*, p. *184*.

V. G. KOHN, M. V. KOVALCHUK, R. M. IMAMOV, B. G. ZAKHAROV, and E. F. LOBANOVICH
The Investigation of Disturbed Surface Layer Structure by the External Photoeffect under X-Ray Diffraction Conditions
(a) Vol. **71**, No. 2, p. 603–610 (1982).
See also *1.6*, p. *43*; *17*, p. *256*; 22.1.1, p. 397; 22.1.2, p. 408.

K. D. GLINCHUK and K. LUKAT
Non-Trivial Annealing-Induced Increase of the Near-Intrinsic Emission Intensity in γ-Irradiated GaAs
(a) Vol. **71**, No. 2, p. K187–K190 (1982).
See also **11**, p. 123; *13.4*, p. *171*; *20.3*, p. *323*.

J. Szczyrbowski
The Exponential Tail of Light Absorption in Disordered Solids
(b) Vol. **113**, No. 2, p. 715–720 (1982).
See also 2, p. 52; *20.1*, p. *312*; 22, p. 390; 22.1.1, p. 398; 22.1.3, p. 417; 22.3, p. 434; 22.6, p. 474.

J. Jimenez, M. A. Gonzalez, L. F. Sanz, and J. Bonnafe
On the Photoconductivity Response of GaAs Doped with Chromium and Oxygen
(a) Vol. **73**, No. 2, p. K189–K192 (1982).
See also 13.4, p. 176; *16*, p. *253*.

K. D. Glinchuk, A. V. Prokhorovich, and N. S. Zayats
Comparison of Irradiation- and Annealing-Induced Changes of the Intrinsic Emission Intensity in Electron-Bombarded Undeformed and Plastically Deformed GaAs
(a) Vol. **73**, No. 2, p. K267–K270 (1982).
See also 11, p. 127; *20.3*, p. *327*.

D. Senguta and Vikram Kumar
Study of n-GaAs MOS Diodes with Spin-on SiO_2 Layer
(a) Vol. **73**, No. 2, p. K279–K282 (1982).
See also *14.3.3*, p. *225*.

S. K. Maksimov, V. L. Egorov, V. V. Kryuk, D. I. Piskunov, E. A. Pitirmova, and V. F. Veselov
Effect of Stoichiometric Disturbances in $A^{III}B^{V}$ Compounds during Ion Implantation
(a) Vol. **73**, No. 2, p. K283–K285 (1982).
See also *11*, p. *127*; 22.2.3, p. 429.

W. Ulrici
Photoinduced Transient Optical Absorption of Cr^{4+} in Semi-Insulating GaAs:Cr
(b) Vol. **114**, No. 1, p. K87–K90 (1982).
See also *13.4*, p. *176*; *20.1*, p. *314*.

J. Wood, M. J. Howes, and D. V. Morgan
The Effect of Dislocations on the Electron Mobility of GaAs
(a) Vol. **74**, No. 2, p. 493–496 (1982).
See also 10.2, p. 117; *14.3.3*, p. *226*.

Sh. G. Suleimanov
Investigation of Semi-Insulating GaAs S-Diodes in Magnetic Field
(a) Vol. **74**, No. 2, p. K97–K100 (1982).
See also *14.3.3*, p. *226*; 14.3.4, p. 233.

22.2.2 Phosphides

Original Papers and Short Notes

E. Kartheuser, A. Babcenco, and R. Evrard
Electron–Hole Correlation Effects on Donor–Acceptor Pairs
(b) Vol. **109**, No. 1, p. 375–386 (1982).
See also *13.4*, p. *163*; 22.4.2, p. 442.

P. Krispin
Thermochemical Model Applied to a Deep-Level Defect in GaP
(a) Vol. **69**, No. 1, p. 193–200 (1982).
See also *13.4*, p. *163*; 14.3.3, p. 219.

G. Oelgart and G. Arnold
Avalanche Multiplication in GaP Studied by Means of the SEM
(a) Vol. **69**, No. 1, p. 207–216 (1982).
See also *11*, p. *118*; *14.3.3*, p. *219*; 14.3.4, p. 227.

V. M. Rubinov and M. Tuychiev
Investigation of the Current Characteristics of Thin-Film Structures Based on GaP under the Action of an Electron Beam
(a) Vol. **69**, No. 1, p. K113–K116 (1982).
See also *11*, p. *118*; *14.3.1*, p. *213*.

A. Kirsch, R. Schwabe, R. Bindemann, R. Dehmlow, and K. Wandel
Studies of Surface Recombination Processes in Photo-Excited VPE-GaP:N, Te
(a) Vol. **69**, No. 2, p. 553–561 (1982).
See also *20.3*, p. *320*.

Tran Chot
Some New Experimental Results of a Study of Metal–n-Type Semiconductor Schottky Barrier Contacts
(a) Vol. **70**, No. 1, p. 311–316 (1982).
See also *14.3.4*, p. *228*; 22.1.2, p. 403; 22.2.1, p. 421; 22.4.1, p. 440; 22.4.2, p. 443; 22.4.3, p. 445.

P. Paufler, O. Boehnke, G. Teichert, and K. Löschke
On the Influence of Plastic Deformation upon the Mass Density of GaP
(a) Vol. **70**, No. 1, p. K81–K83 (1982).
See also *10.2*, p. *107*.

S. Bednarek and U. Rössler
Method of Invariants Applied to Indirect Gap Absorption
(b) Vol. **110**, No. 2, p. 565–570 (1982).
See also 6, p. 73; *20.1*, p. *303*; 22.1.2, p. 404.

T. I. Voronina, Yu. A. Goldberg, and T. V. Lvova
Electrical Properties of GaP Crystals Bombarded with High Energy Protons
(a) Vol. **70**, No. 2, p. K121–K124 (1982).
See also *11*, p. *121*; *14.3*, p. *206*.

G. Oelgart, H. Haefner, R. Reulke, and F. Joachim
Degradation of Green Light-Emitting GaP:N Diodes
(a) Vol. **71**, No. 1, p. 89–98 (1982).
See also *14.3.4*, p. *229*; *20.3*, p. *323*.

W. Lorenz and M. Handschuh
Dynamical Theory of Photocurrent–Voltage Curves and Photocurrent Efficiency on Semiconductor Electrodes
(a) Vol. **71**, No. 1, p. 127–134 (1982).
See also *14.3.2*, p. *217*; *16*, p. *251*.

C. Patel, W. F. Sherman, and G. R. Wilkinson
Lattice Dynamics of GaP under Hydrostatic Pressures
(b) Vol. **111**, No. 2, p. 649–657 (1982).
See also *6*, p. *74*; 20.1, p. 306.

E. Yu. Brailovskii, F. K. Karapetyan, I. G. Megela, and V. P. Tartachnik
Radiation Defects in Electron-Irradiated InP Crystals
(a) Vol. **71**, No. 2, p. 563–568 (1982).
See also *11*, p. *122*; *14.3*, p. *207*.

H. Baránková
Photoluminescence of SSD and LEC Grown InP
(a) Vol. **71**, No. 2, p. K153–K156 (1982).
See also 13.4, p. 170; *20.3*, p. *323*.

A. Vaidyanathan and A. H. Guenther
Comment on "Intense-Field Effects in Solids"
(b) Vol. **112**, No. 1, p. K73–K78 (1982).
See also *13.1*, p. *152*; *20.1*, p. *308*.

N. Erben, P. Paufler, K. Löschke, S. Däbritz, and K. Kleinstück
Kossel Line Broadening near a Scratch on GaP
(a) Vol. **72**, No. 1, p. K15–K18 (1982).
See also *10.2*, p. *111*.

S. V. Bulyarskii, S. I. Radautsan, and A. S. Ambrozhevich
Thermally Stimulated Noise in GaP p–n Junctions
(a) Vol. **72**, No. 1, p. K77–K78 (1982).
See also *14.3.4*, p. *230*.

E. Yu. Brailovskii, I. G. Megela, N. H. Pambuchchyan, and V. V. Teslenko
EPR Study of Electron-Irradiated InP:Fe
(a) Vol. **72**, No. 1, p. K109–K111 (1982).
See also *11*, p. *124*; *19*, p. *288*.

E. Ziegler, W. Siegel, H. Blumtritt, and O. Breitenstein
Electrical and EBIC Investigations of GaP Grain Boundaries
(a) Vol. **72**, No. 2, p. 593–605 (1982).
See also 1.4, p. 35; *14.3.3*, p. *223*.

22.2.3 Antimonides

Original Papers and Short Notes

E. P. Matsas, V. N. Antonyuk, V. F. Bibik, and O. V. Snitko
On the Mechanism of Anodic Oxidation of InSb
(a) Vol. **69**, No. 2, p. K229–K231 (1982).
See also *1.6*, p. *42*.

N. V. Baramidze, V. L. Bonch-Bruevich, M. P. Giorgadze, and N. I. Kurdiani
Electrical Properties of InSb Irradiated with Fast Neutrons
(b) Vol. **110**, No. 1, p. 33–37 (1982).
See also 11, p. 120; *14.3*, p. *204*.

K. P. Ghatak, K. K. Ghosh, H. M. Mukherjee, and A. N. Chakravarti
Effect of Size Quantization on the Electronic Heat Capacity in Ultrathin Films of Non-Parabolic Semiconductors
(b) Vol. **110**, No. 1, p. 323–330 (1982).
See also *8*, p. *85*.

P. M. Fauchet
The Auger Rate in Highly Excited Indium Antimonide
(b) Vol. **110**, No. 1, p. K11–K15 (1982).
See also *14.3*, p. *204*.

E. P. Pokatilov and S. I. Beril
Spatially Extended Optical Interface Modes in Two-Layer Periodical Structure
(b) Vol. **110**, No. 1, p. K75–K78 (1982).
See also *6*, p. *72*.

J. Bracinník and J. Štelina
Electric Field in the Acoustoelectric Stationary Domains of the Second Mode. Simple Model
(a) Vol. **70**, No. 2, p. K135–K139 (1982).
See also *7*, p. *82*; *14.3.4*, p. *228*.

A. R. Beattie
Impact Ionisation and Quantum Efficiency in InSb
(b) Vol. **111**, No. 1, p. 141–153 (1982).
See also *14.3.4*, p. *228*.

V. K. Malyutenko, Yu. M. Malosovskii, A. P. Sashchuk, and A. Chovet
Increase of the Sensitivity of Devices Based on Carrier Concentration Effects. The Optimum Thickness Revisited
(a) Vol. **71**, No. 1, p. 217–222 (1982).
See also *14.3.3*, p. *222*; 22.1.1, p. 397.

B. Escaig, J. L. Farvacque, and D. Ferré
On the Measurement of Thermally Activated Slip Parameters in Tetrahedrally Coordinated Semiconductors
(a) Vol. **71**, No. 2, p. 329–334 (1982).
See also *10.2*, p. *109*.

V. T. Bublik, J. Wilke, and A. T. Pereversev
The Nature of Solid Solutions of the Excess Components in Non-Doped Gallium Antimonide
(a) Vol. **73**, No. 2, p. K271 – K273 (1982).
See also *1.1*, p. *13*; *1.4*, p. *37*.

S. K. Maksimov, V. L. Egorov, V. V. Kryuk, D. I. Piskunov, E. A. Pitirmova, and V. F. Veselov
Effect of Stoichiometric Disturbances in $A^{III}B^{V}$ Compounds during Ion Implantation
(a) Vol. **73**, No. 2, p. K283 – K285 (1982).
See also *11*, p. *127*; 22.2.1, p. 423.

B. B. Tripathi, H. C. Gupta, and M. V. N. Ambika Prasad
Effect of Polarizability in the Phonon Spectrum of Zinc-Blende Compounds
(b) Vol. **114**, No. 1, p. K11 – K14 (1982).
See also 4, p. 65; *6*, p. *80*; 22.2.2, p. 426.

M. Nowak
The Possible Existence of the Interference Photomagnetoelectric Effect
(a) Vol. **74**, No. 1, p. 313 – 322 (1982).
See also 14.3.1, p. 216; *16*, p. *254*.

Al. L. Efros, L. M. Kanskaya, S. I. Kokhanovskii, and R. P. Seisyan
Diamagnetic Excitons and the Indium Antimonide Energy Band Parameters
(b) Vol. **114**, No. 2, p. 373 – 382 (1982).
See also *13.1*, p. *155*; 13.5.1, p. 185; 20.1, p. 315.

E. Litwin-Staszewska, L. Kończewicz, R. Piotrzkowski, and W. Szymańska
On the Deep Impurity Level in n-InSb
(b) Vol. **114**, No. 2, p. K153 – K156 (1982).
See also *13.4*, p. *177*; *14.3*, p. *211*.

22.2.4 Mixed Crystals

Original Papers and Short Notes

S. K. Maksimov and E. N. Nagdaev
Electron Microscopy Images of Composition-Modulated Crystals (II): Dilatational Contrast Images
(a) Vol. **69**, No. 2, p. 505–512 (1982).
See also *1.4*, p. *32*; 10.2, p. 105.

I. V. Ryzhikov, V. P. Sushkov, P. I. Tepaikin, and I. L. Kasatkin
Double Injection at p–n Junctions Based on $Al_xGa_{1-x}As$
(a) Vol. **69**, No. 2, p. 707–718 (1982).
See also *14.3.3*, p. *220*; 14.3.4, p. 227; *20.3*, p. *320*.

K. Ishida, Y. Matsumoto, and K. Taguchi
Lattice Defects in LPE InP–InGaAsP–InGaAs Structure Epitaxial Layers on InP Substrates
(a) Vol. **70**, No. 1, p. 277–286 (1982).
See also *1.5*, p. *39*; *10.2*, p. *106*.

V. Milanović and D. Tjapkin
Energy Band Calculation and Zero Energy Gap Conditions for Semiconductor Superlattices
(b) Vol. **110**, No. 2, p. 687–695 (1982).
See also *13.1*, p. *150*; 14.3.3, p. 221.

R. Stegmann, B. Kloth, and G. Oelgart
The Temperature Dependence of Luminescence Intensity on $GaAs_{1-x}P_x$:N
(a) Vol. **70**, No. 2, p. 423–431 (1982).
See also 13.5.1, p. 183; *20.3*, p. *322*.

E. E. Matyas and A. G. Karoza
Plasma Edge Reflectance Measurements in InAs–AlAs Alloys
(b) Vol. **111**, No. 1, p. K45–K48 (1982).
See also *14.3*, p. *206*; *20.1*, p. *305*.

S. K. Maksimov and E. N. Nagdaev
Electron Microscopy Images of Composition-Modulated Crystals (III): Moire Fringe Images
(a) Vol. **72**, No. 1, p. 135–145 (1982).
See also *1.4*, p. *35*; 10.2, p. 110.

D. V. Morgan, K. Board, C. E. C. Wood, and L. F. Eastman
Current Transport in Al/InAlAs/InGaAs Heterostructures
(a) Vol. **72**, No. 1, p. 251–260 (1982).
See also *14.3.3*, p. *223*.

M. Blasckhe, H. J. Grundmann, and K. Wandel
Current Dependent Electroluminescence Decay and Degradation Behaviour in Green and Yellow LED's
(a) Vol. **72**, No. 2, p. 687–694 (1982).
See also 13.4, p. 173; 14.3.3, p. 224; *20.3*, p. *325*.

P. Swarup, R. K. Jain, S. N. Verma, S. Charan, and D. M. Tandle
LPE Growth of GaInAlAs on GaAs
(a) Vol. **72**, No. 2, p. K189–K192 (1982).
See also *2*, p. *51*; *13.1*, p. *153*; 20.1, p. 310; 20.3, p. 325.

R. Stegmann, B. Kloth, and G. Oelgart
The Influence of the Doping Level on the Cathodoluminescence in $GaAs_{1-x}P_x$:N, Zn Epitaxial Layers
(a) Vol. **73**, No. 1, p. 129–137 (1982).
See also 13.5.1, p. 184; *20.3*, p. *326*.

M. Ludwig and W. Sauer
Hall Measurements along Zinc Diffusion Profiles in n-$GaAs_{1-x}P_x$ ($0.6 \leqq x \leqq 1$)
(a) Vol. **73**, No. 1, p. K83–K86 (1982).
See also *14.3*, p. *210*.

A. Ghosal, D. Chattopadhyay, and N. N. Purkait
High-Frequency Electron Transport in $Ga_xIn_{1-x}As_yP_{1-y}$ Lattice-Matched to InP
(a) Vol. **73**, No. 1, p. K119–K122 (1982).
See also *14.3*, p. *210*.

K. S. Maksimov, L. A. Bondarenko, and G. N. Gaidukov
Stability of Composition Automodulated Complex Epitaxial Films
(a) Vol. **73**, No. 1, p. K129–K132 (1982).
See also *1.5*, p. *41*.

S. Yamazaki, M. Kishi, and T. Katoda
Calculation of Excess Free Energy of Mixing and Analysis of Phonon Spectra with Clustering Parameter in III–V Pseudobinary Alloys
(b) Vol. **113**, No. 2, p. 421–430 (1982).
See also *6*, p. *79*.

V. I. Osinskii, S. A. Malyshev, and M. P. Ryzhkov
Minority Carrier Transport and Radiative Recombination in $Al_xGa_{1-x}As$ Variable-Composition Structures
(a) Vol. **74**, No. 1, p. 43–50 (1982).
See also *14.3.1*, p. *215*; *20.3*, p. *328*.

P. Kleinert
Coherent-Potential Approximation for Vibrations in $Ga_{1-x}In_xSb$ and $Ga_{1-x}In_xAs$ Mixed Crystals
(b) Vol. **114**, No. 2, p. 459–466 (1982).
See also *6*, p. *81*; 20.1, p. 315.

22.3 IV–IV, II–V, and III–VI Compounds

Original Papers and Short Notes

V. RIEDE, H. NEUMANN, F. LÉVY, and H. SOBOTTA
Polarization- and Temperature-Dependent Two-Phonon Absorption Spectra in InSe
(b) Vol. **109**, No. 1, p. 275–281 (1982).
See also *6*, p. *70*; *20.1*, p. *298*.

T. D. DZHAFAROV, A. SH. MEKHTIEV, and M. S. SADIGOV
The Effect of γ-Irradiation on Photoelectric Properties of Sb–GaSe Structures
(a) Vol. **69**, No. 2, p. 643–648 (1982).
See also 13.3, p. 158; *14.3.3*, p. *220*; *16*, p. *250*.

T. SOMA, H. IWANAMI, and H. MATSUO
Equation of State of Si–Ge Solid Solution
(b) Vol. **110**, No. 1, p. 75–79 (1982).
See also *1.3*, p. *26*; 12.2, p. 134; *13*, p. *141*.

Y. HASEGAWA and Y. ABE
Electrical and Optical Characteristics of a Schottky Barrier on a Cleaved Surface of the Layered Semiconductor InSe
(a) Vol. **70**, No. 2, p. 615–621 (1982).
See also *13.3*, p. *159*; 13.4, p. 167; *14.3.3*, p. *221*; 16, p. 250.

J. CISOWSKI
Level Ordering in II_3–V_2 Semiconducting Compounds
(b) Vol. **111**, No. 1, p. 289–293 (1982).
See also *13.1*, p. *150*.

M. S. KUSHWAHA
The Bond-Bending Force Model (BBFM) for Phonons in β-SiC
(b) Vol. **111**, No. 1, p. 337–340 (1982).
See also *6*, p. *73*.

M. M. HAFIZ, M. M. IBRAHIM, and A. S. AHMED
Conduction and Switching in In_xSe_{1-x}
(a) Vol. **71**, No. 1, p. 259–263 (1982).
See also *14.3.1*, p. *214*; *14.3.4*, p. *229*.

B. G. TAGIEV, E. F. BAGIRZADE, and G. M. MAMEDOV
Interference Oscillations in the Temperature Dependence of Photocurrent and Transmission of GaSe, GaTe, and $GaSe_xTe_{1-x}$ Single Crystals
(a) Vol. **71**, No. 1, p. K49–K51 (1982).
See also *16*, p. *251*; *20.1*, p. *305*; 22.3.1, p. 436.

T. IKARI, S. SHIGETOMI, and K. HASHIMOTO
Crystal Structure and Raman Spectra of InSe
(b) Vol. **111**, No. 2, p. 477–481 (1982).
See also 1.3, p. 28; *6*, p. *74*; *20.1*, p. *305*.

C. W. Allen and K. C. Liao
Dislocation Models for Shear Transformations
(a) Vol. **74**, No. 2, p. 673–681 (1982).
See also *1.2*, p. *24*; *10.2*, p. *117*.

22.3.1 Mixed Crystals

Short Note

B. G. Tagiev, E. F. Bagirzade, and G. M. Mamedov
Interference Oscillations in the Temperature Dependence of Photocurrent and Transmission of GaSe, GaTe, and $GaSe_xTe_{1-x}$ Single Crystals
(a) Vol. **71**, No. 1, p. K49 – K51 (1982).
See also *16*, p. *251*; *20.1*, p. *305*; 22.3, p. 433.

22.4 II–VI Compounds

Review Article

S. Chandra and R. K. Pandey
Semiconductor Photoelectrochemical Solar Cells
(a) Vol. **72**, No. 2, p. 415–454 (1982).
See also *14.3.3*, p. *219*; *16*, p. *249*; 22.2, p. 418.

Original Papers and Short Notes

T. Skettrup
Experimental Check of Different Additional Boundary Conditions in Case of Spatial Dispersion
(b) Vol. **109**, No. 2, p. 663–672 (1982).
See also *13.5.1*, p. *181*; *20.1*, p. *300*.

A. Haufe, K. Henneberger, J. Röseler, and H.-J. Wünsche
Local Density Approximation Applied to Bound Excitons in Polar Semiconductors
(b) Vol. **110**, No. 1, p. 157–167 (1982).
See also *13.4*, p. *165*; *13.5.1*, p. *182*; 22.4.1, p. 439; 22.4.2, p. 442; 22.4.3, p. 445.

J. A. Yakimavichius, R. P. Purlys, and A. J. Shirvaitis
Temperature Dependence of Mean-Square Dynamic Displacements and Characteristic Temperatures of $A^{II}B^{VI}$ Compounds
(b) Vol. **110**, No. 1, p. K51–K56 (1982).
See also *6*, p. *72*; 22.4.1, p. 440; 22.4.2, p. 442; 22.4.3, p. 445.

W. Brückner, W. Moldenhauer, and K.-H. Bäther
Time Dependence of Degradation in ZnO Varistors
(a) Vol. **70**, No. 1, p. 167–175 (1982).
See also *14.3.3*, p. *220*.

W. Brückner, S. Lehmann, and G. Pompe
Heat Conductivity in ZnO Varistor Ceramics
(a) Vol. **70**, No. 1, p. K47–K49 (1982).
See also *8*, p. *86*.

E. Ziegler, A. Heinrich, H. Oppermann, and G. Stöver
Growth and Electrical Properties of Non-Stoichiometric ZnO Single Crystals Doped with Co
(a) Vol. **70**, No. 2, p. 563–570 (1982).
See also 3, p. 57; 13.4, p. 167; *14.3*, p. *205*.

U. Pietsch
The Influence of Anharmonic Core Vibrations in the X-Ray Bond Charge Analysis of A^NB^{8-N} Compounds
(b) Vol. **111**, No. 1, p. K7–K12 (1982).
See also *4*, p. *62*; *13*, p. *143*; 22.2, p. 418; 22.5, p. 450.

A. Choudhury and D. Mukhopadhyay
Optimum Second Harmonic Generation Efficiency in II–VI Compound Semiconductors at 77 K
(b) Vol. **113**, No. 1, p. K49–K51 (1982).
See also *14.3.4*, p. *231*.

22.4.1 Sulphides

Review Article

R. Mach and G. O. Müller
Physical Concepts of High-Field, Thin-Film Electroluminescence Devices
(a) Vol. **69**, No. 1, p. 11–66 (1982).
See also 14.3.1, p. 213; *14.3.3*, p. *219*; *20.3*, p. *318*; 22.4.2, p. 442.

Original Papers and Short Notes

P. Palanichamy and S. P. Singh
Change in Electron Mass and Collision Frequency Due to Acoustic Wave Propagation in Piezoelectric Semiconductors
(b) Vol. **109**, No. 1, p. 119–123 (1982).
See also *7*, p. *82*; *13*, p. *139*.

T. Soma, J. Saton, and H. Matsuo
Temperature Dependence of the Linear Thermal Expansion Coefficient and Phonon Curves at the Covalent–Ionic Transition Pressure of ZnS, ZnSe, and ZnTe
(b) Vol. **109**, No. 1, p. K47–K51 (1982).
Erratum
(b) Vol. **111**, No. 1, p. K67 (1982).
See also *6*, p. *70*; *8*, p. *84*; 22.4.2, p. 442; 22.4.3, p. 445.

A. P. Galushka, G. E. Davidyuk, and N. S. Bogdanyuk
Interaction of 1.2 MeV Electron Irradiation Produced Defects with Cu Impurity Atoms in CdS Single Crystals
(a) Vol. **69**, No. 1, p. 71–76 (1982).
See also *11*, p. *118*; *14.3*, p. *202*; 16, p. 249; 20.1, p. 299; 20.3, p. 318.

T. Taguchi and D. W. Palmer
Channeling and Ionoluminescence Spectra of Cubic-Structured ZnS Crystals
(a) Vol. **69**, No. 1, p. K55–K59 (1982).
See also *1*, p. *9*; *20.3*, p. *319*.

E. Gaubas, K. Jarašiunas, and J. Vaitkus
Light Induced Transient Grating Decay in Si and Some $A^{II}B^{VI}$ Compounds
(a) Vol. **69**, No. 1, p. K87–K90 (1982).
See also *20*, p. *293*; 22.4.2, p. 442; 22.4.3, p. 445.

P. L. Kukk and H. A. Aarna
Solubility of Ag in CdS Single Crystals
(a) Vol. **69**, No. 1, p. K109–K112 (1982).
See also *10.2*, p. *104*; *14.3*, p. *203*.

A. Haufe, K. Henneberger, J. Röseler, and H.-J. Wünsche
Local Density Approximation Applied to Bound Excitons in Polar Semiconductors
(b) Vol. **110**, No. 1, p. 157–167 (1982).
See also *13.4*, p. *165*; *13.5.1*, p. *182*; 22.4, p. 437; 22.4.2, p. 442; 22.4.3, p. 445.

M. T. Sebastian, D. Pandey, and P. Krishna
X-Ray Diffraction Study of the 2H to 3C Solid State Transformation in Vapour Grown Single Crystals of ZnS
(a) Vol. **71**, No. 2, p. 633–640 (1982).
See also *1.2*, p. *18*; 10.2, p. 109.

G. K. Padam, B. Singh, S. K. Gupta, and G. D. Sootha
Effect of Impurities on the Oxygen Radicals and F^+ Centres in CdS
(a) Vol. **71**, No. 2, p. K169–K172 (1982).
See also *10.2*, p. *110*; *19*, p. *287*.

J. Echigoya and J. W. Edington
Stacking Faults in Recrystallized Cadmium Sulphide
(a) Vol. **72**, No. 1, p. K1–K4 (1982).
See also *10.2*, p. *111*.

P. Pipinys, A. Proskura, and A. Rimeika
Temperature Dependence of Electroluminescence in ZnS Ceramics
(a) Vol. **72**, No. 2, p. 511–514 (1982).
See also *14.3.4*, p. *230*; *20.3*, p. *324*.

M. A. S. Sweet
The Analysis of Complex Thermoluminescence from Partial Glow Curves
(a) Vol. **72**, No. 2, p. K217–K222 (1982).
See also *13.4*, p. *173*; 20.3, p. 325.

S. Kishida, K. Matsuura, H. Fukuma, F. Takeda, and I. Tsurumi
Optical Absorption Bands in Neutron Irradiated ZnSe and $ZnS_{0.5}Se_{0.5}$ Crystals
(b) Vol. **113**, No. 1, p. K31–K33 (1982).
See also *11*, p. *125*; *20.1*, p. *310*; 22.4.2, p. 443; 22.4.4, p. 448.

T. Lukaszewicz
Studies on the Diffusion of Impurities in II–VI Compounds Based on the Two-Sample Luminescence Method
(a) Vol. **73**, No. 2, p. 611–616 (1982).
See also *9*, p. *93*; 20.3, p. 327; 22.4.2, p. 444.

A. Kreissl, K. Bohnert, V. G. Lyssenko, and C. Klingshirn
Experimental Investigation on the Complex Dielectric Function of CdS in the Exciton and in the Plasma Limit
(b) Vol. **114**, No. 2, p. 537–544 (1982).
See also *13.5*, p. *180*; *20.1*, p. *315*.

V. I. Sokolov, T. P. Surkova, and V. V. Chernyaev
Two Types of Excitons Bound with Ni in ZnSe and ZnS
(b) Vol. **114**, No. 2, p. K195–K199 (1982).
See also *13.5.1*, p. *186*; *14.3.4*, p. *233*; 20.1, p. 316; 22.4.2, p. 444.

22.4.2 Selenides

Review Article

R. Mach and G. O. Müller
Physical Concepts of High-Field, Thin-Film Electroluminescence Devices
(a) Vol. **69**, No. 1, p. 11–66 (1982).
See also **14.3.1**, p. 213; *14.3.3*, p. *219*; *20.3*, p. *318*; 22.4.2, p. 439.

Original Papers and Short Notes

E. Kartheuser, A. Babcenco, and R. Evrard
Electron–Hole Correlation Effects on Donor–Acceptor Pairs
(b) Vol. **109**, No. 1, p. 375–386 (1982).
See also *13.4*, p. *163*; 22.2.2, p. 424.

T. Soma, J. Saton, and H. Matsuo
Temperature Dependence of the Linear Thermal Expansion Coefficient and Phonon Curves at the Covalent–Ionic Transition Pressure of ZnS, ZnSe, and ZnTe
(b) Vol. **109**, No. 1, p. K47–K51 (1982).
Erratum
(b) Vol. **111**, No. 1, p. K67 (1982).
See also *6*, p. *70*; *8*, p. *84*; 22.4.1, p. 439; 22.4.3, p. 445.

E. Gaubas, K. Jarašiūnas, and J. Vaitkus
Light Induced Transient Grating Decay in Si and Some $A^{II}B^{VI}$ Compounds
(a) Vol. **69**, No. 1, p. K87–K90 (1982).
See also *20*, p. *293*; 22.4.1, p. 439; 22.4.3, p. 445.

A. Haufe, K. Henneberger, J. Röseler, and H.-J. Wünsche
Local Density Approximation Applied to Bound Excitons in Polar Semiconductors
(b) Vol. **110**, No. 1, p. 157–167 (1982).
See also *13.4*, p. *165*; *13.5.1*, p. *182*; 22.4, p. 437; 22.4.1, p. 439; 22.4.3, p. 445.

G. I. Tsurkan and G. M. Shmelev
Carrier Mobility in ZnSe in a Wide Temperatur Range
(b) Vol. **110**, No. 1, p. K33–K36 (1982).
See also *13.5.3*, p. *188*; *14.3*, p. *204*.

J. A. Yakimavichius, R. P. Purlys, and A. J. Shirvaitis
Temperature Dependence of Mean-Square Dynamic Displacements and Characteristic Temperatures of $A^{II}B^{VI}$ Compounds
(b) Vol. **110**, No. 1, p. K51–K56 (1982).
See also *6*, p. *72*; 22.4, p. 437; 22.4.1, p. 440; 22.4.3, p. 445.

P. Kukk, Ô. Palmre, and E. Mellikov
The Structure of Recombination Centres in Activated ZnSe Phosphors
(a) Vol. **70**, No. 1, p. 35–42 (1982).
See also *10.2*, p. *106*; *13.4*, p. *166*; 20.3, p. 321.

TRAN CHOT
Some New Experimental Results of a Study of Metal–n-Type Semiconductor Schottky Barrier Contacts
(a) Vol. **70**, No. 1, p. 311–316 (1982).
See also *14.3.4*, p. *228*; 22.1.2, p. 403; 22.2.1, p. 421; 22.2.2, p. 424; 22.4.1, p. 440; 22.4.3, p. 445.

F. SCHAUER, M. JEDLIČKA, and J. POLCER
Amorphous–Crystalline Heterojunction with Built-in Interface Charge (I): Steady-State Transport
(a) Vol. **70**, No. 2, p. 755–762 (1982).
See also 2, p. 48; 13.3, p. 159; *14.3.3*, p. *221*; 22, p. 385.

R. BALTRAMIEYUNAS and E. KUOKSHTIS
Collective Phenomena in Edge Emission of ZnTe and ZnSe Crystals
(b) Vol. **111**, No. 1, p. 187–193 (1982).
See also 13.5.1, p. 183; 20.1, p. 305; *20.3*, p. *322*; 22.4.3, p. 446.

F. SCHAUER, M. JEDLIČKA, and J. MACKŮ
Amorphous–Crystalline Heterojunction with Built-in Interface Charge (II): Non-Steady State Transport
(a) Vol. **71**, No. 2, p. 335–343 (1982).
See also 2, p. 49; 13.3, p. 159; *14.3.3*, p. *222*; 22, p. 387.

L. STAUFFER and G. MUNSCHY
Energy of the Exciton–Ionized Donor Complex in Anisotropic Semiconductors
(b) Vol. **112**, No. 2, p. 501–505 (1982).
See also *13.4*, p. *172*; *13.5.1*, p. *184*.

A. J. WOJTOWICZ and H. J. ŁOŻYKOWSKI
Unknown Excited States of the Self-Activated Center in ZnSe
(b) Vol. **112**, No. 2, p. K103–K106 (1982).
See also *13.4*, p. *172*; *20.3*, p. *324*.

D. SEURET, O. VIGIL, and F. LECCABUE
Electro-Optical Properties of $CuGa_{0.7}In_{0.3}Se_2$/ZnSe Heterojunction
(a) Vol. **72**, No. 2, p. 661–666 (1982).
See also *14.3.3*, p. *223*; *20.3*, p. *324*; 22.8, p. 488.

S. KISHIDA, K. MATSURA, H. FUKUMA, F. TAKEDA, and I. TSURUMI
Optical Absorption Bands in Neutron Irradiated ZnSe and $ZnS_{0.5}Se_{0.5}$ Crystals
(b) Vol. **113**, No. 1, p. K31–K33 (1982).
See also *11*, p. *125*; *20.1*, p. *310*; 22.4.1, p. 441; 22.4.4, p. 448.

J. DAUNAY, JAC. DAUNAY, and P. BUGNET
A Possible Origin of the Internal Electric Field Responsible of the Anomalous Photovoltage in Twinned ZnSe Single Crystals
(a) Vol. **73**., No. 1, p. 199–203 (1982).
See also *16*, p. *253*.

G. RIVAUD, M. F. DENANOT, H. GAREM, and J. C. DESOYER
Dissociated Dislocations in ZnSe
(a) Vol. **73**, No. 2, p. 401–408 (1982).
See also *10.2*, p. *114*.

T. Łukaszewicz
Studies on the Diffusion of Impurities in II–VI Compounds Based on the Two-Sample Luminescence Method
(a) Vol. **73**, No. 2, p. 611–616 (1982).
See also *9*, p. *93*; 20.3, p. 327; 22.4.1, p. 441.

H. L. Oczkowski
Thermoluminescence Model with Intra-Pair Metastable State
(a) Vol. **74**, No. 1, p. 65–74 (1982).
See also *13.4*, p. *176*; *20.3*, p. *328*.

T. Łukaszewicz
The Diffusion of Atoms and Luminescence Centres in II–VI Compounds
(a) Vol. **74**, No. 1, p. 307–312 (1982).
See also *9*, p. *93*; 20.3, p. 328.

V. I. Sokolov, T. P. Surkova, and V. V. Chernyaev
Two Types of Excitons Bound with Ni in ZnSe and ZnS
(b) Vol. **114**, No. 2, p. K195–K199 (1982).
See also *13.5.1*, p. *186*; *14.3.4*, p. *233*; 20.1, p. 316; 22.4.1, p. 441.

A. I. Öpik and J. A. Varvas
Defect Structure of Cu-Doped Cadmium Selenide
(a) Vol. **74**, No. 2, p. 467–473 (1982).
See also *10.2*, p. *117*; *14.3*, p. *212*.

Yu. A. Vashpanov, V. V. Serdyuk, and V. A. Smyntyna
Current Instabilities in Thin Cadmium Selenide Films
(a) Vol. **74**, No. 2, p. K131–K135 (1982).
See also 14.3.1, p. 216; *14.3.4*, p. *233*.

22.4.3 Tellurides

Original Papers and Short Notes

T. Soma, J. Saton, and H. Matsuo
Temperature Dependence of the Linear Thermal Expansion Coefficient and Phonon Curves at the Covalent–Ionic Transition Pressure of ZnS, ZnSe, and ZnTe
(b) Vol. **109**, No. 1, p. K47 – K51 (1982).
Erratum
(b) Vol. **111**, No. 1, p. K67 (1982).
See also *6*, p. *70*; *8*, p. *84*; 22.4.1, p. 439; 22.4.2, p. 442.

E. Gaubas, K. Jarašiūnas, and J. Vaitkus
Light Induced Transient Grating Decay in Si and Some $A^{II}B^{VI}$ Compounds
(a) Vol. **69**, No. 1, p. K87 – K90 (1982).
See also *20*, p. *293*; 22.4.1, p. 439; 22.4.2, p. 442.

E. Molva, J. P. Chamonal, and J. L. Pautrat
Shallow Acceptors in Cadmium Telluride
(b) Vol. **109**, No. 2, p. 635 – 644 (1982).
See also *13.4*, p. *164*; 20.3, p. 319.

A. Haufe, K. Henneberger, J. Röseler, and H.-J. Wünsche
Local Density Approximation Applied to Bound Excitons in Polar Semiconductors
(b) Vol. **110**, No. 1, p. 157 – 167 (1982).
See also *13.4*, p. *165*; *13.5.1*, p. *182*; 22.4, p. 437; 22.4.1, p. 439; 22.4.2, p. 442.

J. A. Yakimavichius, R. P. Purlys, and A. J. Shirvaitis
Temperature Dependence of Mean-Square Dynamic Displacements and Characteristic Temperatures of $A^{II}B^{VI}$ Compounds
(b) Vol. **110**, No. 1, p. K51 – K56 (1982).
See also *6*, p. *72*; 22.4, p. 437; 22.4.1, p. 440; 22.4.2, p. 442.

Tran Chot
Some New Experimental Results of a Study of Metal–n-Type Semiconductor Schottky Barrier Contacts
(a) Vol. **70**, No. 1, p. 311 – 316 (1982).
See also *14.3.4*, p. *228*; 22.1.2, p. 403; 22.2.1, p. 421; 22.2.2, p. 424; 22.4.1, p. 440; 22.4.2, p. 443.

W. Schmid and P. J. Dean
Bound Exciton Time Decay and the Auger Effect in Zinc Telluride
(b) Vol. **110**, No. 2, p. 591 – 600 (1982).
See also *13.4*, p. *167*; *13.5.1*, p. *183*; 20.3, p. 321.

W. G. Proctor, P. Lawaetz, Y. Marfaing, and R. Triboulet
On the S-Shaped Characteristic of Semiconductors at Low Temperature
(b) Vol. **110**, No. 2, p. 637 – 641 (1982).
See also *13.4*, p. *167*; *14.3.4*, p. *228*.

22.4.4 Mixed Crystals

Original Papers and Short Notes

R. MACH, P. FLÖGEL, L. G. SUSLINA, A. G. ARESHKIN, J. MAEGE, and G. VOIGT
The Influence of Compositional Disorder on Electrical and Optical Properties of ZnS_xSe_{1-x} Single Crystals
(b) Vol. **109**, No. 2, p. 607–615 (1982).
See also 13.5.1, p. 181; *14.3*, p. *203*; 20.1, p. 300; *20.3*, p. *319*.

B. S. SUNDERSHESHU and T. KENDELEWICZ
Temperature Dependence of the Absorption Edge of $Cd_{1-x}Mn_xTe$ Mixed Crystals
(a) Vol. **69**, No. 2, p. 467–474 (1982).
See also 13.1, p. 149; *20.1*, p. *300*.

J. CALAS, C. FAU, J. R. VELLAS, and M. ROYER
Shubnikov-de-Haas Effect in EDRI – Type Epitaxial HgCdTe Thin Films
(a) Vol. **69**, No. 2, p. 751–756 (1982).
See also 13.1, p. 149; *14.3.1*, p. *214*.

A. TWARDOWSKI and J. GINTER
Simple Exciton in Intermediate Magnetic Field. Application to $Cd_{1-x}Mn_xTe$ Mixed Crystals
(b) Vol. **110**, No. 1, p. 47–55 (1982).
Erratum
(b) Vol. **112**, No. 2, p. 747 (1982).
See also *13.5.1*, p. *182*.

O. GOEDE and W. HEIMBORDT
Isoelectronic Impurity Te in $CdS_{1-x}Se_x$ Mixed Crystals
(b) Vol. **110**, No. 1, p. 175–182 (1982).
See also *13.5.1*, p. *182*; 20.1, p. 301; *20.3*, p. *320*.

X. W. FAN and J. WOODS
Blue and Violet Electroluminescence Emission in $ZnSe_xS_{1-x}$ MIS Diodes
(a) Vol. **70**, No. 1, p. K27–K30 (1982).
See also 13.5.1, p. 182; *14.3.3*, p. *221*; *20.3*, p. *321*.

R. J. IWANOWSKI
The Γ-Point Band Structure Parameters of $Cd_xHg_{1-x}Se$ ($x \leqq 0.1$) Mixed Crystals from Thermomagnetic Measurements
(b) Vol. **110**, No. 2, p. 461–470 (1982).
See also *13.1*, p. *150*; 15, p. 246.

YU. G. ARAPOV, A. B. DAVYDOV, I. M. TSIDILKOVSKII, and N. G. SHELUSHININA
On the Anomalies in the Temperature Dependence of Conductivity in $Hg_{1-x}Cd_xTe$
(b) Vol. **110**, No. 2, p. 619–630 (1982).
See also *14.3*, p. *205*.

S. KATSURA, S. FUJIKI, T. SUENAGA, and I. NAGAHARA
Spin Glass and Superantiferromagnet in the Site Diluted Ising Model on the Frustrated Square and F.C.C. Lattices with First- and Second-Neighbor Interactions
(b) Vol. **111**, No. 1, p. 83–88 (1982).
See also *18.2*, p. *270*; *18.4*, p. *280*.

M. YA. VALAKH, M. P. LISITSA, G. S. PEKAR, G. N. POLYSSKII, V. I. SIDORENKO, and A. M. YAREMKO
Anharmonic Coupling of Phonon Modes in Mixed $Zn_xCd_{1-x}Se$ Crystals
(b) Vol. **113**, No. 2, p. 635–645 (1982).
See also *6*, p. *79*; *20.1*, p. *312*.

D. HENNIG, G. OEDE, and W. HEIMBRODT
Koster-Slater Model for Isolated and Paired Isoelectronic Te Impurities in II–VI Semiconductors
(b) Vol. **113**, No. 2, p. K163–K167 (1982).
See also *13.4*, p. *175*; *13.5.1*, p. *185*.

A. TWARDOWSKI and J. GINTER
The Influence of Exchange Spin Splitting on Exciton in $Cd_{1-x}Mn_xTe$ Mixed Crystals
(b) Vol. **114**, No. 2, p. 331–336 (1982).
See also *13.5.1*, p. *185*; 18, p. 264; 20.1; p. 315.

22.5 Halides

Original Papers and Short Notes

P. GLUCK
Transitions in Fast Ionic Conductors
(b) Vol. **109**, No. 1, p. 437–439 (1982).
See also *1.2*, p. *14*; *14.4*, p. *234*.

P. J. ALONSO, V. M. ORERA, F. PALACIO, and R. ALCALA
Optical and EPR Studies of NH_4MnF_3 above 80 K
(b) Vol. **109**, No. 1, p. K81–K83 (1982).
See also *19*, p. *283*; *20.1*, p. *299*.

K. S. ARNTSON and M. B. LANGLEY
A Phenomenological Theory for Hysteresis in the Ionic Conductivity of Solid Electrolytes
(a) Vol. **69**, No. 1, p. 185–192 (1982).
See also 8, p. 85; *10.2*, p. *103*; *14.4*, p. *234*.

M. YOSHIDA, N. OHNO, K. NAKAMURA, and Y. NAKAI
Exciton Spectra of Orthorhombic Indium Bromide
(b) Vol. **109**, No. 2, p. 503–509 (1982).
See also *13.5.1*, p. *181*; 20.1, p. 300.

F. HENNEBERGER and V. MAY
Non-Linear Optical Response and Optical Bistability Due to Excitonic Molecules
(b) Vol. **109**, No. 2, p. K139–K143 (1982).
See also *13.5.1*, p. *181*; *20.1*, p. *300*.

K. TENNAKONE and W. G. D. DHARMARATNE
Electrical Conduction in Lead Bromide
(a) Vol. **70**, No. 1, p. K5–K6 (1982).
See also *14.3*, p. *204*.

V. P. GUPTA and S. K. SIPANI
Cohesive Energy of Ionic Crystals within the Framework of Compressible Ion Theory
(b) Vol. **111**, No. 1, p. 295–301 (1982).
See also *4*, p. *62*; 22, p. 386.

U. PIETSCH
The Influence of Anharmonic Core Vibrations in the X-Ray Bond Charge Analysis of A^NB^{8-N} Compounds
(b) Vol. **111**, No. 1, p. K7–K12 (1982).
See also *4*, p. *62*; *13*, p. *143*; 22.2, p. 418; 22.4, p. 437.

W. KORCZAK and M. SUBOTOWICZ
Analysis of Temperature Variation of EPR Linewidths for Gd^{3+} Ion in CeF_3 Single Crystals
(b) Vol. **111**, No. 2, p. 455–460 (1982).
See also 13.4, p. 169; *19*, p. *287*.

K. M. Häussler, A. Lehmeyer, and L. Merten
Far-Infrared Reflectivity of Optical Magnons in FeF_2 and CoF_2
(b) Vol. **111**, No. 2, p. 513–518 (1982).
Erratum
(b) Vol. **114**, No. 1, p. 277 (1982).
See also *13.5.4*, p. *189*; 18, p. 260; *20.1*, p. *306*.

M. S. Brodin, I. V. Blonskii, B. M. Nitsovich, A. S. Krochuk, and A. V. Franiv
Dynamical Properties of Excitons in Layer Crystals of PbI_2
(b) Vol. **111**, No. 2, p. 625–630 (1982).
See also *13.5.1*, p. *184*; 20.1, p. 306.

A. J. McEvoy
ESCA Spectrum and Band Structure of Ruthenium Chloride
(a) Vol. **71**, No. 2, p. 569–574 (1982).
See also 13.1, p. 151; *17*, p. *256*.

B. S. Umarov, V. S. Gorelik, M. M. Sushinskii, and J. F. Vetelino
Raman Scattering from Nonfundamental States in Ammonium Chloride Crystals at Low Temperatures
(b) Vol. **112**, No. 1, p. 69–73 (1982).
See also *6*, p. *75*; *20.1*, p. *307*.

N. A. Kulagin and D. T. Sviridov
Change of the Effective Occupation Numbers of 3d-Shells of Mn^{2+} Ions in Manganese Halide Crystals
(b) Vol. **112**, No. 1, p. K61–K66 (1982).
See also *4*, p. *63*; *13.4*, p. *171*.

B. Kuhlow
Magnetic Ordering in $CrCl_3$ at the Phase Transition
(a) Vol. **72**, No. 1, p. 161–168 (1982).
See also 18.2, p. 271; *18.4*, p. *280*; *20*, p. *296*.

Wee Lam Ng and P. W. M. Jacobs
Electric Field- and Self-Induced Polarization in Ammonium Chloride Crystals
(a) Vol. **72**, No. 1, p. 281–287 (1982).
See also *14.4*, p. *236*.

B. P. Chandra, R. D. Verma, A. K. Jaiswal, B. Majumdar, T. R. Chandraker, and S. V. Deshpande
Mechanoluminescence of Halides and Other Inorganic Crystals
(a) Vol. **72**, No. 1, p. 361–367 (1982).
See also *20.3*, p. *324*.

T. M. Batirov, V. M. Fridkin, R. Nitsche, and K. A. Verkhovskaya
Bulk Photovoltaic and Illumination Effects in the Argyrodite-Type Ionic Conductors Cu_6PS_5Br and Cu_6PS_5I
(a) Vol. **72**, No. 1, p. K105–K108 (1982).
See also 14.4, p. 237; *16*, p. *252*.

22.5.1 Silver Halides

Original Papers and Short Notes

R. J. DERI, CH. HAYNAM, and J. P. SPOONHOWER
Photoconductivity of Lead(II)- and Cadmium(II)-Doped Silver Bromide
(b) Vol. **109**, No. 2, p. 645–653 (1982).
See also *10.2*, p. *104*; *16*, p. *250*.

F. CALLENS, W. MAENHOUT-VAN DER VORST, and L. E. KETELLAPPER
The Effect of the Solution pAg and pH on the Space Charge Characteristics of Silver Bromide Emulsion Grains
(a) Vol. **70**, No. 1, p. 189–195 (1982).
See also *10.2*, p. *106*; *14.4*, p. *235*.

J. SHANKER and J. P. SINGH
Third- and Fourth-Order Elastic Constants for Silver Halides, Alkaline Earth Oxides, and Chalcogenides of Pb and Sn
(a) Vol. **70**, No. 2, p. 677–681 (1982).
See also *12.2*, p. *135*; 22, p. 385; 22.6, p. 470.

I. Z. KOSTADINOV and VU THI BINH
Temperature Dependence of the Phonon Spectra of β-AgI
(b) Vol. **111**, No. 1, p. 53–58 (1982).
See also *6*, p. *73*; 20.1, p. 304.

F. CALLENS and W. MAENHOUT-VAN DER VORST
The Influence of EDTA on the Space Charge Characteristics of AgBr and β-AgI Emulsion Grains
(a) Vol. **71**, No. 2, p. K161–K163 (1982).
See also *10.2*, p. *110*; *14.4*, p. *236*.

D. A. GOLOPENTIA and L. M. SLIFKIN
Dielectric Relaxation in Manganese-Doped Silver Chloride
(a) Vol. **72**, No. 1, p. 123–129 (1982).
See also *10.2*, p. *110*; *14.4*, p. *236*.

F. VÁVRA
Formation of Wavy Slip Bands in AgCl Crystals
(a) Vol. **72**, No. 2, p. K125–K128 (1982).
See also *12.2*, p. *136*.

J. SHANKER and K. SINGH
Analysis of Crystal Binding and Grüneisen Parameter in Cesium Halides, Silver Halides and Alkaline Earth Oxides
(b) Vol. **113**, No. 2, p. 737–745 (1982).
See also *4*, p. *64*; *6*, p. *79*; 22.5.2, p. 461; 22.6, p. 474.

22.5.2 Alkali Halides

Review Article

E. A. Kotomin and A. B. Doktorov
Theory of Tunneling Recombination of Defects Stimulated by Their Motion (II): Three Recombination Mechanisms
(b) Vol. **114**, No. 2, p. 287–318 (1982).
See also *9*, p. *91*; *10.2*, p. *103*.

Original Papers and Short Notes

E. A. Kotomin and A. L. Shluger
Semiempirical Calculations of Defect Properties in LiF Crystal (II): Electron and Hole Centres and Their Recombination
(b) Vol. **109**, No. 1, p. 75–81 (1982).
See also *10.2*, p. *103*; *13.4*, p. *162*.

Zh. Egemberdiev, S. Zazubovich, and V. Nagirnyi
Polarized Luminescence of Lead Ions Associated with Interstitial Radiation Defects in $KCl:PbCl_2$ Crystals
(b) Vol. **109**, No. 1, p. 173–182 (1982).
See also *10.2*, p. *103*; *20.3*, p. *318*.

A. A. Alybakov, N. Toichiev, and Sh. Akchalov
Investigation of EPR and Thermoluminescence Glow Curves of X-Irradiated $KCl.CrO_4^{2-}$, Ca^{2+} Crystals
(b) Vol. **109**, No. 1, p. 295–301 (1982).
See also *19*, p. *282*; *20.3*, p. *318*.

Y. V. G. S. Murti and N. Sucheta
Thermoluminescence of Coloured Alkali Halides
(b) Vol. **109**, No. 1, p. 325–334 (1982).
See also 13.4, p. 163; *20.3*, p. *318*.

J. Shanker and K. Singh
Evaluation of the Grüneisen Parameter for NaCl-Structure Alkali Halide Crystals
(b) Vol. **109**, No. 1, p. 411–416 (1982).
See also *6*, p. *70*.

J. Shanker, D. P. Agrawal, and A. K. Rajauria
Analysis of the Relative Stability of the NaCl, CsCl, and Cubic ZnS Structures in Alkali Halide Crystals
(b) Vol. **109**, No. 1, p. K23–K26 (1982).
See also *1*, p. *9*; *4*, p. *61*.

S. Bhan
Thermoluminescence of NaF Harshaw Single Crystals Irradiated to Medium Doses
(a) Vol. **69**, No. 1, p. 367–376 (1982).
See also *10.2*, p. *104*; *20.3*, p. *319*.

G. A. ANDREEV, T. S. ORLOVA, and B. I. SMIRNOV
A Difference between Tension and Compression in the Density Change of Strained LiF Crystals
(a) Vol. **69**, No. 1, p. 419–423 (1982).
See also *10.2*, p. *104*.

Y. S. HOROWITZ
Criticism of the Z-Center Model in LiF-TLD
(a) Vol. **69**, No. 1, p. K29–K32 (1982).
See also *10.2*, p. *104*.

ZH. EGEMBERDIEV, S. ZAZUBOVICH, V. SEEMAN, and V. NAGIRNYI
Luminescence Accompanying the Recombination of Electrons with V_F Centres in KCl Crystals
(b) Vol. **109**, No. 2, p. 473–482 (1982).
See also *10.2*, p. *104*; 19, p. 283; *20.3*, p. *319*.

S. BENCI, F. FERMI, and R. REVERBERI
Symmetries of Luminescent Centers in NaCl:Pb^{2+} Excited in the A-Absorption Band
(b) Vol. **109**, No. 2, p. 543–550 (1982).
See also *10.2*, p. *104*; *20.3*, p. *319*.

I. I. GEGUSIN, V. N. DATSYUK, and R. V. VEDRINSKII
Electron Structure and X-Ray Absorption Spectra of Potassium Chloride (I): Band Structure Approach
(b) Vol. **109**, No. 2, p. 563–572 (1982).
See also *13.1*, p. *149*; *20*, p. *293*.

C. BANSAL and K. N. SHRIVASTAVA
Mössbauer Isomer Shifts of $^{40}K^+$ in KF, KCl, and KBr Lattices
(b) Vol. **109**, No. 2, p. 743–748 (1982).
See also *5*, p. *66*.

V. A. STRELTSOV, V. I. BARBASHOV, and YU. B. TKACHENKO
Effect of Hydrostatic Pressure on Dislocation Mobility in KCl Crystals
(a) Vol. **69**, No. 2, p. 497–503 (1982).
See also *10.2*, p. *104*; 12.2, p. 134.

T. M. BULATOVA and E. M. NADGORNYI
Effect of the Polycrystalline Structure on the Relaxation Processes and Dislocation Motion in NaCl Crystals
(a) Vol. **69**, No. 2, p. 527–533 (1982).
See also *10.2*, p. *105*.

B. S. ACHARYA and M. L. MUKHERJEE
Role of Tin Centres on the Thermoluminescence of KCl:$SnCl_2$ Single Crystals
(a) Vol. **69**, No. 2, p. 649–655 (1982).
See also *10.2*, p. *105*; 20.1, p. 301; *20.3*, p. *320*.

R. R. Mijangos, J. L. Marín, A. Clark, A. Navarro, and C. Ruíz Mejía
Effect of Surface Polarization on the Absorption Spectrum of an F-Center
(b) Vol. **110**, No. 2, p. 703–708 (1982).
See also *10.2*, p. *107*; *13.4*, p. *167*.

S. Bhan
F_{Z_1} Centres in NaF:Ca^{2+} Single Crystals
(b) Vol. **110**, No. 2, p. K125–K127 (1982).
See also *10.2*, p. *107*; *20.1*, p. *304*.

Z. G. Liu and W. Skrotzki
Slip on {100} Planes in LiF
(a) Vol. **70**, No. 2, p. 433–440 (1982).
See also *10.2*, p. *107*.

J. Prakash
On the Occurrence of Photo-Induced Polarization and Photo-Induced Depolarization
(a) Vol. **70**, No. 2, p. 747–753 (1982).
See also *10.2*, p. *108*; *14.4*, p. *235*.

K. Narasimha Reddy and V. Hari Babu
Study of Hardness Anisotropy in Doped NaCl Crystals
(a) Vol. **70**, No. 2, p. K97–K100 (1982).
See also *12.2*, p. *135*.

S. P. Kathuria and C. M. Sunta
TL Response of LiF TLD-100 at High Irradiation Temperatures
(a) Vol. **70**, No. 2, p. K125–K128 (1982).
See also 11, p. 121; *20.3*, p. *322*.

S. Bhan
Thermoluminescence of NaF Harshaw Single Crystals Irradiated to Very Low and High Doses
(a) Vol. **71**, No. 1, p. 73–78 (1982).
See also 11, p. 121; *20.3*, p. *323*.

M. Georgiev
Nucleation Characteristics of the Suzuki Phase in LiF (TLD-100)
(a) Vol. **71**, No. 1, p. K89–K92 (1982).
See also *1.4*, p. *34*.

B. R. Sood
Suitability of NaF as a Phaser Material
(a) Vol. **71**, No. 1, p. K117–K119 (1982).
See also *6*, p. *74*; 10.2, p. 109; *20.2*, p. *317*.

R. V. Vedrinskii, I. I. Gegusin, V. N. Datsyuk, A. A. Novakovich, and V. L. Kraizman
Electron Structure and X-Ray Absorption Spectra of Potassium Chloride (II): Effect of a Core Hole in the Cluster Approach
(b) Vol. **111**, No. 2, p. 433–442 (1982).
See also *13.1*, p. *151*; *20*, p. *295*.

M. IKEYA
ESR Studies of Manganese Centres in LiH X-Rayed at Low Temperature
(b) Vol. **111**, No. 2, p. 525–530 (1982).
See also *13.4*, p. *169*; *19*, p. *287*.

P. VAROTSOS, K. ALEXOPOULOS, and K. NOMICOS
Comments on the Pressure Variation of the Gibbs Energy for Bound and Unbound Defects
(b) Vol. **111**, No. 2, p. 581–590 (1982).
See also *8*, p. *87*; **9**, p. **92**; *10*, p. *94*; **14.4**, p. **236**; **21**, p. **335**; **21.1**, p. **349**; **21.6**, p. **376**.

V. P. IZVEKOV, A. WATTERICH, T. GYÖRGYI, and E. PUNGOR
Infrared Properties of Doped and Irradiated NaCl Crystals
(b) Vol. **111**, No. 2, p. K153–K156 (1982).
See also *6*, p. *75*; *20.1*, p. *307*.

K. RAMANI and A. M. GHODGAONKAR
Lattice Parameters of Alkali Halides by the Simons-Parr-Finlan Technique
(a) Vol. **71**, No. 2, p. 345–350 (1982).
See also 8, p. 87; *12.2*, p. *135*.

V. B. DUDNIKOVA and V. S. URUSOV
Solvus of $KCl:Sr^{2+}$ Solid Solution by Density Measurements of Single Crystals with the Flotation Method
(a) Vol. **71**, No. 2, p. 409–414 (1982).
See also *1.2*, p. *18*; *10.2*, p. *109*.

G. A. ANDREEV, M. HARTMANOVÁ, and V. A. KLIMOV
Precipitation-Induced Dislocation Loops and Helices in NaCl–Ni Crystals
(a) Vol. **71**, No. 2, p. K137–K138 (1982).
See also *10.2*, p. *109*.

M. CHINNUSAMY and T. NAGARAJAN
Positron Annihilation in Mn^{2+}-Doped KCl Single Crystals
(b) Vol. **112**, No. 1, p. 297–301 (1982).
See also *10.2*, p. *110*.

V. S. URUSOV and V. B. DUDNIKOVA
Energy of Anion Polarisation in Layered Structures
(b) Vol. **112**, No. 1, p. 303–310 (1982).
See also *1.3*, p. *29*; *4*, p. *63*; 22.5.3, p. 465.

I. SH. AKHMADULLIN and R. V. SABUROVA
Features of the Molecular Field Distribution Function of Electric Dipoles in Crystals
(b) Vol. **112**, No. 1, p. K15–K18 (1982).
See also *14.4*, p. *236*.

B. M. LOGINOV and A. A. PREDVODITELEV
Computer Simulation of Dislocation Motion through a Flexible and Reactionable Dislocation Forest of Different Density in NaCl and Mg Crystals
(a) Vol. **72**, No. 1, p. 69–77 (1982).
See also *10.1*, p. *98*; *10.2*, p. *110*; 21.3, p. 367.

J. Kikas and M. Rätsep
Correlation of Transition Frequencies in Impurity Spectra of Inhomogeneous Solids
(b) Vol. **112**, No. 2, p. 409–415 (1982).
See also *10.2*, p. *111*; *20.1*, p. *309*.

H. Opyrchał, K. D. Nierzewski, and B. Macalik
Effect of γ-Irradiation on Eu^{2+} Ions in KCl Crystals
(b) Vol. **112**, No. 2, p. 429–434 (1982).
See also *10.2*, p. *111*; 11, p. 124; 19, p. 288; 20.1, p. 309; 20.3, p. 324.

A. Dauletbekova, A. Akilbekov, and A. Elango
Thermo- and Photostimulated Recombinations of $F{-}H_A$ and $\alpha{-}I_A$ Centres in KBr with Large Na Concentration
(b) Vol. **112**, No. 2, p. 445–452 (1982).
See also *10.2*, p. *111*; 20.3, p. 324.

S. Bhan
Thermoluminescence of $NaF{:}Ca^{2+}$ Single Crystals
(b) Vol. **112**, No. 2, p. 507–514 (1982).
See also *10.2*, p. *112*; *20.3*, p. *324*.

B. Pfalzgraf, O. Roder, and J.-M. Spaeth
Optical Investigation of Ag^+ Centers in Cesium Halides
(b) Vol. **112**, No. 2, p. 565–574 (1982).
See also *10.2*, p. *112*; 13.5.1, p. 184; *20.1*, p. *309*.

J. Shanker and K. Singh
Analysis of the Grüneisen Parameter and Its Volume Dependence in Alkali Halide Crystals
(b) Vol. **112**, No. 2, p. 615–620 (1982).
See also 4, p. 63; *6*, p. *77*; 12.2, p. 136.

O. H. Prasad and M. Suryanarayana
Fourth-Order Elastic Constant, C_{1111}, of Cubic Crystals
(b) Vol. **112**, No. 2, p. 627–631 (1982).
See also *12.1*, p. *132*; *12.2*, p. *136*; 21, p. 337 ; 21.1, p. 351; 21.6, p. 377; 22.1, p. 393.

M. Siu Li, M. de Souza, and S. E. Kapphan
Ionic Thermal Currents under Uniaxial Stress. A New Method to Determine Electric and Elastic Dipole Properties
(b) Vol. **112**, No. 2, p. 685–693 (1982).
See also *10.2*, p. *112*; *14.4*, p. *237*.

W. van Puymbroeck, N. Schrijvers, A. Bouwen, and D. Schoemaker
Electron Spin Resonance Study of H_D-Type $BrCl^-$ Defects in KCl Doped with Divalent Cation Impurities
(b) Vol. **112**, No. 2, p. 725–733 (1982).
See also *10.2*, p. *112*; *19*, p. *289*.

F. Mariani and J. Vignolo
Effect of γ-Irradiation on I–V Dipoles in KCl Crystals
(b) Vol. **112**, No. 2, p. K85–K87 (1982).
See also *10.2*, p. *112*; *11*, p. *124*.

N. D. Vieira, Jr., I. M. Ranieri, and S. P. Morato
Room Temperature Visible Laser Action of F Aggregated Centers in LiF:Mg, OH Crystals
(a) Vol. **73**, No. 1, p. K115–K117 (1982).
See also 10.2, p. 114; *20.2*, p. *317*.

V. V. Hiznyakov and A. V. Sherman
Self-Trapping of Polaritons
(b) Vol. **113**, No. 2, p. 471–479 (1982).
See also *13.5*, p. *179*; *13.5.1*, p. *185*; 20, p. 296.

J. Katriel and R. Bar-Adon
Spectrum of the F_2^+-Center
(b) Vol. **113**, No. 2, p. 481–486 (1982).
See also 10.2, p. 114; *13.4*, p. *174*.

K. Schmitt
Absorption and Stress-Induced Linear Dichroism of Pb^{++} Centres in KCl and RbCl
(b) Vol. **113**, No. 2, p. 559–568 (1982).
See also 13.4, p. 175; *20.1*, p. *311*.

J. Provoost, P. Debergh, and G. Hoebeeck
Formation of Lattice Defects by Two-Photon Excitation in Alkali Halides
(b) Vol. **113**, No. 2, p. 657–663 (1982).
See also 10.2, p. 114; *20.1*, p. *312*.

J. Grammatikakis
Measurements of the Ionic Conductivity of NaI and Calculation of the Bulk Modulus and Specific Heat of Vacancies
(b) Vol. **113**, No. 2, p. 721–726 (1982).
See also 8, p. 89; *10.2*, p. *114*; 12.2, p. 137; *14.4*, p. *238*.

J. Shanker and K. Singh
Analysis of Crystal Binding and Grüneisen Parameter in Cesium Halides, Silver Halides, and Alkaline Earth Oxides
(b) Vol. **113**, No. 2, p. 737–745 (1982).
See also *4*, p. *64*; *6*, p. *79*; 22.5.1, p. 453; 22.6, p. 474.

M. R. Sen, A. V. Krishna Rao, and K. V. Rao
Dielectric Properties of Bismuth-Doped NaCl Single Crystals
(a) Vol. **73**, No. 2, p. 395–399 (1982).
See also *14.4*, p. *238*.

S. V. Moharil
On the General-Order Kinetics in Thermoluminescence
(a) Vol. **73**, No. 2, p. 509–514 (1982).
See also *20.3*, p. *326*.

P. Veeresham, U. V. Subba Rao, and V. Hari Babu
Thermoluminescence of Z_1-Centres in KCl_xBr_{1-x}:Ca^{2+} Crystals
(a) Vol. **73**, No. 2, p. 565–572 (1982).
See also *10.2*, p. *115*; 20.1, p. 313; *20.3*, p. *326*.

N. Nishimaki and S. Shimanuki
New Luminescence from M and N Centers Perturbed by Divalent and Monovalent Impurities in γ-Irradiated NaCl
(b) Vol. **114**, No. 1, p. K59 – K62 (1982).
See also *10.2*, p. *116*; 11, p. 127; *20.3*, p. *328*.

J. Shanker and R. K. Varshney
Temperature Dependence of Elastic Constants of Alkali Halide Crystals
(b) Vol. **114**, No. 1, p. K71 – K74 (1982).
See also 6, p. 80; *12.2*, p. *137*.

S. P. Kuila and V. V. Ratnam
Monochromatic Thermoluminescence Glow Curve Method for Evaluating Trapping Parameters
(a) Vol. **74**, No. 1, p. 255 – 263 (1982).
See also 10.2, p. 116; *13.4*, p. *177*; *20.3*, p. *328*.

M. Bucher
Core Energy and Peierls Stress of Edge Dislocations on {110} and {001} Slip Planes in NaCl
(b) Vol. **114**, No. 2, p. 383 – 392 (1982).
See also *10.2*, p. *116*.

J. Z. Damm and Z. Chvoj
Optimization of Tosi-Fumi Ionic Radii for F.C.C. Alkali Halide Crystals
(b) Vol. **114**, No. 2, p. 413 – 418 (1982).
See also *4*, p. *65*.

M. Elango, A. Kikas, E. Nômmiste, J. Pruulmann, and A. Saar
The Quantum Yield Spectra of Electron Emission of Solids in XUV Region. Alkali Chlorides at 100 to 250 eV
(b) Vol. **114**, No. 2, p. 487 – 493 (1982).
See also *17*, p. *257*.

M. Kucharczyk and S. Olszewski
Repulsive Energy and the Grüneisen Parameter of Alkali Halides Calculated on the Basis of a Quantum-Statistical ab Initio Theory
(b) Vol. **114**, No. 2, p. 589 – 598 (1982).
See also *4*, p. *65*; 6, p. 81; 12.2, p. 138.

J. Shanker and R. Sundaraj
Dielectric and Anharmonic Behaviour of Alkali Halide Crystals
(b) Vol. **114**, No. 2, p. 659 – 665 (1982).
See also *4*, p. *65*; 12.2, p. 138; 14.4, p. 239.

V. Kumar Srivastava
Plasmon Approach to the Band Gap of Alkali Halides
(b) Vol. **114**, No. 2, p. 667 – 671 (1982).
See also *13.1*, p. *156*; 13.5.2, p. 187; 14.4, p. 239.

V. V. Hizhnyakov
Vibrational Relaxation in the Excited Electronic State
(b) Vol. **114**, No. 2, p. 721 – 730 (1982).
See also *6*, p. *81*; 20.3, p. 328.

22.5.3 Alkali-Earth Halides

Original Papers and Short Notes

G. Jeandel, A. Warin, G. Villermain-Lecolier, and G. Morlot
Long-Wavelength Infrared Absorption Due to Y^{3+} Ion in BaF_2
(b) Vol. **109**, No. 1, p. K39–K40 (1982).
See also *6*, p. *70*; *20.1*, p. *299*.

H. J. Fuchs, W. Karthe, and R. Böttcher
EPR and ENDOR Investigations of an Oxygen Centre in CaF_2
(b) Vol. **110**, No. 1, p. 203–209 (1982).
See also **13.4**, p. 165; *19*, p. *284*.

C. E. Bielschowsky and B. Maffeo
Cluster Calculations of Alkaline-Earth Fluorides by the MSXα Method
(b) Vol. **110**, No. 1, p. 263–271 (1982).
See also *13*, p. *141*.

R. Kh. Sabirov and I. A. Vinukhov
On the Nature of the Oscillating Behaviour of the Free Induction Decay
(b) Vol. **110**, No. 1, p. 281–284 (1982).
See also *19*, p. *284*.

J. Shanker and J. P. Singh
Third-Order Elastic Constants and Pressure Derivatives of Second-Order Elastic Constants of Fluorite-Structure Crystals
(b) Vol. **110**, No. 2, p. 697–702 (1982).
See also *4*, p. *62*; *12.2*, p. *135*.

W. Pastusiak and J. Kuriata
A New Model for the Interpretation of Spin–Lattice Coefficients for S-State ($4f^7$) Ions
(b) Vol. **111**, No. 2, p. K147–K151 (1982).
See also *13.4*, p. *170*; *19*, p. *287*.

Y. M. M. Hornos, H. Panepucci, C. J. Magon, and M. C. Terrile
Discrete Saturation Effects of the ESR Spectrum of H^0 in CaF_2
(b) Vol. **112**, No. 1, p. 91–96 (1982).
See also *10.2*, p. *110*; *19*, p. *288*.

V. S. Urusov and V. B. Dudnikova
Energy of Anion Polarisation in Layered Structures
(b) Vol. **112**, No. 1, p. 303–310 (1982).
See also *1.3*, p. *29*; *4*, p. *63*; 22.5.2, p. 458.

A. K. Gerasyuk, V. F. Pisarenko, and V. V. Popov
On the Theory of Dipolar Thermally Stimulated Depolarization in Doped Ionic Crystals
(b) Vol. **112**, No. 1, p. K11–K14 (1982).
See also *10.2*, p. *110*; *14.4*, p. *236*.

22.5.4 Mixed Crystals

Original Papers and Short Notes

J. W. BRIGHTWELL, C. N. BUCKLEY, and B. RAY
Polymorphism in AgI and $Ag_{1-x}Cu_xI$
(a) Vol. **69**, No. 1, p. K11–K13 (1982).
See also *1.2*, p. *15*; *1.3*, p. *26*.

J. SHANKER and G. D. JAIN
Cohesive Energies and Bulk Modulus for Alkali Halide Mixed Crystals
(b) Vol. **110**, No. 1, p. 257–262 (1982).
See also *4*, p. *61*; *12.2*, p. *134*.

P. GÜNTER, R. SANCTUARY, and F. ROHNER
Changes of Optical Properties at the Successive Phase Transitions in Rb_2ZnCl_4
(a) Vol. **70**, No. 2, p. 583–589 (1982).
See also *1.2*, p. *16*; *20.1*, p. *304*.

R. K. SINHA and M. L. MUKHERJEE
Lattice Parameter and F-Band Position in $KCl_{1-x}Br_x$ Crystals
(b) Vol. **111**, No. 2, p. K117–K121 (1982).
See also *1.1*, p. *12*; *10.2*, p. *109*; 20.1, p. 306.

D. VERGNAT-GRANDJEAN, P. VERGNAT, and J.-F. BRICE
Lattice Vibrations of Alkaline-Earth Hydride-Fluoride Mixed Crystals Determined from Infrared Spectroscopy
(b) Vol. **113**, No. 1, p. 353–360 (1982).
See also *6*, p. *78*; *20.1*, p. *310*.

C. BASU, A. KANRAR, and U. S. GHOSH
Elastic Behaviour of Mixed Alkali Halide Crystals of Rocksalt Structure
(b) Vol. **114**, No. 1, p. 221–225 (1982).
See also *12.2*, p. *137*.

G. GRANGE, R. LANDERS, and B. MUTAFTSCHIEV
The Bubble Method: A Tool for Growth Morphology Studies of the Crystal–Melt Interface
(a) Vol. **74**, No. 1, p. 233–245 (1982).
See also *1.6*, p. *44*; *3*, p. *59*.

22.6 Simple Oxides (CdO, ZnO See 22.4)

Review Article

Z. M. Jarzębski
Preparation and Physical Properties of Transparent Conducting Oxide Films
(a) Vol. **71**, No. 1, p. 13–41 (1982).
See also 1.5, p. 38; 3, p. 57; 13.1, p. 148; *14.3.1*, p. *213*; *20.1*, p. *298*; 22.6.1, p. 477.

Original Papers and Short Notes

W. Tews and R. Gründler
Electron-Energy-Loss Spectroscopy of Different Al_2O_3 Modifications (I): Energy Loss Function, Dielectric Function, Oscillar Strength Sum Rule and the Quantity $\varepsilon_2 E$
(b) Vol. **109**, No. 1, p. 255–264 (1982).
See also 13, p. 139; *14.4*, p. *234*; *20.1*, p. *298*.

J. P. Lascaray, J. Diouri, J. P. Nadai, and J. Ferre
Infrared Absorption Spectra of Eu-Rich EuO. Excitations of the Bound Magnetic Polaron
(b) Vol. **109**, No. 1, p. 417–422 (1982).
See also 13.5.3, p. 188; *20.1*, p. *299*.

R. Strehlow and W. H. Lamfried
A New Approach to the Electronic Structure of SiO_x
(b) Vol. **109**, No. 1, p. K75–K79 (1982).
See also *13*, p. *140*.

E. I. Terukov, N. Mattern, W. Reichelt, H. Oppermann, and M. Wolf
The Behaviour of Resistance and Thermopower and the Coexistence of Two Phases at Metal–Insulator Transitions
(a) Vol. **69**, No. 1, p. 97–101 (1982).
See also 1.2, p. 14; *14.1*, p. *193*; *14.3*, p. *202*; 15, p. 245.

J. Szczyrbowski, A. Dietrich, and H. Hoffmann
Optical Properties of RF-Sputtered Indium Oxide Films
(a) Vol. **69**, No. 1, p. 217–226 (1982).
See also 14.3.1, p. 213; *20.1*, p. *299*.

H. Gruber and E. Krautz
Magnetoresistance and Conductivity in the Binary System Titanium-Oxygen (I): Titanium Oxides with Metallic Conductivity
(a) Vol. **69**, No. 1, p. 287–295 (1982).
See also 8, p. 85; *14.1*, p. *193*; *14.3*, p. *203*; 15, p. 245; 18, p. 259.

H.-J. Fitting and J.-U. Friemann
Monte-Carlo Studies of the Electron Mobility in SiO_2
(a) Vol. **69**, No. 1, p. 349–358 (1982).
See also 6, p. 71; *13*, p. *140*; *14.3.4*, p. *227*.

G. Gergely and H.-J. Müssig
Notes on Quantitative AES of the Fe–O System
(a) Vol. **69**, No. 1, p. K69–K72 (1982).
See also *1.6*, p. *42*; *17*, p. *255*.

A. J. McEvoy and W. Gissler
ESCA Spectra and Electronic Properties of Some Ruthenium Compounds
(a) Vol. **69**, No. 1, p. K91–K96 (1982).
See also **13**, p. 140; *17*, p. *255*; 21.4, p. 369.

H. Fischer, H.-D. Geiler, G. Götz, A. V. Dvurechenskii, B. P. Kashnikov, and L. S. Smirnov
Antiannealing Effects in Ion-Implanted and Electron Pulse Irradiated Quartz Crystals
(a) Vol. **69**, No. 2, p. K139–K142 (1982).
See also 1.2, p. 15; *11*, p. *119*.

H. Näfe, G. Boden, H. Ullmann, and D. Rettig
On the Luminescence Behaviour of Yttria Doped Thoria
(a) Vol. **69**, No. 2, p. K223–K227 (1982).
See also 11, p. 120; *20.3*, p. *320*.

P. Clippe and A. Ausloos
Size Effect on the Infrared Absorption Spectrum of Clustered Ionic Spheres
(b) Vol. **110**, No. 1, p. 211–218 (1982).
See also 1, p. 9; *20.1*, p. *302*.

S. Mochizuki
Infrared Spectra of NiO Microcrystals at High Temperatures
(b) Vol. **110**, No. 1, p. 219–225 (1982).
See also 6, p. 72; *20.1*, p. *302*.

S. Mochizuki and S. Tateyama
Reststrahlen Spectra of Cerium Dioxide
(b) Vol. **110**, No. 1, p. K1–K4 (1982).
See also *6*, p. *72*; *20.1*, p. *302*.

L. Schumann, A. Lehmann, H. Sobotta, V. Riede, U. Teschner, and K. Hübner
Infrared Studies of Reactively Sputtered SiO_x Films in the Composition Range $0.2 \leqq x \leqq 1.9$
(b) Vol. **110**, No. 1, p. K69–K73 (1982).
See also 2, p. 47; *6*, p. *72*; *20.1*, p. *302*.

L. N. Skuja and A. R. Silin
A Model for the Non-Bridging Oxygen Center in Fused Silica. The Dynamic Jahn-Teller Effect
(a) Vol. **70**, No. 1, p. 43–49 (1982).
See also *10.2*, p. *106*; *13.4*, p. *166*; 20.1, p. 302; 20.3, p. 321.

W. Gräfe
Description of Internal Friction Background of Glasses by Means of Migration Processes
(a) Vol. **70**, No. 1, p. 101–108 (1982).
See also *2*, p. *47*; *12.2*, p. *134*.

K. B. Kadyrakunov, E. V. Nidaev, and L. S. Smirnov
Behaviour of a Built-in Charge in a Dielectric at Pulsed Annealing of Si–SiO_2 Structures
(a) Vol. **70**, No. 1, p. K15–K17 (1982).
See also *13.3*, p. *158*; *14.4*, p. *235*; 22.1.2, p. 403.

A. HAYDAR, A. CORET, and N. OUDJEHANE
Dark Current–Voltage Characteristics of Cu_2O Crystals and Their Relation to Structural Defects
(a) Vol. **70**, No. 2, p. 683–690 (1982).
See also *14.3*, p. *205*.

J. LLOPIS, J. PIQUERAS, and C. BALLESTEROS
Influence of Purity on Cathodoluminescence from Dislocations in MgO
(a) Vol. **70**, No. 2, p. 739–746 (1982).
See also *10.2*, p. *108*; *20.3*, p. *322*.

E. I. SHTYRKOV, V. S. LOBKOV, S. A. MOISEEV, and N. G. YARMUKHAMETOV
Forced Deformation of the Reversed Three-Pulsed Photon-Echo Signal on the Ruby R_1-Line
(a) Vol. **70**, No. 2, p. K93–K95 (1982).
See also *20*, p. *295*.

K. P. KOVAL, E. V. BABKIN, and V. G. PYNKO
Growth-Induced Anisotropy in Epitaxial Crystals of γ-Fe_2O_3
(a) Vol. **70**, No. 2, p. K117–K120 (1982).
See also *18.3*, p. *276*.

J. GARBARCZYK, M. WASIUCIONEK, W. JAKUBOWSKI, and A. GRODZIŃSKI
TGA and DTA Studies on Moist Na, Li/Na, and Ag β''-Aluminas
(a) Vol. **70**, No. 2, p. K163–K166 (1982).
See also *8*, p. *86*; *10.2*, p. *108*.

H.-J. FITTING, J. J. PINNIS, and A. R. LUSIS
Exoelectron Emission of Thin WO_3 Films after Electron Beam Excitation
(a) Vol. **70**, No. 2, p. K185–K187 (1982).
See also *17*, p. *256*.

D. FRÖHLICH and R. KENKLIES
Polarization Dependence of Two-Photon Magnetoabsorption of the 1S Exciton in Cu_2O
(b) Vol. **111**, No. 1, p. 247–253 (1982).
See also *13.5.1*, p. *183*; *20.1*, p. *305*.

A. A. BEREZIN and F. L. WEICHMAN
Cuprous Oxide–Copper Photovoltaic Cells Formed by High-Temperature Reduction
(a) Vol. **71**, No. 1, p. 265–274 (1982).
See also *16*, p. *251*; 20.1, p. 305.

J. BULLOT, O. GALLAIS, M. GAUTHIER, and J. LIVAGE
Threshold Switching in V_2O_5 Layers Deposited from Gels
(a) Vol. **71**, No. 1, p. K1–K4 (1982).
See also *2*, p. *48*; *14.3.4*, p. *229*.

K. ATOBE
The Effect of Radiation Type on Defect Production in Li_2O
(a) Vol. **71**, No. 1, p. K25–K28 (1982).
See also *11*, p. *122*.

R. GÖRING, K. KNEIPP, and H. NASS
Investigation of Glasses in the $BaO-B_2O_3-GeO_2$ System with Low BaO Content by NMR and Raman Spectroscopy
(a) Vol. **72**, No. 2, p. 623–630 (1982).
See also *2*, p. *51*; 6, p. 78; 19, p. 289; 20.1, p. 309.

B. YANGUI
Electron Microscopic Evidence of New Twins in Monoclinic Rare-Earth Sesquioxides ($B-Ln_2O_3$)
(a) Vol. **72**, No. 2, p. 631–643 (1982).
See also *1.2*, p. *20*; *1.4*, p. *35*.

M. WOLF, E. I. TERUKOV, W. REICHELT, and H. OPPERMANN
Influence of ^{18}O Substitution on the High-Temperature Phase Transition of Cr-Doped V_2O_3 and Its Relation to Low-Temperature Phase Transition
(a) Vol. **72**, No. 2, p. 655–659 (1982).
See also *1.2*, p. *20*; 6, p. 78; 14.1, p. 196; 14.3, p. 209; 18, p. 262.

H. BIALAS and G. SCHAUER
Circular Acoustic Birefringence in Quartz Observed by Time-of-Flight Measurement
(a) Vol. **72**, No. 2, p. 679–686 (1982).
See also *7*, p. *83*.

C. PARACCHINI and G. SCHIANCHI
Luminescence of WO_3
(a) Vol. **72**, No. 2, p. K129–K132 (1982).
See also *20.3*, p. *325*.

U. PIETSCH
A Proof of Aspherical Valence Charge Parts in the Rock-Salt Structure Compounds by Means of an X-Ray Bond Charge Model
(b) Vol. **113**, No. 1, p. 203–207 (1982).
See also *4*, p. *64*; 13.1, p. 154; 22.5.2, p. 460.

B. HENDERSON and A. H. GARRISON
Ion Implantation in Alkaline Earth Oxides
(b) Vol. **113**, No. 1, p. 267–276 (1982).
See also *11*, p. *125*; 19, p. 289; 20.1, p. 310; 20.3, p. 325.

M. SAINT-PAUL and G. FROSSATI
Dielectric Relaxation in Smoky Quartz at Very Low Temperatures
(b) Vol. **113**, No. 1, p. 373–377 (1982).
See also *14.4*, p. *237*.

G. DRÄGER, F. WERFEL, aand O. BRÜMMER
Polarized OKα Valence Band Spectra of TiO_2 (Rutile)
(b) Vol. **113**, No. 1, p. K15–K17 (1982).
See also *13.1*, p. *154*; *20*, p. *296*.

D. SENDOREK, M. LUBECKA, and A. WĘGRZYN
Carrier Mobility in EuO
(b) Vol. **113**, No. 1, p. K19–K22 (1982).
See also *14.3*, p. *209*; *18.2*, p. *272*.

B. Yangui and C. Boulesteix
Domains and Interfaces Created during Cooparative Crystallographic Transformation – Application to Hexagonal-Monoclinic Transformation of Rare Earth Sesquioxides (II): Relative Position and Connection of the Variants of Rotation and Translation Domains: Connection into Stars
(a) Vol. **73**, No. 2, p. 515–527 (1982).
See also *1.2*, p. *22*; *1.4*, p. *37*.

F. Delannay
On the Reduction of Orthorhombic MoO_3 to MoO_2
(a) Vol. **73**, No. 2, p. 529–537 (1982).
See also *1.2*, p. *22*; *1.4*, p. *37*.

A. J. Cohen and L. N. Makar
Models for Color Centers in Smoky Quartz
(a) Vol. **73**, No. 2, p. 593–596 (1982).
See also *10.2*, p. *115*; *20.1*, p. *313*.

B. P. Chandra
Uncommon Mechanoluminescence Spectra of Certain Organic and Inorganic Crystals
(a) Vol. **73**, No. 2, p. K205–K208 (1982).
See also *20.3*, p. *327*; 22.9, p. 507.

G. Rudlof and H. Glaefeke
Comparison of Isochronal and Time-Linear Thermal Annealing Curves and Conclusions about the Correlation Glow-Peak Nature of Traps and the Mechanism of TSEE for α-Al_2O_3
(a) Vol. **73**, No. 2, p. K253–K258 (1982).
See also *13.4*, p. *176*; *17*, p. *257*.

S. Mochizuki
Infrared Optical Properties of Cerium Dioxide
(b) Vol. **114**, No. 1, p. 189–199 (1982).
See also *6*, p. *80*; *20.1*, p. 314.

V. P. Gupta and C. P. Agrawal
Rydberg Potential in Diatomic Ionic Crystals
(b) Vol. **114**, No. 1, p. 265–270 (1982).
See also *4*, p. *65*; *6*, p. *80*; 8, p. 89; 12.2, p. 137; 22.5.2, p. 462.

E. Hartmann and L. Kovács
Electrical Conductivity of Paratellurite (TeO_2) Crystals
(a) Vol. **74**, No. 1, p. 59–64 (1982).
See also *14.3*, p. *211*; 16, p. 253.

T. Jung and W. Titel
Application of Auger Depth Profiling on the Al–SiO_2 Interface
(a) Vol. **74**, No. 1, p. 85–90 (1982).
See also *1.6*, p. *44*; *13.3*, p. *161*.

Hj. Bernhardt
A Contribution to High-Temperature Thermoluminescence Investigations in Quartz Crystals
(a) Vol. **74**, No. 1, p. 159–165 (1982).
See also 10.2, p. 116; *20.3*, p. *328*.

22.6.1 Mixed Crystals

Review Article

Z. M. JARZĘBSKI
Preparation and Physical Properties of Transparent Conducting Oxide Films
(a) Vol. **71**, No. 1, p. 13–41 (1982).
See also **1.5**, p. 38; 3, p. 57; **13.1**, p. 148; *14.3.1*, p. *213*; *20.1*, p. *298*; 22.6, p. 468.

Original Papers and Short Notes

A. WITEK
The Effect of Illumination on Anomalous Ultrasonic Attenuation in Doped Single Crystals of $Bi_{12}GeO_{20}$
(a) Vol. **69**, No. 1, p. 67–70 (1982).
See also *7*, p. *82*; 16, p. 249.

W. RADLOFF, E. BELOW, H. WAGNER, and P. KLEINERT
Laser Absorption Measurement of OH Ion Distribution Profiles in Silica Glasses
(a) Vol. **69**, No. 1, p. K21–K24 (1982).
See also *2*, p. *46*; *20*, p. *293*.

H. BIŃCZYCKA, O. GZOWSKI, L. MURAWSKI, and J. SAWICKI
Mössbauer Effect and Electrical Conductivity in TeO_2–Fe_2O_3 Glasses
(a) Vol. **70**, No. 1, p. 51–55 (1982).
See also 2, p. 47; *5*, p. *66*; *14.3*, p. *204*.

H. TAGUCHI
Effect of CO Adsorption on $(La_{1-x}Sr_x)FeO_3$ $(0 \leqq x \leqq 0.5)$
(a) Vol. **70**, No. 1, p. K31–K33 (1982).
See also *14.3.1*, p. *214*.

H. KUWAMOTO
Domain Effects Accompanying the Metal–Insulator Transition in $(V_{0.99}Cr_{0.01})_2O_3$
(b) Vol. **110**, No. 2, p. K175–K178 (1982).
See also *1.4*, p. *33*; *14*, p. *191*.

M. Y. NADEEM and C. A. HOGARTH
DC Conduction in Thin Films of GeO_2/BaO Complexes prior to Electroforming
(a) Vol. **72**, No. 2, p. K203–K206 (1982).
See also *14.3.1*, p. *215*.

H. HENNIG, K.-H. HECKNER, D. HIRSCH, and H. LADWIG
The Influence of the Preparation on Layer Properties of Transparent Indium–Tin Oxide Electrodes
(a) Vol. **74**, No. 1, p. 133–139 (1982).
See also *1.5*, p. *41*; *14.3.1*, p. *216*.

C. MIYAKE, H. TAKEUCHI, H. OHYA-NISHIGUCHI, and S. IMOTO
Electron Spin Resonance and Magnetic Susceptibility of M–U–O Ternary Mixed Oxides (I): Anomalies of Electron Spin Resonance and Magnetic Susceptibility of $Li_{2n-5}UO_n$ $(n = 3, 4,$ and 6)
(a) Vol. **74**, No. 1, p. 173–180 (1982).
See also *18.1*, p. *266*; 18.4, p. 281; *19*, p. *290*.

22.7 Semiconducting Intermetallic Compounds

Original Papers and Short Notes

H. T. Langhammer, M. Stordeur, H. Sobotta, and V. Riede
Optical and Electrical Investigations of the Anisotropy of Sb_2Te_3 Single Crystals
(b) Vol. **109**, No. 2, p. 673–681 (1982).
See also *14.3*, p. *203*; 15, p. 245; *20.1*, p. *300*.

L. G. Khvostantsev, A. I. Orlov, N. Kh. Abrikosov, T. E. Svechnikova, and S. N. Chizhevskaya
Thermoelectric Properties and Phase Transitions in Bi_2Te_3 under Hydrostatic Pressure up to 9 GPa and Temperature up to 300 °C
(a) Vol. **71**, No. 1, p. 49–53 (1982).
See also *1.2*, p. *17*; 12.2, p. 135; 14.3, p. 206; *15*, p. *246*.

J. Horák, L. Koudelka, J. Klikorka, and L. Šiška
Transport Properties and Plasma Resonance of Bismuth Selenide Crystals Doped with Cadmium
(b) Vol. **111**, No. 2, p. 575–580 (1982).
See also *13.4*, p. *169*; *14.3*, p. *207*; 15, p. 247.

T. Satow, O. Uemura, and Y. Sagara
The Neutron Diffraction Study of Liquid Antimony and Bismuth Chalcogenides
(a) Vol. **71**, No. 2, p. 555–561 (1982).
See also *2*, p. *49*; 14.3, p. 207.

M. A. Korzhuev
On the Fundamental Energy Gap in GeTe
(b) Vol. **112**, No. 1, p. K39–K41 (1982).
See also 1.2, p. 19; *13.1*, p. *152*.

S. Santhanam, B. K. Samantharay, and A. K. Chaudhuri
X-Ray Studies on the Microstructure of Vacuum Evaporated SnTe Thin Films
(a) Vol. **72**, No. 2, p. 521–527 (1982).
See also *1.5*, p. *40*; 2, p. 51; 16, p. 252.

W. Hoyer, A. Müller, E. Thomas, and M. Wobst
X-Ray-Diffraction Investigation on Molten In–Te Alloys
(a) Vol. **72**, No. 2, p. 585–592 (1982).
See also 1.3, p. 30; *2*, p. *51*.

B. V. Kobrin, R. M. Kupriyanova, V. S. Minaev, E. P. Prokopiev, and V. P. Shantarovich
Study of Imperfections in Ge–Te Semiconductors by the Positron Annihilation Technique
(a) Vol. **73**, No. 2, p. 321–324 (1982).
See also *2*, p. *53*; *10.2*, p. *114*.

J. Horák, P. Lošťák, L. Šiška, and M. Stordeur
The Nature of Silver Impurity Atoms in Antimony Telluride
(b) Vol. **114**, No. 1, p. 39–45 (1982).
See also 13.1, p. 155; *14.3*, p. *210*; 15, p. 248; *20.1*, p. *314*.

22.8 Tertiary and Higher Compounds

Original Papers and Short Notes

S. K. Misra and P. Mikolajczak
EPR Linewidth Studies of Gd^{3+}-Doped $Yb_xY_{1-x}Cl_3 \cdot 6\,H_2O$ Single Crystals. Yb^{3+} Spin-Lattice Relaxation Times
(b) Vol. **109**, No. 1, p. 59–66 (1982).
See also *18.1*, p. *265*; *19*, p. *282*.

A. M. Sandauer and P. Byszewski
Evidence of the Conduction–Valence Band Overlap Induced by a Magnetic Field in Zero-Gap $Hg_{1-x}Mn_xTe$
(b) Vol. **109**, No. 1, p. 167–171 (1982).
See also *13.1*, p. *148*; *14.3*, p. *202*; 18, p. 259.

O. Uemura, M. Itoh, and T. Satow
Electronic Properties of the Tl–$AgTlTe_2$ System in the Liquid State
(b) Vol. **109**, No. 1, p. 361–368 (1982).
See also *2*, p. *45*; *14.3*, p. *202*; 15, p. 245; 18, p. 259.

I. V. Bodnar, A. P. Bologa, and B. V. Korzun
Composition Dependence of the Band Gap of $CuGa_xIn_{1-x}Se_2$ Solid Solutions
(b) Vol. **109**, No. 1, p. K31–K33 (1982).
See also *13.1*, p. *148*; 20.1, p. 299.

A. Leblé, J. J. Rousseau, J. C. Fayet, J. Pannetier, J. L. Fourquet, and R. de Pape
Order–Disorder Transition of NH_4AlF_4 through EPR Investigations
(a) Vol. **69**, No. 1, p. 249–256 (1982).
See also 1.2, p. 14; *19*, p. *283*.

A. H. Abou El Ela, N. Abdelmohsen, and H. H. A. Labib
Dielectric Relaxation Study of $CuTlSe_2$ in the Solid and Liquid Phases
(a) Vol. **69**, No. 1, p. 377–384 (1982).
See also 2, p. 45; 13.4, p. 163; *14.4*, p. *234*.

B. Capelle and C. Malgrange
X-Ray Topographic Study of the Phase Transition in $GdDy(MoO_4)_3$ Crystals
(a) Vol. **69**, No. 1, p. K1–K2 (1982).
See also *14.4.1*, p. *240*.

S. G. Abdullaeva and V. A. Aliev
Special Features of Negative Photoconductivity in $TlGaSe_2$
(a) Vol. **69**, No. 1, p. K33–K36 (1982).
See also 13.4, p. 163; *16*, p. *249*.

J. González, J. Alberto Torres, and G. Sanchez Pérez
Photoconductivity Spectrum of p-Type $CuInS_2$ Single Crystals
(a) Vol. **69**, No. 1, p. K37–K41 (1982).
See also 13.4, p. 163; *16*, p. *250*.

A. N. Georgobiani, S. I. Radautsan, and I. M. Tiginyanu
Electroabsorption and Non-Equilibrium Carrier Recombination in $CdGa_2S_4$ Single Crystals
(a) Vol. **69**, No. 2, p. 513–520 (1982).
See also *13.4*, p. *165*; 16, p. 250; 20.1, p. 301; *20.3*, p. *320*.

M. N. Avasthi, M. T. El-Gemal, and M. Saleem
High Ionic Conductivity of $Ag_{19}I_{15}P_2O_7$ Solid Electrolyte
(a) Vol. **69**, No. 2, p. 535–539 (1982).
See also *10.2*, p. *105*; *14.4*, p. *234*.

D. Kostopoulos, S. Kouremenou, P. Varotsos, and S. Mourikis
Thermocurrents in Ferroelectric Materials
(a) Vol. **70**, No. 1, p. K19–K21 (1982).
See also *14.4.1*, p. *241*.

C. Laflamme, M. Aubin, R. Provencher, and S. Jandl
Resistivity Measurements of $ZrS_{3-x}Se_x$ Alloys
(a) Vol. **70**, No. 1, p. K67–K69 (1982).
See also *14.3*, p. *205*.

R. Durný, P. Macko, and R. D. Tomlinson
Electrical Properties of Neutron-Irradiated $CuInTe_2$
(a) Vol. **70**, No. 1, p. K89–K91 (1982).
See also *11*, p. *120*; 13.4, p. 166; *14.3*, p. *205*.

G. Blasse and G. J. Dirksen
The Luminescence of Broad-Band Emitters in $LiLaP_4O_{12}$
(b) Vol. **110**, No. 2, p. 487–494 (1982).
See also *20.3*, p. *321*.

M. Sekita, M. Ishii, and Y. Miyazawa
Raman Scattering in $Bi_2Ge_3O_9$
(b) Vol. **110**, No. 2, p. K141–K144 (1982).
See also *6*, p. *73*; *20.1*, p. *304*.

C. Rincoń and J. González
Influence of Impurities on the Optical Properties of $CuInSe_2$ near the Fundamental Absorption Edge
(b) Vol. **110**, No. 2, p. K171–K174 (1982).
See also 13.1, p. 150; 13.4, p. 167; *20.1*, p. *304*.

A. A. Kaminskii, N. R. Agamalyan, G. A. Denisenko, S. E. Sarkisov, and P. P. Fedorov
Spectroscopy and Laser Emission of Disordered GdF_3–CaF_2:Nd^{3+} Trigonal Crystals
(a) Vol. **70**, No. 2, p. 397–406 (1982).
See also 20.1, p. 304; *20.2*, p. *317*; *20.3*, p. *322*.

G. A. Petrakovskii, K. A. Sablina, V. P. Ikonnikov, I. A. Volkov, and A. G. Klimenko
The Effect of Amorphization on the Magnetic Properties of $KFeS_2$
(a) Vol. **70**, No. 2, p. 507–511 (1982).
See also *2*, p. *47*; *18*, p. *260*.

A. N. ABBASOV, K. R. ALLAKHVERDIEV, F. G. ALIEV, and R. M. ZAMANOVA
Far-Infrared Absorption Spectra of $CdInAlS_4$ Single Crystals
(b) Vol. **111**, No. 2, p. K129–K131 (1982).
See also *6*, p. *75*; *20.1*, p. *306*.

H. NEUMANN, B. PERLT, B. SCHUMANN, and G. KÜHN
Optical Properties of $CuIn_2Se_{3.5}$ near the Fundamental Absorption Edge
(b) Vol. **111**, No. 2, p. K133–K136 (1982).
See also *13.1*, p. *151*; *20.1*, p. *306*.

A. A. KAMINSKII, A. G. PETROSYAN, G. A. DENISENKO, T. I. BUTAEVA, V. A. FEDOROV, and S. E. SARKISOV
Spectroscopic Properties and 3 μm Stimulated Emission of Er^{3+} Ions in the $(Y_{1-x}Er_x)_3Al_5O_{12}$ and $(Lu_{1-x}Er_x)_3Al_5O_{12}$ Garnet Crystal Systems
(a) Vol. **71**, No. 2, p. 291–312 (1982).
See also 20.1, p. 307; *20.2*, p. *317*; *20.3*, p. *323*.

Y. SEKI, S. ENDO, and T. IRIE
Negative Resistance in $CdIn_2S_4$ Single Crystals
(a) Vol. **71**, No. 2, p. 365–370 (1982).
See also *14.3.4*, p. *229*.

X.-Z. DONG, B. GRÖGER, T. JENDRYSIK, and H. KRONMÜLLER
Annealing Effect of Domain Patterns of the Nearly Non-Magnetostrictive Amorphous Alloy $Co_{58}Ni_{10}Fe_5Si_{11}B_{16}$
(a) Vol. **71**, No. 2, p. 441–449 (1982).
See also 2, p. 49; *18*, p. *261*; *20*, p. *295*.

K. BOHMHAMMEL, P. DEUS, G. KÜHN, and W. MÖLLER
Specific Heat, Debye Temperature, and Related Properties of Chalcopyrite Semiconducting Compounds $CuGaSe_2$, $CuGaTe_2$, and $CuInTe_2$
(a) Vol. **71**, No. 2, p. 505–510 (1982).
See also *6*, p. *75*; *8*, p. *87*.

S. M. WASIM, G. S. PORRAS, and R. D. TOMLINSON
Some Electrical Characteristics of Cu- and In-Doped $CuInTe_2$
(a) Vol. **71**, No. 2, p. 523–530 (1982).
See also *13.4*, p. *170*; *14.3*, p. *207*.

G. A. SCHOLZ, R. F. FRINDT, and A. E. CURZON
Electron Diffraction Investigation of the Ag_xTaS_2 System (I): Superlattices in $Ag_{1/3}TaS_2$ and $Ag_{2/3}TaS_2$
(a) Vol. **71**, No. 2, p. 531–542 (1982).
See also *1.3*, p. *28*.

A. N. ANAGNOSTOPOULOS and A. KAROUTIS
Electron Trap Distribution Determined by Means of Conductivity Measurements on Highly Anisotropic $ZnIn_2S_4$ Crystals
(a) Vol. **71**, No. 2, p. 543–548 (1982).
See also *13.4*, p. *170*; *14.3.4*, p. *229*.

M. D. KAPLAN
Metamagnetoelastic and Metamagnetoelectric Behaviour of the Jahn-Teller Antiferroelastics
(b) Vol. **112**, No. 1, p. 351–356 (1982).
See also *1.2*, p. *19*; 18.4, p. 280.

H. NEUMANN, W. KISSINGER, R. D. TOMLINSON, and N. AVGERINOS
Two-Phonon Absorption Spectra in $CuInTe_2$
(b) Vol. **112**, No. 1, p. K19–K23 (1982).
See also *6*, p. *76*; *20.1*, p. *308*.

G. A. SCHOLZ, R. F. FRINDT, and A. E. CURZON
Electron Diffraction Investigation of the Ag_xTaS_2 System (II): Superlattices, Structure, and Charge Density Waves in Ag_xTaS_2
(a) Vol. **72**, No. 1, p. 375–390 (1982).
See also *1.3*, p. *29*.

A. KIDAWA and R. WITULA
The Influence of Surface Perfection on Photoconductivity in SbSI Single Crystals
(a) Vol. **72**, No. 1, p. K9–K10 (1982).
See also *16*, p. *252*.

A. E. MOSKALEV, I. A. BELOBROVA, L. I. ZHEREBTSOVA, and I. P. ALEKSANDROVA
Ferroelectric Phase Transition in $(NH_4)_2ZnBr_4$
(a) Vol. **72**, No. 1, p. K19–K21 (1982).
See also *1.2*, p. *19*; *14.4.1*, p. *242*; 19, p. 288.

K. G. NIKIFOROV, A. G. GUREVICH, S. I. RADAUTSAN, L. M. EMIRYAN, and V. E. TEZLEVAN
Effect of Nonstoichiometry on Ferromagnetic Resonance in $CdCr_2S_4$ Crystals
(a) Vol. **72**, No. 1, p. K37–K39 (1982).
See also 10.2, p. 111; *19*, p. *288*.

H. ROSENBERGER, G. SCHELER, and YU. N. MOSKVICH
^1H Chemical Shifts in Polycrystalline $RbHSeO_4$ and NH_4HSeO_4
(a) Vol. **72**, No. 1, p. K49–K51 (1982).
See also *19*, p. *288*.

K. ZÁVĚTA, J. SCHNEIDER, A. HANDSTEIN, and Z. KALVA
The Influence of Annealing on the Magnetization of Amorphous $(Ni_{0.93}Fe_{0.07})_{80}P_{10}B_{10}$ Alloy
(a) Vol. **72**, No. 1, p. K79–K82 (1982).
See also 2, p. 50; *18*, p. *261*.

D. V. GITSU, I. N. GRINCHESHEN, and N. S. POPOVICH
Dependence of Electrophysical Parameters of $TlSbSe_2$ Layer Crystals on Growth Conditions
(a) Vol. **72**, No. 1, p. K113–K116 (1982).
See also 3, p. 58; *14.3*, p. *208*; *16*, p. *252*.

K. BANGE, A. LIEBEGALL, and B. LORENTZ
Interband Magnetoreflectivity at $Pb_{1-x}Sn_xSe$ Single Crystals
(b) Vol. **112**, No. 2, p. 543–547 (1982).
See also 13.1, p. 153; *18*, p. *262*; *20.1*, p. *309*.

K. R. ALLAKHVERDIEV, M. M. GODZHAEV, A. I. NADZHAFOV, and R. M. SARDARLY
Light Scattering Spectra of the Solid Solutions $TlGa_xIn_{1-x}Se_{2x}S_{2(1-x)}$
(b) Vol. **112**, No. 2, p. K93–K97 (1982).
See also *6*, p. *77*; *20.1*, p. *309*.

R. A. ALIEV, K. R. ALLAKHVERDIEV, R. M. SARDARLY, and V. E. SHTEINSHREIBER
Angular Behaviour of the Polar Optical Phonons in $A^{III}B^{III}C_2^{VI}$ Layered Semiconductors
(b) Vol. **112**, No. 2, p. K153–K156 (1982).
See also *6*, p. *78*.

D. SEURET, O. VIGIL, and F. LECCABUE
Electro-Optical Properties of $CuGa_{0.7}In_{0.3}Se_2/ZnSe$ Heterojunction
(a) Vol. **72**, No. 2, p. 661–666 (1982).
See also *14.3.3*, p. *223*; *20.3*, p. *324*; 22.4.2, p. 443.

E. WIESER, G. KRABBES, and E. I. TERUKOV
Detection of the Metastable Simultaneous Occurrence of High- and Low-Temperature States in $Co_yFe_{1-y}S$ by Mössbauer Spectroscopy
(a) Vol. **72**, No. 2, p. 695–699 (1982).
See also *1.2*, p. *20*; *5*, p. *68*; 14, p. 191; 18, p. 262.

A. N. ANAGNOSTOPOULOS, C. MANOLIKAS, D. PAPADOPOULOS, and J. SPYRIDELIS
Composition Faults in $ZnIn_2S_4$ (III) Layered Crystals and Their Influence on the Anisotropic Conductivity of This Compound
(a) Vol. **72**, No. 2, p. 731–736 (1982).
See also 1.4, p. 35; *14.3*, p. *209*.

M. HIDAKA, K. INOUE, B. J. GARRARD, and B. M. WANKLYN
Study of Structural Phase Transitions in the Layer Compound $RbVF_4$
(a) Vol. **72**, No. 2, p. 809–816 (1982).
See also *1.2*, p. *20*; 6, p. 78; 8, p. 88.

M. BACMANN and B. LAMBERT-ANDRON
Crystalline Structure of $Co_3U_2O_8$ from 300 to 110 K
(a) Vol. **72**, No. 2, p. 833–837 (1982).
See also *1.2*. p. *20*.

K. K. P. SRIVASTAVA and S. N. MISHRA
Magnetic Relaxation Effects in the Mössbauer Spectrum of Fe^{2+} in Gillespite ($BaFeSi_4O_{10}$)
(b) Vol. **113**, No. 1, p. 93–104 (1982).
See also *5*, p. *68*.

V. M. VOLKOV, E. F. KUSTOV, T. K. MAKETOV, and G. STECZKO
Invariant Expansion of Energy of the Induced Magnetic Anisotropy
(b) Vol. **113**, No. 1, p. 301–305 (1982).
See also *13*, p. *145*; *18*, p. *262*.

K. R. ALLAKHVERDIEV, T. G. MAMEDOV, E. YU. SALAEV, and I. K. EFENDIEVA
The Fundamental Absorption Edge of $TlInSe_2$
(b) Vol. **113**, No. 1, p. K43–K47 (1982).
See also *13.1*, p. *154*; *20.1*, p. *311*.

CH. KLEINT, S. SCHULZE, and R. D. TOMLINSON
Low Energy Electron Loss Peaks of $CuInTe_2$
(b) Vol. **113**, No. 1, p. K53–K56 (1982).
See also *13.1*, p. *154*; 13.5.2, p. 187.

A. N. ANAGNOSTOPOULOS and B. PLOSS
Influence of Composition Faults on the Anisotropic Conductivity of Layered Semiconductors
(a) Vol. **73**, No. 1, p. 91–94 (1982).
See also *14.3*, p. *210*.

G. BLASSE
Energy Transfer Phenomena in the System (Y, Ce, Gd, Tb)F_3
(a) Vol. **73**, No. 1, p. 205–208 (1982).
See also *13.4*, p. *174*; *20.3*, p. *326*.

G. SORGE, H. BEIGE, A. ALMEIDA, and L. A. SHUVALOV
Elastic Nonlinearity of $KH_3(SeO_3)_2$ and $KD_3(SeO_3)_2$ Crystals
(a) Vol. **73**, No. 1, p. K63–K66 (1982).
See also *1.2*, p. *21*; 12.2, p. 136; *14.4.1*, p. *243*.

A. D. AVESOV, E. M. GASANOV, A. I. NOVOZHILOV, and V. I. SAMOILOVICH
Some Structure Defects in Neutron Irradiated Fluorphlogopite
(a) Vol. **73**, No. 1, p. K67–K68 (1982).
See also *11*, p. *126*; *20.3*, p. *326*.

I. HENNING, M. MERTIG, R. PLATH, G. POMPE, E. HEGENBARTH, and R. SCHALGE
Glass-Like Behaviour of $Sr_{1-x}Ba_xNb_2O_6$ (SBN) Single Crystals Demonstrated by Heat Capacity Measurements
(a) Vol. **73**, No. 1, p. K105–K109 (1982).
See also *8*, p. *89*; *14.4.1*, p. *243*.

K. SURENDRANATH and C. BANSAL
Mössbauer Effect Study of Polymetallic Nodules from Indian Ocean Bed
(a) Vol. **73**, No. 1, p. K133–K136 (1982).
See also *5*, p. *68*.

T. G. KERIMOVA, E. YU. SALAEV, A. SH. KHIDIROV, N. G. DERVISHOV, and SH. M. EFENDIEV
Effect of Cationic Ordering on Vibrational Spectra of $CdIn_2S_4$
(b) Vol. **113**, No. 2, p. K107–K109 (1982).
See also *6*, p. *79*; *20.1*, p. *312*.

R. CHAVES, H. AMARAL, A. LEVELUT, S. ZIOLKIEWICZ, M. BALKANSKI, M. K. TENG, J. F. VITTORI, and H. STONE
Tricritical Point Induced by Atomic Substitution in $SbSe_xS_{1-x}I$
(a) Vol. **73**, No. 2, p. 367–376 (1982).
See also *14.4*, p. *238*.

H. AMARAL, R. CHAVES, A. LEVELUT, and S. ZIOLKIEWICZ
Diffuse Phase Transition in Ferroelectric $SbSe_{0.40}S_{0.60}I$
(a) Vol. **73**, No. 2, p. 421–427 (1982).
See also *14.4.1*, p. *243*.

U. Straube, G. Sorge, and N. R. Ivanov
Temperature-Dependent Electrostriction Coefficient and Pseudoproper Phase Transition in TSCC
(a) Vol. **73**, No. 2, p. 467–474 (1982).
See also **1.2**, p. **22**; *12.2*, p. *137*; *14.4.1*, p. *243*.

Tran Chot
On the Location of the Interface Fermi Level in Metal–Semiconductor Schottky-Barrier Contacts
(a) Vol. **73**, No. 2, p. 551–558 (1982).
See also *13.3*, p. *160*; *14.3.2*, p. *218*; 14.3.4, p. 232; 22.1.1, p. 398; 22.1.2, p. 413; 22.2, p. 419; 22.4, p. 438.

E. Bărbulescu, A. Bărbulescu, and D. Barb
On the Ferroelectric Structural Phase Transitions in Solid Solutions
(a) Vol. **73**, No. 2, p. 603–610 (1982).
See also *14.4.1*, p. *243*.

A. D. Avesov and E. M. Gasanov
Some Radiation-Optical Properties of Fluorphlogopite with Admixture of Vanadium
(a) Vol. **73**, No. 2, p. K259–K261 (1982).
See also **11**, p. 127; *20.1*, p. *313*.

H. Cofta and P. Pawlicki
The Model of Metamagnetic Phase Transition in Mn_3GaC
(b) Vol. **114**, No. 1, p. 53–60 (1982).
See also *18.3*, p. *273*; *18.4*, p. *281*.

Sh. M. Efendiev, T. Z. Kulieva, V. A. Lomonov, M. I. Chiragov, M. Grandolfo, and P. Vecchia
Crystal Structure of Bismuth Titanium Oxide $Bi_{12}TiO_{20}$
(a) Vol. **74**, No. 1, p. K17–K21 (1982).
See also *1.1*, p. *13*.

G. M. Ramāns, J. V. Gabrusenoks, and A. A. Veispāls
Structure of Tungstic Acids and Amorphous and Crystalline WO_3 Thin Films
(a) Vol. **74**, No. 1, p. K41–K44 (1982).
See also **2**, p. 54; **6**, p. 80; *8*, p. *90*; *20.1*, p. *315*; 22.6, p. 476.

J. Pirnat, J. Lužnik, and Z. Trontelj
Temperature Dependence of ^{81}Br NQR in Rb_3ZnBr_5
(a) Vol. **74**, No. 1, p. K55–K57 (1982).
See also *19*, p. *291*.

L. A. Makovetskaya, I. V. Bodnar, B. V. Korzun, and G. P. Yaroshevich
Thermal Conductivity, Thermoelectric Power, and Thermal Expansion of $CuInS_{2x}Se_{2(1-x)}$
(a) Vol. **74**, No. 1, p. K59–K62 (1982).
See also *8*, p. *90*; *15*, p. *248*.

G. M. Guseinov, G. M. Niftiev, and M. G. Shakhtakhtinskii
Temperature and Infrared Quenching of Photoconductivity in $GdIn_3S_6$ Single Crystals
(a) Vol. **74**, No. 1, p. K63–K67 (1982).
See also *16*, p. *254*.

J. A. Tuszyński and H. Cofta
On Thermodynamics of Spontaneous Metamagnetism
(b) Vol. **114**, No. 2, p. 359–364 (1982).
See also *18.2*, p. *274*; 18.4, p. 281.

D. Fröhlich, K. Reimann, and P. Koidl
Investigation of Mixed-Mode Polariton Dispersion in $AgGaS_2$
(b) Vol. **114**, No. 2, p. 553–559 (1982).
See also *13.5.1*, p. *185*; *20.1*, p. *315*.

G. Heber, D. Żymierska, and H. Beyer
Dynamics of Two-Spin Cluster in Spin Glass (III): Magnetization Process at Finite, but Low Temperature
(b) Vol. **114**, No. 2, p. 571–580 (1982).
See also *13*, p. *147*; *18*, p. *264*.

A. Antonini, V. Fano, I. Ortalli, and A. Vera
Mössbauer Study of the $Pb_xSn_{1-x}{}^{125}Te$ System
(b) Vol. **114**, No. 2, p. 585–588 (1982).
See also *5*, p. *69*.

A. Selmi, R. Le Toullec, and R. Faymonville
Influence of the Spin Disorder on the Plasma Edge in n-Type $HgCr_2Se_4$
(b) Vol. **114**, No. 2, p. K97–K99 (1982).
See also *13.5.2*, p. *187*; *20.1*, p. *315*.

V. K. Jain and S. K. Yadav
EPR of Mn^{2+} in $K_2Zn(XO_4)_2 \cdot 6\,H_2O$ (X = S, Se) Single Crystals
(b) Vol. **114**, No. 2, p. K131–K134 (1982).
See also *19*, p. *291*.

K. Bohmhammel, P. Deus, and H. A. Schneider
Investigation of Acoustic Phonon Modes of GaP and Compound Semiconductors $A^{II}B^{IV}C_2^{V}$ by Specific Heat Measurements
(a) Vol. **74**, No. 2, p. 485–492 (1982).
See also *6*, p. *81*; *8*, p. *90*; 22.2.2, p. 426.

T. H. O'Dell
Measurement of Magnetomechanical Coupling Factor in Amorphous Ribbons
(a) Vol. **74**, No. 2, p. 565–572 (1982).
See also 2, p. 56; *12.2*, p. *138*; *18.2*, p. *274*.

L. V. Rao, B. S. Naidu, and P. J. Reddy
Structure and Optical Energy Gap Measurements in As–Te–Cd Chalcogenide Thin Films
(a) Vol. **74**, No. 2, p. K137–K140 (1982).
See also *1.5*, p. *41*; 2, p. 56; *20.1*, p. *316*.

22.8.1 Salts of Oxy-Acids (Carbonates, Phosphates, Silicates Including Mica etc.)

Original Papers and Short Notes

K. N. Kalaidjiev, M. P. Mikhailov, G. I. Bozhanov, and St. R. Stoyanov
Influence of a Constant Electric Field on the Heat-Conductivity Coefficient of TGS near the Phase Transition Temperature
(a) Vol. **69**, No. 2, p. K163–K166 (1982).
See also *8*, p. *85*; *14.4.1*, p. *240*.

G. I. Rosenmann, E. I. Boikova, and Yu. L. Chepelov
Emission Study of Surface States in Ferroelectrics
(a) Vol. **69**, No. 2, p. K173–K177 (1982).
See also *13.3*, p. *158*; 14.4.1, p. 240; *17*, p. *255*.

R. Parrot and B. Canny
ESR Study of the Monoclinic–Orthorhombic Phase Transition in Neodymium Pentaphosphates
(b) Vol. **110**, No. 1, p. 123–131 (1982).
See also *1.2*, p. *16*; *19*, p. *284*.

V. N. Denisov, B. N. Mavrin, V. B. Podobedov, and Kh. E. Sterin
Hyper-Raman Scattering by Phonons and Mixed Polaritons in a Calcite Crystal
(b) Vol. **110**, No. 1, p. 183–189 (1982).
See also *6*, p. *72*; *20.1*, p. *302*.

Sh. M. Efendiev, N. G. Darvishov, V. M. Nagiev, N. M. Gasanly, V. T. Gabrielyan, and N. S. Nikogosyan
Raman Scattering in $Pb(MoO_4)_x(WO_4)_{1-x}$ Mixed Crystals
(b) Vol. **110**, No. 1, p. K21–K26 (1982).
See also *6*, p. *72*; *20.1*, p. *302*.

O. Hudák
Critical Properties of CsH_2PO_4 and CsD_2PO_4
(b) Vol. **110**, No. 1, p. K85–K88 (1982).
See also *14.4.1*, p. *240*.

Z. Czapla, L. Sobczyk, and J. Mróz
Isotopic Effect in Ferroelectric Rubidium Hydrogen Selenate Crystals
(a) Vol. **70**, No. 1, p. 73–78 (1982).
See also *1.2*, p. *16*; *14.4.1*, p. *240*.

A. D. Reddy, S. M. D. Rao, and G. S. Sastry
Electrical Conductivity and Stimulated Thermo-Current Studies on VO^{2+} Doped $Li(N_2H_5)SO_4$ Single Crystals
(a) Vol. **70**, No. 1, p. 269–275 (1982).
See also *14.4*, p. *235*; 14.4.1, p. 240.

I. Henning, P. Frach, E. Hegenbarth, and V. I. Fritsberg
Glass-Like Behaviour of PLZT Demonstrated by Heat Capacity Measurements
(a) Vol. **70**, No. 1, p. K7–K9 (1982).
See also *8*, p. *86*; 14.4.1, p. 240.

R. Bharati and R. A. Singh
Electrical Transport in Chromium(III) Tungstate
(a) Vol. **70**, No. 1, p. K11–K14 (1982).
See also *14.3*, p. *204*; *15*, p. *245*.

N. R. Ivanov, J. Kroupa, J. Fousek, and B. Březina
Birefringence of $PbHPO_4$-Family Ferroelectrics
(a) Vol. **70**, No. 1, p. K85 – K87 (1982).
See also *14.4.1*, p. *241*; *20.1*, p. *303*.

B. Z. Malkin, A. A. Kaminskii, N. R. Agamalyan, L. A. Bumagina, and T. I. Butaeva
Spectra of Rare-Earth Ions in the Crystal Fields of Double Tungstates and Molybdates
(b) Vol. **110**, No. 2, p. 417 – 422 (1982).
See also *13.4*, p. *166*; 20.1, p. 303; 20.3, p. 321.

Fat Duen Ho
The F^+ Center in Lithium Tantalate
(b) Vol. **110**, No. 2, p. 437 – 443 (1982).
See also *10.2*, p. *107*; *13.4*, p. *166*; 20.1, p. 303.

F. Brehat, B. Wyncke, and M. El Sherif
Lattice Mode Anharmonicity in Sodium Nitrite
(b) Vol. **110**, No. 2, p. 499 – 506 (1982).
See also 1.2, p. 16; *6*, p. *73*; 14.4.1, p. 241; *20.1*, p. *303*.

R. Huzimura and K. Atarashi
The Role of Sulphur-Oxy Radicals in Thermoluminescence and Exoelectron Emission of $CaSO_4$ Phosphors
(a) Vol. **70**, No. 2, p. 649 – 657 (1982).
See also 10.2, p. 108; *17*, p. *256*; *20.3*, p. *322*.

J. Torrès, C. Roucau, and R. Ayroles
Investigation of the Interactions between Ferroelastic Domain Walls and of the Structural Transition in Lead Phosphate Observed by Electron Microscopy (I): Experimental Results
(a) Vol. **70**, No. 2, p. 659 – 669 (1982).
See also *1.2*, p. *16*; *1.4*, p. *34*.

V. Narasimhan, K. Seshan, D. K. Chakrabarty, and H. V. Keer
Magnetic Properties of the $LaCoO_3$–$LaMnO_3$ System
(a) Vol. **70**, No. 2, p. K155 – K157 (1982).
See also *18.2*, p. *270*.

D. Ball
The Paramagnetic Resonance of Nd^{3+} and Yb^{3+} in Zircon Structure Silicates
(b) Vol. **111**, No. 1, p. 311 – 320 (1982).
See also *13.4*, p. *168*; *19*, p. *286*.

R. J. Matthews and T. G. Stoebe
Cathodoluminescence in $CaSO_4$:Dy
(a) Vol. **71**, No. 1, p. 55 – 58 (1982).
See also *20.3*, p. *322*.

J. Torrès, C. Roucau, and R. Ayroles
Investigation of the Interactions between Ferroelastic Domain Walls and of the Structural Transition in Lead Phosphate Observed by Electron Microscopy (II): Interpretation of the Interactions between Ferroelastic Domain Walls
(a) Vol. **71**, No. 1, p. 193 – 205 (1982).
See also *1.2*, p. *17*; *1.4*, p. *34*.

M. SAUVAGE, F. LEFAUCHEUX, M. C. ROBERT, and M. RIBET
Synchrotron Radiation Section Topographs in $Sr(NO_3)_2$
(a) Vol. **71**, No. 1, p. K5–K7 (1982).
See also *1.4*, p. *34*.

P. E. TOMASZEWSKI and K. ŁUKASZEWICZ
Orthorhombic Low-Temperature Phase of $LiKSO_4$
(a) Vol. **71**, No. 1, p. K53–K55 (1982).
See also *1.2*, p. *17*.

J. ROSA, K. POLÁK, and J. KUBÁTOVÁ
ESR and Optical Studies of Impurity Centres in γ- and X-Irradiated $LiNbO_3$
(b) Vol. **111**, No. 2, p. K85–K87 (1982).
See also *13.4*, p. *170*; 19, p. 287; 20.1, p. 306.

D. L. SASTRY and K. V. S. RAMA RAO
ESR and Optical Absorption Studies of ClO_3 in $Ba(ClO_4)_2 \cdot 3\,H_2O$ Single Crystals
(b) Vol. **112**, No. 1, p. 133–142 (1982).
See also 11, p. 123; *19*, p. *288*; *20.1*, p. *307*.

A. NATH and U. S. GHOSH
Ligand Field Theory of the Magnetic Behaviours of Neodymium Ethylsulphate Nonahydrate
(b) Vol. **112**, No. 1, p. 187–192 (1982).
See also *13*, p. *144*; *18*, p. *261*.

A. A. KAMINSKII, S. E. SARKISOV, H. D. KÜRSTEN, and D. SCHULTZE
Crystal Growth and Spectroscopic Properties of Nd^{3+} Ions in Ferroelectric $Pb_5Ge_3O_{11}$ Crystals
(a) Vol. **72**, No. 1, p. 207–213 (1982).
See also 3, p. 58; 14.4.1, p. 242; *20.1*, p. *308*; *20.3*, p. *324*.

R. A. RUPP and E. KRÄTZIG
Investigation of Refractive Index Gratings in Electro-Optic Crystals by a Microscope Technique
(a) Vol. **72**, No. 1, p. K5–K8 (1982).
See also *20.1*, p. *308*.

J. DEL CERRO, S. RAMOS, and J. ISIDORO
Thermal Study of α-$LiIO_3$ Single Crystals below Room Temperature
(a) Vol. **72**, No. 1, p. K23–K26 (1982).
See also *8*, p. *87*; 14.4, p. 237.

Y. A. BADR and S. AWAD
On the Low Temperature Phase Transition in Ammonium Sulfate
(a) Vol. **72**, No. 1, p. K27–K32 (1982).
See also *1.2*, p. *19*; 8, p. 87; *14.4.1*, p. *242*.

A. SPROGIS and V. DIMZA
Investigation of Dark Conductivity and Optical Absorption in Reduced PLZT
(a) Vol. **72**, No. 1, p. K57–K59 (1982).
See also *14.4.1*, p. *242*; *20.1*, p. *308*.

M. Korkmaz, Ö. Korkmaz, and S. Kospançali
EPR of Copper-Doped $(NH_4)_2C_2O_4 \cdot H_2O$ Single Crystals
(b) Vol. **112**, No. 2, p. 423–427 (1982).
See also *13.4*, p. *172*; *19*, p. *288*.

H. Presting and R. von Baltz
Bulk Photovoltaic Effect in a Ferroelectric Crystal. A Model Calculation
(b) Vol. **112**, No. 2, p. 559–564 (1982).
See also 13.1, p. 153; *14.4.1*, p. *242*; *16*, p. *252*.

R. K. Singh and S. L. Chaplot
Lattice Mechanics of Molecular Solid Sodium Nitrate
(b) Vol. **112**, No. 2, p. 717–724 (1982).
See also *6*, p. *77*.

P. Orlandi, A. Magistris, and A. Rigamonti
Mössbauer and Conductivity Study of Ferroelastic Ferric Molybdate
(a) Vol. **72**, No. 2, p. 543–549 (1982).
See also *5*, p. *67*; *14*, p. *191*.

S. Jerzak, S. Waplak, aned L. A. Shuvalov
Influence of an External Electric Field and Uniaxial Stress on the Domain Structure of $RbH_3(SeO_3)_2$ Crystals Studied by EPR Spectra of the Cr^{3+} Ion
(a) Vol. **72**, No. 2, p. 783–787 (1982).
See also *14.4.1*, p. *242*; 19, p. 289.

E. Fischer, W. Hässler, and E. Hegenbarth
Glass-Like Behaviour in the Thermal Conductivity of $Sr_{1-x}Ba_x(Nb_2O_6)$ Single Crystal
(a) Vol. **72**, No. 2, p. K169–K171 (1982).
See also *8*, p. *88*; *14.4.1*, p. *242*.

O. S. Kolotov, V. A. Pogozhev, R. V. Telesnin, G. V. Smirnov, Yu. V. Shvydko, S. Kadeĉkova, M. Kotrbova, and J. Novak
Pulse Magnetization Reversal in $FeBO_3$ Crystals
(a) Vol. **72**, No. 2, p. K197–K201 (1982).
See also 3, p. 58; *18.2*, p. *271*.

H. G. Festl, P. Hertel, E. Krätzig, and R. von Baltz
Investigations of the Photovoltaic Tensor in Doped $LiNbO_3$
(b) Vol. **113**, No. 1, p. 157–164 (1982).
See also *14.4.1*, p. *242*; *16*, p. *253*.

F. Prokert
Neutron Scattering Studies on Strontium Barium Niobate
(b) Vol. **113**, No. 1, p. 239–252 (1982).
See also *6*, p. *78*; 12.2, p. 136; 14.4.1, p. 242.

M. Sekita and T. Kobayashi
Oscillator Strengths of Transitions between Stark Levels of $Nd_xGd_{1-x}P_5O_{14}$
(a) Vol. **73**, No. 1, p. 61–64 (1982).
See also 13.4, p. 174; *20.1*, p. *311*.

A. D. REDDY, S. G. SATHYANARAYAN, and G. S. SASTRY
Proton Conduction in $(NH_4)HSO_4$ Single Crystals
(a) Vol. **73**, No. 1, p. K41–K45 (1982).
See also *14.4.1*, p. *243*.

A. S. LITOVCHENKO, A. V. BRODOVOI, and A. A. MELNIKOV
Study of the Temperature Dependences of Magnetic Susceptibilities and ^7Li NMR Spectra of Ferriferrous Micas
(a) Vol. **73**, No. 1, p. K79–K82 (1982).
See also 18.1, p. 266; *19*, p. *289*.

D. GOURIER, J. P. PARANT, and D. VIVIEN
ESR Investigation of a $[Cr(V)O_4Pb_8]$ Center in $PbMoO_4$. Origin of the Yellow Coloration of Lead Molybdate
(b) Vol. **113**, No. 2, p. 509–518 (1982).
See also *13.4*, p. *174*; *19*, p. *290*.

J. KROUPA, J. FOUSEK, F. SMUTNÝ, and B. BŘEZINA
Birefringence and Electrooptic Effect in K_2SeO_4
(b) Vol. **113**, No. 2, p. K153–K157 (1982).
See also *14.4.1*, p. *243*; *20.1*, p. *313*.

T. KALA
Study of the Influence of Manganese and Lanthanum Oxides on the Band Structure of $Pb(Zr, Ti)O_3$
(a) Vol. **73**, No. 2, p. 573–578 (1982).
See also *13.1*, p. *155*; *20.1*, p. *313*.

V. A. ZUIKOV and V. V. SAMARTSEV
Reverse Photon Echo as a Method of Investigation of Resonant Medium Parameters
(a) Vol. **73**, No. 2, p. 625–632 (1982).
See also 13.4, p. 175; *20.1*, p. *313*; *20.3*, p. *327*.

B. V. R. CHOWDARI, A. RAJARATNAM, and Y. RAVI SEKHAR
Radiation Damage Studies of Dichromate Doped Alums
(a) Vol. **73**, No. 2, p. K231–K235 (1982).
See also *11*, p. *127*; 19, p. 290; 20.1, p. 313.

FAT DUEN HO
Point-Ion-Lattice Calculations on F^+ Centers in Lithium Niobate and Lithium Tantalate
(b) Vol. **114**, No. 1, p. 209–212 (1982).
See also *13.4*, p. *176*.

S. BARB, E. BĂRBULESCU, and A. BĂRBULESCU
Diffuse Phase Transitions and Ferroelectric–Paraelectric Diagram for the $BaTiO_3$–$SrTiO_3$ System
(a) Vol. **74**, No. 1, p. 79–83 (1982).
See also 1.2, p. 23; *14.4.1*, p. *244*.

G. SORGE, A. ALMEIDA, and L. A. SHUVALOV
Influence of Deuteration and Defects on Elastic Properties of $KH_3(SeO_3)_2$
(a) Vol. **74**, No. 1, p. 91–96 (1982).
See also *12.2*, p. *137*; 14.4.1, p. 244.

22.8.2 Spinels, Garnets, Ferrites

Review Article

J. C. Joshi and A. L. Dawar
Pyroelectric Materials, Their Properties and Applications
(a) Vol. **70**, No. 2, p. 353–369 (1982).
See also *8*, p. *84*; *14.4*, p. *234*; 22.9, p. 503.

Original Papers and Short Notes

V. Nekvasil
Crystal-Field Analysis of Electronic Raman Scattering Data in YbAG and YbGG
(b) Vol. **109**, No. 1, p. 67–74 (1982).
See also *13.4*, p. *162*; *20.1*, p. *298*.

V. A. M. Brabers and J. C. J. M. Terhell
Electrical Conductivity and Cation Valencies in Nickel Manganite
(a) Vol. **69**, No. 1, p. 325–332 (1982).
See also *1*, p. *9*; *14.3*, p. *203*; 15, p. 245.

A. Meenakshisundaram, N. Gunasekaran, and V. Srinivasan
Distribution of Metal Ions in Transition Metal Manganites AMn_2O_4(A–Co, Ni, Cu, or Zn)
(a) Vol. **69**, No. 1, p. K15–K19 (1982).
See also *1*, p. *9*.

T. Duczmal, S. P. Kunsevich, V. P. Palekhin and J. Pietrzak
Temperature Dependence of Magnetocrystalline Anisotropy Constant of $BaCo_2Ti_2Fe_8O_{19}$ Single Crystals
(a) Vol. **69**, No. 1, p. K125–K127 (1982).
See also *18.3*, p. *276*; 19, p. 283.

B. Gillot
Thermopower–Composition Dependence in Iron–Chromium, Iron–Zinc, and Iron–Cobalt Ferrites
(a) Vol. **69**, No. 2, p. 719–725 (1982).
See also *15*, p. *245*.

H. K. D. H. Bhadeshia
Diffusional Transformations: A Theory for the Formation of Superledges
(a) Vol. **69**, No. 2, p. 745–750 (1982).
See also *1.3*, p. *26*; 1.4, p. 32.

A. V. Muromtsev and E. A. Raitman
Relaxation of Complex Susceptibility of Ferrites in AC Magnetic Field
(a) Vol. **70**, No. 1, p. 29–34 (1982).
See also *18.3*, p. *276*.

S. A. Mazen, M. A. Ahmed, and B. A. Sabrah
Thermal Studies on the Electrical Conduction Mechanism of $CuFe_2O_4$
(a) Vol. **70**, No. 1, p. K71–K75 (1982).
See also *14.4*, p. *235*; 18.3, p. 276.

M. Wójcik, H. Le Gall, and J. M. Desvignes
Two Types of Tetrahedral Sites in the Mixed $\{Ca_2Na\}$ $[Mg_2]$ $(V_3)O_{12}$ Garnet
(b) Vol. **112**, No. 2, p. 483–488 (1982).
See also *19*, p. *289*.

A. S. Parshin and N. S. Chistyakov
Spin Wave Resonance in Thin Monocrystalline Iron–Manganese Ferrite Films
(b) Vol. **112**, No. 2, p. K89–K92 (1982).
See also *18.3*, p. *277*.

Yu. E. Perlin, A. A. Kaminskii, M. G. Blazha, and V. N. Enakii
Multiphonon Nonradiative Transitions of TR^{3+} Ions in $Y_3Al_5O_{12}$ and $Lu_3Al_5O_{12}$ Garnets
(b) Vol. **112**, No. 2, p. K125–K130 (1982).
See also 6, p. 77; *13.4*, p. *172*; *20*, p. *296*.

L. N. Bezmaternykh, V. G. Mashchenko, N. A. Sokolova, and V. L. Temerov
Interaction of Magnetostatic and Magnetoelastic Waves with Growth Inhomogeneities in Iron Garnet Single Crystals
(a) Vol. **72**, No. 2, p. 469–474 (1982).
See also *1.4*, p. *35*; *18*, p. *262*.

K. Takada and K. Goto
A Numerical Explanation for a Relation of the Domain Width to the Thickness in Barium Ferrite Crystals
(a) Vol. **73**, No. 1, p. 209–213 (1982).
See also *18.3*, p. *277*.

R. A. Jago and P. L. Rossiter
A Re-Examination of Long-Range Ordering at Fe–25 at% Ni
(a) Vol. **73**, No. 2, p. 497–502 (1982).
See also *1.3*, p. *31*; 1.4, p. 36; 21.1, p. 355; 21.1.1, p. 362.

I. Onyszkiewicz and J. Pietrzak
Magnetocrystalline Anisotropy of Tetragonal Copper Ferrite
(a) Vol. **73**, No. 2, p. 641–650 (1982).
See also *18.3*, p. *277*; 19, p. 290.

G. Albanese, M. Carbucicchio, L. Pareti, and S. Rinaldi
Cation Distribution and Magnetic Properties of $BaZn_2Fe_{16-x}In_xO_{27}$ and $BaZn_2Fe_{16-x}Sc_2O_{27}$ Hexagonal Ferrites
(a) Vol. **73**, No. 2, p. K193–K197 (1982).
See also *5*, p. *68*; *18*, p. *263*.

R. Z. Sadikhov, I. G. Kerimov, L. M. Valiev, and A. D. Namazov
Temperature Dependence of Spontaneous Magnetization of the $FeCr_{2-x}In_xS_4$ $(0 \leqq x \leqq 1)$ System
(a) Vol. **73**, No. 2, p. K209–K212 (1982).
See also *18.3*, p. *277*.

I. Onyszkiewicz, N. T. Malafaev, A. A. Murakhovskii, and J. Pietrzak
Magnetic and FMR Investigation of the Structural Phase Transition in Cu-Ferrite with Cooperative Jahn-Teller Effect
(a) Vol. **73**, No. 2, p. K243–K247 (1982).
See also *1.2*, p. *22*; 18.3, p. 277; 19, p. 290.

22.9 Organic Semiconductors

Review Article

J. C. JOSHI and A. L. DAWAR
Pyroelectric Materials, Their Properties and Applications
(a) Vol. **70**, No. 2, p. 353–369 (1982).
See also *8*, p. *84*; *14.4*, p. *234*; 22.8.2, p. 499.

Original Papers and Short Notes

U. SCHMID and P. REINEKER
Coherent Motion of Triplet Excitons in Molecular Crystals with Two Differently Oriented Molecules per Unit Cell. Application to the Charge Transfer Crystal Anthracene/Tetracyanobenzene
(b) Vol. **109**, No. 1, p. 265–273 (1982).
See also *13.5.1*, p. *181*; 19, p. 282.

P. P. REDDY and Y. P. REDDY
Absorption Spectrum of Fe^{3+} in Strontium Tartrate Tetrahydrate Single Crystals
(b) Vol. **109**, No. 1, p. K27–K29 (1982).
See also *20.1*, p. *299*.

G. J. ASHWELL
Subtle Structural Variations in (DEPE) $(TCNQ)_{4.5}(H_2O)_x$
(b) Vol. **109**, No. 2, p. K89–K91 (1982).
See also *1*, p. *9*.

YU. S. BULYSHEV, I. M. KASHIRSKII, V. V. SINITSKII, G. F. MYATCHINA, T. G. ERMAKOVA, and V. A. LOPYREV
Photoinjection of Charge Carriers in Poly-1-Vinyl-1,2,4-Triazole
(a) Vol. **69**, No. 2, p. 637–641 (1982).
See also *14.3.3*, p. *220*; 16, p. 250; 20.1, p. 301.

M. ALDISSI, M. ROLLAND, and F. SCHUE
Middle Infrared Study of the Cis Trans Isomerisation of $(CH)_x$ Induced by Oxidation
(a) Vol. **69**, No. 2, p. 733–737 (1982).
See also 14.3, p. 204; *20.1*, p. *301*.

A. SIRCAR, B. MALLIK, and T. N. MISRA
Compensation Effect in the Electrical Conduction Process in a Series of Naphthyl Polyenes
(a) Vol. **69**, No. 2, p. 767–771 (1982).
See also *14.3*, p. *204*.

A. GRAJA, R. SWIETLIK, J. PETZELT, and L. DOBIÁŠOVÁ
Phase Transition in $MTPP(TCNQ)_2$ Single Crystals by Infrared Reflection Measurements
(a) Vol. **69**, No. 2, p. K205–K208 (1982).
See also *1.2*, p. *15*; 6, p. 71; 20.1, p. 301.

J. AAVIKSOO, G. LIIDJA, and P. SAARI
Luminescence Spectra of Anthracene Crystals at 0.4 K
(b) Vol. **110**, No. 1, p. 69–73 (1982).
See also 13.5.1, p. 182; *20.3*, p. *320*.

J. L. Miane, F. Carmona, and P. Delhaes
Microwave Measurements of the Thermal Expansion of Organic Metals
(b) Vol. **111**, No. 1, p. 235–245 (1982).
See also **6**, p. 73; *8*, p. *86*; **14.1**, p. 195.

K. Parliński
Phase Transitions in KND_2
(b) Vol. **111**, No. 1, p. K55–K59 (1982).
See also *1.2*, p. *17*.

P. Ranson, P. Peretti, and Y. Rousset
Crystal Heating and Non-Radiative Relaxation Processes of Triplet Excitons in 4–4′ Dichlorobenzophenone
(a) Vol. **71**, No. 1, p. K69–K71 (1982).
See also *13.5.1*, p. *183*; **20.3**, p. 323.

L. Komorowski
Numerical Approach to the Band Structure in TCNQ Stack
(b) Vol. **111**, No. 2, p. 443–447 (1982).
See also *13.1*, p. *151*.

B. Schügerl and H. Kuzmany
Isomerization and Disorder of Polyacetylene Induced by Doping
(b) Vol. **111**, No. 2, p. 607–617 (1982).
See also *6*, p. *74*; **20.1**, p. 306.

J. F. J. Jordan, A. Axmann, H. Egger, and J. Kalus
Temperature Dependence of the Compressibility of d_8-Naphthalene
(a) Vol. **71**, No. 2, p. 457–462 (1982).
See also *1*, p. *10*; **6**, p. 75; **12.2**, p. 135.

W. J. Kuhn, J. Petersson, and H. E. Müser
Phase Transition Fronts in Single Crystals of the Antiferroelectric Squaric Acid
(a) Vol. **71**, No. 2, p. 483–492 (1982).
See also *1.2*, p. *18*; *14.4*, p. *236*.

J. K. Krüger, M. Pietralla, and H.-G. Unruh
The Elastic Properties of Hexatriacontane Single Crystals at Their Various Phase Transitions
(a) Vol. **71**, No. 2, p. 493–504 (1982).
See also *1.2*, p. *18*; *6*, p. *75*; **12.2**, p. 135.

B. L. Sharma and P. K. C. Pillai
Photoconduction in Kapton Polyimide Film at High Electric Fields
(a) Vol. **71**, No. 2, p. 583–588 (1982).
See also *16*, p. *252*.

M. G. Kaplunov, K. I. Pokhodniya, and Yu. G. Borodko
Temperature Dependence of Reflection Spectra of Conductive TCNQ Salts
(a) Vol. **71**, No. 2, p. 649–656 (1982).
See also **6**, p. 75; **14.3.4**, p. 229; *20.1*, p. *307*.

V. M. Yartsev
Electron–Molecular Vibration Coupling in Trimerized Organic Ion-Radical Semiconductors
(b) Vol. **112**, No. 1, p. 279–287 (1982).
See also *6*, p. *76*; 20.1, p. 308.

A. Buka and F. Leyvraz
Dielectric Relaxation in Polar Liquid Crystals
(b) Vol. **112**, No. 1, p. 289–296 (1982).
See also *2*, p. *50*; *14.4*, p. *236*.

T. G. Abdel-Malik, A. M. Abdeen, H. M. El-Labany, and A. A. Aly
Bulk Trapping States in β-Zinc Phthalocyanine Single Crystals
(a) Vol. **72**, No. 1, p. 99–104 (1982).
See also *13.4*, p. *171*; *14.3.4*, p. *230*.

V. I. Melnik, K. I. Nelipovich, A. N. Faidysh, M. T. Shpak, and L. B. Yankovskaya
Influence of Secondary Excitons on Intensity and Decay of Phosphorescence in Doped Benzophenone Single Crystals
(b) Vol. **112**, No. 2, p. 435–443 (1982).
See also *13.5.1*, p. *184*; *20.3*, p. *324*.

H. Grassi, A. Brau, and J. P. Farges
Heat Pulse Diffusion in TEA $(TCNQ)_2$
(b) Vol. **112**, No. 2, p. 633–640 (1982).
See also 8, p. 88; *14.3*, p. *208*.

J. Tyczkowski, G. Czeremuszkin, and M. Kryszewski
On Differentiation of Electrical Conduction Processes in Polymers
(a) Vol. **72**, No. 2, p. 751–762 (1982).
See also *14.3.1*, p. *215*; *14.3.4*, p. *230*.

J.-P. Crine
Influence of Electric Field and Low Pressure on the Activation Energy for Conduction in Polyethylene
(a) Vol. **72**, No. 2, p. 789–797 (1982).
See also *14*, p. *191*.

E. A. Silinsh, V. A. Kolesnikov, I. J. Muzikante, and D. R. Balode
On Charge Carrier Photogeneration Mechanisms in Organic Molecular Crystals
(b) Vol. **113**, No. 1, p. 379–393 (1982).
See also 14.3.1, p. 215; *16*, p. *253*.

T. Goworek, C. Rybka, R. Wasiewicz, and J. Wawryszczuk
Positronium Trapping at Thermally Produced Defects in Naphthalene
(b) Vol. **113**, No. 1, p. K9–K13 (1982).
See also *10.2*, p. *113*.

M. Godlewska, M. Massalska-Arodź, and S. Urban
Dielectric Relaxation in Liquid and Plastic Phases of 2,2-Dinitropropane
(a) Vol. **73**, No. 1, p. 65–69 (1982).
See also *1.2*, p. *20*; 14.4, p. 237.

A. P. Tyutnev, V. S. Saenko, E. D. Pozhidaev, and A. F. Akkerman
Transient Radiation-Induced Conductivity in Polymers
(a) Vol. **73**, No. 1, p. 81 – 89 (1982).
See also *11*, p. *126*; *14.3*, p. *210*.

B. Pura, J. Przedmojski, A. Jabłonka, and R. Dąbrowski
Anisotropy of X-Ray Critical Scattering in 4,4'-Bi-Pentylobiphenyl Liquid Crystal
(b) Vol. **113**, No. 2, p. K119 – K122 (1982).
See also 1.2, p. 21; *2*, p. *53*.

A. P. Tyutnev, G. S. Mingaleev, V. S. Saenko, E. D. Pozhidaev, and A. F. Akkerman
Bulk Charging of Dielectric Films by Low Energy Electrons
(a) Vol. **73**, No. 2, p. 361 – 366 (1982).
See also *11*, p. *127*; *14.4*, p. *238*.

M. Januszczyk, J. Pietrzak, and A. Stecki
Electrical Conductivity of $Fe(PY)_3Cl_3 \cdot PY$ Complex Single Crystals near the Structural Phase Transition
(a) Vol. **73**, No. 2, p. K185 – K188 (1982).
See also *1.2*, p. *22*; *14.3*, p. *210*.

B. P. Chandra
Uncommon Mechanoluminescence Spectra of Certain Organic and Inorganic Crystals
(a) Vol. **73**, No. 2, p. K205 – K208 (1982).
See also *20.3*, p. *327*; 22.6, p. 475.

E. G. Sharoyan and H. A. Samuelyan
ESR and Dielectric Anomalies in Amorphous Cu-Phthalocyanine–Iodine Compounds
(a) Vol. **73**, No. 2, p. K213 – K216 (1982).
See also 2, p. 53; *14.4*, p. *238*; *19*, p. *290*.

A. Graja, E. Sopa, and L. Dobiášová
Infrared Study of the Structural Phase Transition in a Ternary 1 D Salt $(TMA^+)\ (TCNQ)^{2/3-}(I_3^-)_{1/3}$
(a) Vol. **73**, No. 2, p. K237 – K241 (1982).
See also *1.2*, p. *22*; *20.1*, p. *313*.

G. B. Talapatra and T. N. Misra
Delayed Luminescence of Organic Mixed Crystals. Delayed Excimer Fluorescence of Pyrene in p-Terphenyl
(b) Vol. **114**, No. 1, p. 73 – 83 (1982).
See also *13.5.1*, p. *185*; *20.3*, p. *327*.

G. Völkel
The Importance of Anharmonic Effects in the Electron Spin–Lattice Relaxation at Ferroelectric Phase Transitions
(b) Vol. **114**, No. 1, p. K67 – K69 (1982).
See also *14.4.1*, p. *243*; *19*, p. *290*.

J. S. Brown
Some Improved Lennard-Jones 6–12 Potentials for Methane
(b) Vol. **114**, No. 1, p. K91 – K96 (1982).
See also *6*, p. *80*.

23. Solidified Gases

Review Article

F. BECHSTEDT
Electronic Relaxation Effects in the Core Level Spectra of Solids
(b) Vol. **112**, No. 1, p. 9–49 (1982).
See also **4**, p. 61; *13.1*, p. *148*; **17**, p. 255; 20, p. 293; 21, p. 330; 22, p. 382.

Original Papers and Short Notes

YU. YA. BOGUSLAVSKII
Equation for the Melting Curve of Solids under High Pressure
(b) Vol. **109**, No. 1, p. 145–151 (1982).
See also *1.2*, p. *14*; *8*, p. *84*; 12.1, p. 130; 21.2, p. 365.

U. HAHN, R. HAENSEL, and N. SCHWENTNER
Kr and Xe Guest Atoms in Ar Matrix-: Emission Spectra, Excitation Spectra, and Lifetimes
(b) Vol. **109**, No. 1, p. 233–244 (1982).
See also *13.4*, p. *162*; 20.1, p. 298; *20.3*, p. *318*.

J. SHANKER and A. K. RAJAURIA
Theory of Interatomic Forces in Rare Gas Crystals
(b) Vol. **109**, No. 2, p. 711–716 (1982).
See also *4*, p. *61*; 12.2, p. 134.

S. BARONI, G. GROSSO, and G. PASTORI PARRAVICINI
Energy Bands of Krypton and Xenon with the GOPW Method
(b) Vol. **110**, No. 1, p. 227–232 (1982).
See also *13.1*, p. *149*.

I. YA. FUGOL, O. N. GRIGORASHCHENKO, and E. V. SAVCHENKO
Polariton and Spatial Dispersion Effects in Exciton Spectra of Xenon Cryocrystals
(b) Vol. **111**, No. 1, p. 397–406 (1982).
See also *6*, p. *74*; *13.5.1*, p. *183*; 20.1, p. 305; 20.3, p. 322.

R. K. SINGH and D. K. NEB
Role of Zero-Point Motion and Three-Body Effects in the Study of Crystal Properties of Rare Gas Solids
(b) Vol. **112**, No. 2, p. 735–746 (1982).
See also *4*, p. *63*; 6, p. 77; 8, p. 88; 12.2, p. 136.

S. I. KOVALENKO and N. V. KRAINYUKOVA
Effect of Sample Size upon Mean-Square Displacements of Atoms in Solidified Noble Gases
(a) Vol. **73**, No. 1, p. 49–53 (1982).
See also *1.5*, p. *40*.

N. P. TISHCHENKO
Self-Diffusion in Rare-Gas Solids
(a) Vol. **73**, No. 1, p. 279–286 (1982).
See also *9*, p. *92*.

PREM PAUL and S. P. PURI
Impurity–Host to Host–Host Coupling Force Constant Ratio for ^{57}Fe in Rare Gas Solids
(b) Vol. **114**, No. 1, p. 213–220 (1982).
See also *4*, p. *64*; 5, p. 68; *6*, p. *80*; 10.2, p. 116.

Author Index

19. Magnetic Resonance

20. Optical Properties
    20.1 Spectra. Optical Constants (X-Ray Spectra See 20)
    20.2 Lasers
    20.3 Luminescence (X-Ray Spectra See 20)

21. Metals. Alloys
    21.1 Transition Metals (Group Sc to Zn) and Their Alloys
        21.1.1 Fe and Fe Alloys
    21.2 Alkali-Metals
    21.3 Alkali-Earth Metals
    21.4 Rare-Earth Metals
    21.5 Actinides
    21.6 Noble Metals
    21.7 Semi-Metals (Semiconducting Alloys See 22.7)

22. Semiconductors and Ionic Crystals
    22.1 Elements
        22.1.1 Germanium
        22.1.2 Silicon
        22.1.3 Group VI Elements
    22.2 III–V Compounds
        22.2.1 Arsenides
        22.2.2 Phosphides
        22.2.3 Antimonides
        22.2.4 Mixed Crystals
    22.3 IV–IV, II–V, and III–VI Compounds
        22.3.1 Mixed Crystals
    22.4 II–VI Compounds
        22.4.1 Sulphides
        22.4.2 Selenides
        22.4.3 Tellurides
        22.4.4 Mixed Crystals
    22.5 Halides
        22.5.1 Silver Halides
        22.5.2 Alkali Halides
        22.5.3 Alkali-Earth Halides
        22.5.4 Mixed Crystals
    22.6 Simple Oxides (CdO, ZnO See 22.4)
        22.6.1 Mixed Crystals
    22.7 Semiconducting Intermetallic Compounds
        22.7.1 Mixed Crystals
    22.8 Tertiary and Higher Compounds
        22.8.1 Salts of Oxy-Acids (Carbonates, Phosphates, Silicates Including Mica etc.)
        22.8.2 Spinels, Garnets, Ferrites
    22.9 Organic Semiconductors

23. Solidified Gases

For comments on the use of the classification scheme see p. 6.

32913/32914

Manuscripts and letters for physica status solidi (b) — basic research and physica status solidi (a) — applied research

should be sent to

DDR-1020 Berlin, Neue Schönhauser Str. 20

or

Internationaler Buchversand GmbH, 1000 Berlin 19, Eschenallee 21.

The object of "physica status solidi (b)" and "physica status solidi (a)" is to publish rapidly important and topical results concerning fundamental research in the field of solid state physics and aspects of application and experimental techniques of fundamental solid state research, respectively. The issues of "physica status solidi (b)" and "physica status solidi (a)" will be published on the 1st and the 16th day of each month, respectively. Either series will contain the following categories:

1. Review articles written by arrangement with the editors. They will be remunerated.
2. Original papers selected from those which have not been submitted for publication elsewhere, after being reviewed. (The latest date for acceptance of contributions is the 8th and the 23rd day of each month for "physica status solidi (b)" and "physica status solidi (a)", respectively 50 days before the delivery of the journal.)
3. Short notes published in the off-set part of the journal. They are, if necessary after being reviewed, selected by the editorial staff. (The latest date for acceptance of manuscripts for this part is the 12th and the 27th day of each month for "physica status solidi (b)" and "physica status solidi (a)", respectively, 16 days before the delivery of the journal.)
4. Pre-printed titles of papers to be published in the next issue of both series.

Editorial Information

Manuscripts should be submitted preferentially in English. Besides, papers in German, French, and Russian are admitted. The final decision on the incorporation of the papers in the series (b) or (a) is taken by the editorial staff after consultation of the referee. Authors are asked to observe the following rules:

1. Manuscripts have to be written as concisely as possible. As most, 20 and 4 pages (28 lines per page; abstracts, tables, references, figures, and captions of figures included, where 3 figures are counted as about one page) are admitted for original papers and short notes, respectively. All manuscripts should be type-written on one side only. Two copies are required. Badly arranged manuscripts are returned.
2. The author's name, initials, and institution should be given.
3. One abstract of about ten lines in English and one in German, French, or Russian are required for original papers.
4. Because of lack of space conclusions (if any) must not have the character of a summary (i. e. substantially a repetition of the contents of the abstract).
5. Formulae should not be written too small and not with pencil. Separate lines for formulae are desirable. Indistinct writing will lead to a return of the manuscript.
6. Drawings should be enclosed on separate sheets of paper, not exceeding the maximum and minimum sizes of 20×30 and 10×15 cm^2, respectively. Photographs should be submitted, on white paper, only for review articles and original papers. Duplicates are required.
7. Captions for the figures should be submitted on separate sheets of paper, and special instructions with regard to the size of illustrations should be given on these.
8. References should be numbered in order of appearance and should be given, on a separate sheet, in the form used in physica status solidi.
9. SI-units should be used.
10. Each paper should contain a proposal for the subject classification according to the classification scheme appearing on the second and third cover of each issue.

The authors will be sent galley proofs of review articles and original papers. Proof reading of short notes and pre-printed titles will be done by the editorial staff.

For each original paper, and short note, a total of 75 reprints is available free of charge. The number of reprints free of charge per review article is 25. Requests for additional reprints (against payment) should be given together with the submitted manuscript. **The deadline for orders of additional reprints is the 15th and the 30th day of each month** for "physica status solidi (b)" and "physica status solidi (a)", respectively, two weeks before the delivery of the journal.

Bei Fragen zur Produktsicherheit wenden Sie sich bitte an:
If you have any questions regarding product safety,
please contact:

Walter de Gruyter GmbH
Genthiner Straße 13
10785 Berlin
productsafety@degruyterbrill.com